AF684820

State-of-the-Art Avian Viruses Research in Asia

State-of-the-Art Avian Viruses Research in Asia

Editor

Chi-Young Wang

MDPI • Basel • Beijing • Wuhan • Barcelona • Belgrade • Manchester • Tokyo • Cluj • Tianjin

Editor
Chi-Young Wang
Veterinary Medicine
National Chung
Hsing University
Taichung
Taiwan

Editorial Office
MDPI
St. Alban-Anlage 66
4052 Basel, Switzerland

This is a reprint of articles from the Special Issue published online in the open access journal *Viruses* (ISSN 1999-4915) (available at: www.mdpi.com/journal/viruses/special_issues/avian_viruses_asia).

For citation purposes, cite each article independently as indicated on the article page online and as indicated below:

LastName, A.A.; LastName, B.B.; LastName, C.C. Article Title. *Journal Name* **Year**, *Volume Number*, Page Range.

ISBN 978-3-0365-6623-8 (Hbk)
ISBN 978-3-0365-6622-1 (PDF)

© 2023 by the authors. Articles in this book are Open Access and distributed under the Creative Commons Attribution (CC BY) license, which allows users to download, copy and build upon published articles, as long as the author and publisher are properly credited, which ensures maximum dissemination and a wider impact of our publications.
The book as a whole is distributed by MDPI under the terms and conditions of the Creative Commons license CC BY-NC-ND.

Contents

About the Editor

Chi-Young Wang

Dr. Chi-Young Wang received his bachelor's degree in veterinary medicine from National Chung Hsing University, Taiwan. He earned his Ph.D. in animal virology and avian diseases at Auburn University, then spent a year doing postdoctoral research at the University of Alabama at Birmingham, where he studied genetics. He joined the faculty of National Chung Hsing University as an assistant professor of veterinary virology and avian diseases, and was promoted to full professor in 2015.

Preface to "State-of-the-Art Avian Viruses Research in Asia"

To date, the rapid progression in avian virus research in Asia has resulted in abundant achievements. This has mainly contributed to the development of some innovative techniques, advanced approaches, and the timely resolution of urgent issues. More and more new research topics focusing on emerging novel viruses have been found in various avian species. As the source of an important food supply, retaining the health of the birds remains the first priority for the poultry industry and research teams around Asia. Therefore, the accumulated knowledge about the fundamental biology, pathogenicity, epidemiology, antiviral strategies, vaccination schemes, and exceptional applications of avian viruses built up by expert research groups in Asia has accelerated the overall understanding of those key pathogens. Thus, the research articles on all research aspects of avian viruses from Asia were collected in this Special Issue. These valuable studies surely demonstrated some parts of advances in the research on modern avian virology.

Chi-Young Wang
Editor

viruses

MDPI

Article

Phylodynamics of Highly Pathogenic Avian Influenza A(H5N1) Virus Circulating in Indonesian Poultry

Desniwaty Karo-karo [1,2], Rogier Bodewes [3], Restuadi Restuadi [4], Alex Bossers [1,5], Agustiningsih Agustiningsih [6], Jan Arend Stegeman [1], Guus Koch [7] and David Handojo Muljono [8,9,10,*]

1 Department Population Health Sciences, Faculty of Veterinary Medicine, Utrecht University, 3584 CL Utrecht, The Netherlands
2 Centre of Diagnostic Standard Indonesian Agricultural Quarantine Agency, Ministry of Agriculture, Jakarta 13220, Indonesia
3 National Institute for Public Health and the Environment, 3720 BA Bilthoven, The Netherlands
4 Great Ormond Street Institute of Child Health, University College London, London WC1N 1EH, UK
5 Institute for Risk Assessment Sciences (IRAS), Department Population Health Sciences, Faculty of Veterinary Medicine, Utrecht University, 3584 CL Utrecht, The Netherlands
6 National Agency for Research and Innovation of The Republic of Indonesia, Jakarta 10340, Indonesia
7 Wageningen Bioveterinary Research, 8221 RA Lelystad, The Netherlands
8 Faculty of Medicine, Universitas Hasanuddin, Makassar 90245, Indonesia
9 Faculty of Medicine and Health, University of Sydney, Camperdown, NSW 2006, Australia
10 Eijkman Institute for Molecular Biology, Jakarta 10430, Indonesia
* Correspondence: davidmuljono@med.unhas.ac.id; Tel.: +62-8161-923-563

Abstract: After its first detection in 1996, the highly pathogenic avian influenza A(H5Nx) virus has spread extensively worldwide. HPAIv A(H5N1) was first detected in Indonesia in 2003 and has been endemic in poultry in this country ever since. However, Indonesia has limited information related to the phylodynamics of HPAIv A(H5N1) in poultry. The present study aimed to increase the understanding of the evolution and temporal dynamics of HPAIv H5N1 in Indonesian poultry between 2003 and 2016. To this end, HPAIv A(H5N1) hemagglutinin sequences of viruses collected from 2003 to 2016 were analyzed using Bayesian evolutionary analysis sampling trees. Results indicated that the common ancestor of Indonesian poultry HPAIv H5N1 arose approximately five years after the common ancestor worldwide of HPAI A(H5Nx). In addition, this study indicated that only two introductions of HPAIv A(H5N1) occurred, after which these viruses continued to evolve due to extensive spread among poultry. Furthermore, this study revealed the divergence of H5N1 clade 2.3.2.1c from H5N1 clade 2.3.2.1b. Both clades 2.3.2.1c and 2.3.2.1b share a common ancestor, clade 1, suggesting that clade 2.3.2.1 originated and diverged from China and other Asian countries. Since there was limited sequence and surveillance data for the HPAIv A(H5N1) from wild birds in Indonesia, the exact role of wild birds in the spread of HPAIv in Indonesia is currently unknown. The evolutionary dynamics of the Indonesian HPAIv A(H5N1) highlight the importance of continuing and improved genomic surveillance and adequate control measures in the different regions of both the poultry and wild birds. Spatial genomic surveillance is useful to take adequate control measures. Therefore, it will help to prevent the future evolution of HPAI A(H5N1) and pandemic threats.

Keywords: HPAI; H5N1; Indonesia; phylodynamic; Bayesian evolutionary analysis

check for updates

Citation: Karo-karo, D.; Bodewes, R.; Restuadi, R.; Bossers, A.; Agustiningsih, A.; Stegeman, J.A.; Koch, G.; Muljono, D.H. Phylodynamics of Highly Pathogenic Avian Influenza A(H5N1) Virus Circulating in Indonesian Poultry. *Viruses* **2022**, *14*, 2216. https://doi.org/10.3390/v14102216

Academic Editor: Chi-Young Wang

Received: 13 August 2022
Accepted: 5 October 2022
Published: 8 October 2022

Publisher's Note: MDPI stays neutral with regard to jurisdictional claims in published maps and institutional affiliations.

Copyright: © 2022 by the authors. Licensee MDPI, Basel, Switzerland. This article is an open access article distributed under the terms and conditions of the Creative Commons Attribution (CC BY) license (https://creativecommons.org/licenses/by/4.0/).

1. Introduction

In 1996, the first outbreak of highly pathogenic avian influenza virus (HPAIv) A(H5N1) occurred in China. Subsequently, this virus from the goose/Guangdong (Gs/Gd) lineage spread to multiple other countries. Nowadays, outbreaks of HPAIv A(H5N1) and related HPAIv have caused economic losses due to the deaths and culling of millions of chickens and other poultry worldwide. In addition, 865 human cases of HPAIv A(H5N1) infections were reported with a case-fatality rate of 53% from 2003 to 2022 [1].

The HPAIv A(H5N1) virus was first reported in Indonesia in 2003 and became endemic in multiple regions afterward. The introduction to and spread of HPAIv A(H5N1) within Indonesia was facilitated by several factors [2]. First, Indonesia is located at the crossroads of international trade between two continents (Asia and Australia) and two oceans (Peace, the Indian Oceans). Second, two wild bird migratory flyways, the East Asian–Australasian (EAAF) and the West Pacific (WPF) flyways include Indonesia. Third, the high contact rate between poultry from different locations [3] and between domestic ducks and wild birds due to poor biosecurity, particularly for backyard and moving or scavenging ducks [4]. Virus transmission between farms was facilitated by poultry trade and live bird markets and by human–animal interaction from inbound and outbound visits to poultry farms and live bird markets. Humans, via contact with poultry, could act as a vector of HPAIv A(H5N1) and facilitate transmission between poultry flocks [3,5].

Molecular surveillance is an important tool to support the control of HPAIv A(H5N1). HPAI genome sequence data obtained from avian and human cases can be used to understand transmission pathways [6], identify molecular markers for disease [7,8], expand host coverage [9], and detect variants associated with vaccine escape [10]. Molecular surveillance can also help to identify possible genetic drift and reassortments of HPAIv A(H5N1) with other influenza A viruses that may result in newly emerging viruses with possible increased transmission in poultry and wild birds, different pathogenicity which may also result in a wider host range [11,12].

Based on the global analysis of genomic data of HPAIv A(H5N1) detected in Indonesia, HPAIv A (H5N1) were classified into various clades, starting with clade 2.1, which subsequently branched into clades 2.1.1, 2.1.2, 2.1.3.2, and 2.1.3.2a; most clades have been reported to affect poultry [13–15]. In 2012, a new clade, 2.3.2.1c, was isolated from a duck farm and live bird markets in Java with high mortality among duck and amino changes such as a Ser deletion at position 325 in the multibasic amino acid cleavage site, and a K328R substitution [16]. The detection of HPAIv from this new clade was thought to be the result of a new incursion from other parts of South East Asia to Indonesia [11,17], as the clades 2.3.2.1, 2.3.2.1a, and 2.3.2.1b have been reported in other South-East countries such as China, Vietnam and Bangladesh [13,14,18,19]. Clade 2.3.2.1c subsequently circulated in poultry, while HPAIv from clade 2.1.3.2a was only detected in Sumatra [11,16,20–22]. A molecular study of HPAIv A(H5N1) carried out in 2015 and 2016 suggested that this new clade had diverged into two putative subgroups, clades 2.3.2.1c (A) and 2.3.2.1c (B) [11].

Although the major clades of HPAIv A(H5N1) in Indonesia are known, there is limited understanding of the evolution of HPAIv A(H5N1) in Indonesia. This knowledge can be useful to help focus surveillance and strengthen control measures aiming to reduce future reassortments and transmission of HPAIv among poultry and humans. The present study aimed to increase the knowledge of HPAIv A(H5N1) evolution in Indonesia from 2003–2016, with a particular focus on the HA gene segment and the jump of the clades of H5N1v.

To this end, we analyzed the available sequences of hemagglutinin (HA) in the genome database to improve the understanding of the phylodynamics of HPAIv A(H5N1) in Indonesia.

2. Materials and Methods

2.1. Dataset Preparation

Complete sequences of HA genes obtained from HPAIV A(H5Nx) detected in poultry in Indonesia from 2003 to 2016 were downloaded from the genome database, GISAID, and GENBANK and compiled as Indonesian H5N1 (HA). Another data compilation was downloaded from all available global sequences including Indonesia from 1966 to 2022 and separated as Global H5 (HA). Additional separated data for clades 2.3.2.1c, 2.3.2.1a, and 2.3.2.1 were also downloaded from the database. The HA gene was chosen because the HA protein is located on the outer surface of the virus particle, has a role in the virus–host cell interaction and is the main target for the protective antibody response [23]. Additionally, HA genes are published most frequently in the genome database, indicating that a worldwide phylodynamic analysis of H5N1v using HA genes will provide the most information.

The sequences were then aligned using MUSCLE [24] and the HA clades of the virus were phylogenetically analyzed using MEGA 7 [25] as described in a previous study [11]. The clade of HA was confirmed using the Highly Pathogenic H5N1 Clade Classification Tool of the Influenza Research Database (https://www.fludb.org/brc/home.spg?decorator=influenza, last accessed on 13 September 2022).

2.2. Clustering HA Gene Segments

The dataset of HA genome sequences was processed with cd-hit-est software of the CD-HIT Suite (http://weizhong-lab.ucsd.edu/cdhit_suite/cgi-bin/index.cgi?cmd=cd-hit-est, last accessed on 13 September 2022) to cluster sequences that shared 100% nucleotide identity [26–28]. The CD-HIT_EST test was performed on the globally available 12,018 HA genome sequences (1966–2022) irrespective of the accompanying NA. To condense the global taxa of the full genomes of HA genes, 80 to 99% identity thresholds were examined to obtain the cluster representative sequences. Maximum-likelihood analysis with bootstrapping was performed at different thresholds, and clusters of representative taxa were selected from taxa that share a larger identity than 98%. The representative sequences were used as a dataset for time-scale phylogeny analysis and demography reconstruction.

2.3. Time-Scale Phylogeny of Indonesian HPAIv A(H5N1) Sequences

Divergence times and evolutionary analysis were estimated simultaneously with Bayesian phylogenetic inference (BI) implemented in BEAST v.2.6.7 [29] (http://www.beast2.org/). The optimal substitution model was selected by the BEAST-ModelTest (bModelTest) v.1.2.1 package implemented in BEAST using transdimensional Markov chain Monte Carlo (MCMC) methods [30]. The best substitution model from bModelTest was also compared to the best substitution model selected by the Modeltest in the phangorn package implemented in the R (version R-4.0.3) environment for statistical analysis. bModelTest was also used to infer the gamma-distributed rate of heterogeneity, invariable site proportions, and unequal base frequencies [30].

Tree and clock priors were set on a coalescent Bayesian skyline tree and a relaxed molecular clock (assuming an uncorrelated lognormal distribution clock model) which was calibrated by using the sample collection dates. The Bayesian MCMC analysis was performed for 150–300 million generations sampled every 1000–3000 generations.

The parameter convergences were viewed and evaluated using Tracer v.1.7.1 [31] (http://tree.bio.ed.ac.uk/software/tracer/). The maximum clade credibility (MCC) phylogenetic trees were constructed by TreeAnnotator v.2.6.7 (BEAST package) by removing the initial 10–25% (burn-in) trees (burn-in settings depend on convergence). Then, phylogenetic trees were visualized by using FigTree version 1.4.4 (http://tree.bio.ed.ac.uk/software/figtree/). The calendar date of origin of tMRCA of Indonesian HPAIv A(H5N1) estimated in the BEAST analysis was converted using the lubridate package (https://lubridate.tidyverse.org/) implemented in R (version R-4.0.3).

The first step of BEAST was to analyze the Indonesian HA HPAIv A(H5N1) gene segments from viruses collected from 2010 to 2016. Then, HA gene segment analysis was performed separately on viral sequences of three different HA clades, clades 2.1.3.2, 2.1.3.2a, and 2.3.2.1c, collected from 2005 to 2016. To confirm the evolution in 2010–2016, MCMC analysis (BEAST) of the Indonesian HA HPAIv A(H5N1) gene was performed over the extended period of 2003–2016. The BEAST analysis over the period 2003–2016 is displayed in the results of Indonesian HA (H5).

In the final stage, we performed BEAST to analyze the worldwide HA of all available avian influenza viral sequences (2005–2021). The full length 12,018 HA sequences were downloaded from GISAID and clustered using CD-HIT-EST as described above. the Reference sequences closely related to Indonesian HPAIv H5N1 according to the maximum likelihood tree were selected and aligned using MEGA 7 [25] before proceeding to the BEAST analysis.

3. Results

3.1. Bayesian Evolutionary Analysis of HA of Indonesian H5N1

The number of taxa used for the BEAST analysis performed on HA sequences from Indonesia and worldwide with different times of collection and the number of sites is presented in Table 1.

Table 1. The substitution rates of Indonesian HPAIv A(H5N1) 2003–2016 and representative worldwide HA of H5N1v with the estimation of root height.

Hemagglutinin Gene	Collection Time (Year)	Taxa (n)	Sites (Character)	Evolutionary Models	Substitution Rate (10^{-3} s/s/y)		Number of Substituted Sequences (Subst/Genome/Year)
					Mean	95% HPD Interval	
Geographical Zone							
Indonesian H5N1 (HA)	2003–2016	502	1559	TVM + Γ + I	0.0042	0.0038–0.0046	6
Global H5 (HA)	2005–2021	284	1723	TVM+ Γ + I	0.0065	0.0061–0.0070	11
HA clade (Indonesian HA (H5N1))							
Clade 2.1.3.2	2005–2010	203	1559	TIM1 + Γ + I	0.0046	0.0036–0.0056	7
Clade 2.1.3.2a	2008–2016	73	1730	TIM1 + Γ + I	0.0049	0.0032–0.0069	7
Clade 2.3.2.1c	2012–2016	94	1707	TIM1 + Γ + I	0.0036	0.0026–0.0046	6

The mean of rates is posteriorly estimated based on Bayesian MCMC analysis using evolutionary models. The number of sequences is labeled as a taxon (taxa). The character of the number of differed sites is normalized from the length of a sequence to get the proportion of differences between two sequences [24]. Abbreviations: TVM (transversion model), TIM (transition model), Γ (gamma), I (Invariant), bp (base pair), s/s/y (substitution/site/years).

Time-measured phylogenetic analysis of 1707 sites from 94 taxa using the substitution model TIM1 + Γ + I showed the evolution of various clades HPAIv A(H5N1) in Indonesia (Table 1).

Time-measured phylogenetic analysis (Figures 1 and A1) estimated that HPAIv A(H5N1) clade 2.1-like, 2.1.1, 2.1.2, 2.1.3, 2.1.3.1, 2.1.3.3, 2.1.3.2 and 2.1.3.2a evolved from a common ancestor in the year 2002. In addition, the analysis indicated that some of the HA clades 2.1.1 shared a common ancestor with 2.3.2.1c in the year 2001. Subsequently, the HPAIv A(H5N1) clade 2.1.1 in 2003 diverged into HPAIv A(H5N1) HA clade 2.3.2.1c, which was detected mostly in 2015/16. A significant divergence in 2011 was also observed between HA subgroup 2.3.2.1c (A) and 2.3.2.1c (B) (Figure A1) in the phylogenetic analysis, with posterior values of more than 0.7 (Figure A2).

In 2015–2016, HPAIv A(H5N1) clade 2.1.3.2a was still detected. In contrast, the clades 2.1-like 2.1.2, 2.1.3, 2.1.3.1, 2.1.3.2, and 2.1.3.3 were no longer detected in Indonesia after 2012. Detection of clades of HPAIv A (H5N1) in Indonesian poultry varied between years. From 2005 to 2007 (2 years), the HA clades 2.1.1, 2.1.2, and 2.1.3 were detected. Between 2005 and 2012 (7 years), the HPAIv H5N1 clade 2.1.3.1 was detected. From 2005 to 2010 (5 years), the virus HA sequences were classified as clades 2.1.3.2 and 2.1.3.3. Of the subclades of HPAIv A(H5N1) clade 2.1, only 2.1.3.2a was still detected in 2016, while subclade 2.3.2.1c was mostly detected after 2010.

Spatiotemporal analysis indicated that various HA clades of Indonesian HPAIv A(H5N1) were detected in different areas (Figure 2). Most viruses were detected on Java Island. HA clades 2.1.3.2 and 2.1.3.2a were detected in most of the regions of Indonesia, while some clades were only detected in specific regions. For example, HPAIv A(H5N1) clade 2.1-like viruses were only detected in Jakarta and Yogyakarta, and HPAIv (AH5N1) clade 2.1.3 was only detected in Central Java, East Java, Yogyakarta, and Bali.

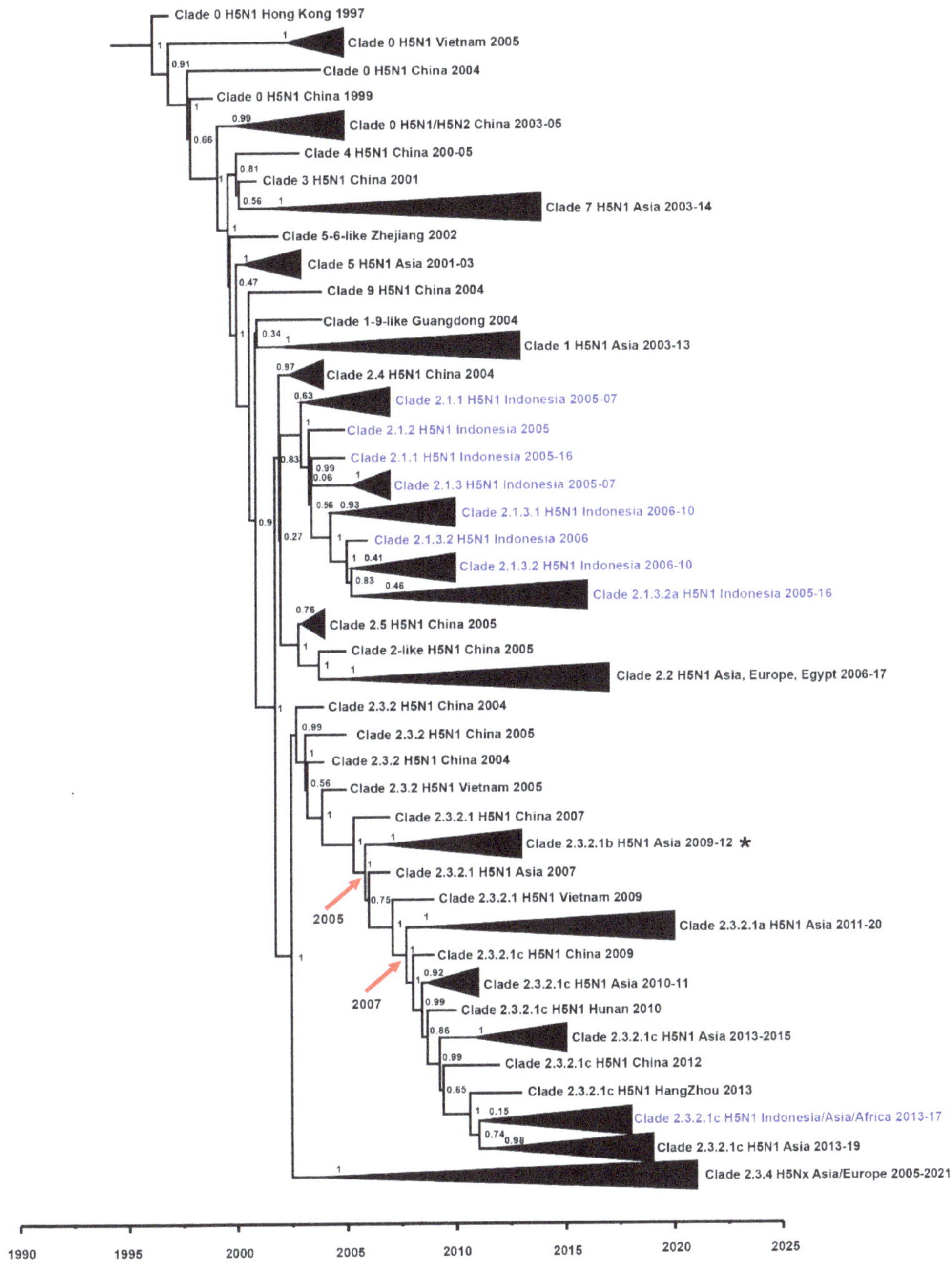

Figure 1. Time-scale phylogeny of selected worldwide HA of H5N1v. The estimated origin of the divergence of the HA 2.3.2.1c clade is highlighted in the asterisk symbol. The tMRCA of HA clades 2.3.2.1b and 2.2.3.1a are pointed out by the arrow. The blue colour highlights the HA of H5N1v from Indonesia. The node labels display the posterior value. The original sequences (GISAID ID) for worldwide HA of H5N1v phylogeny are displayed in Supplementary Table S1.

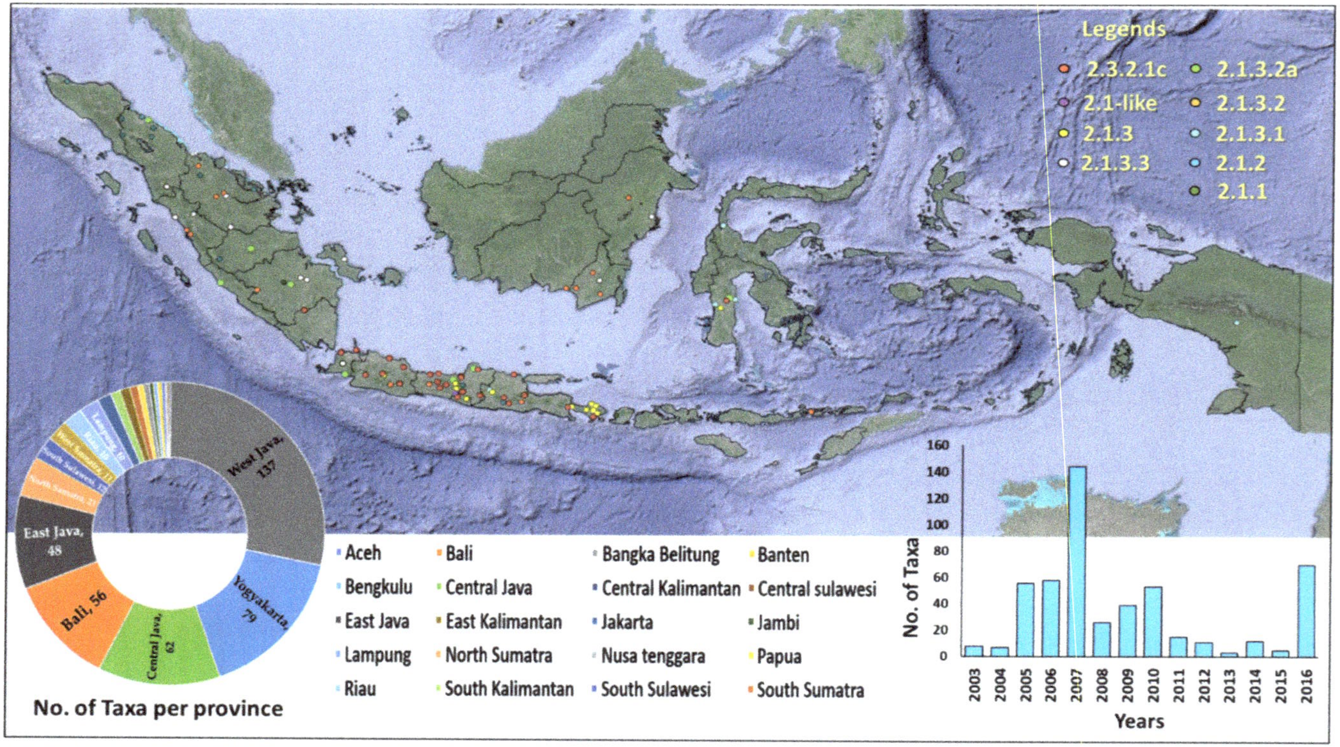

Figure 2. The distribution of HA of HPAIv A(H5N1) was detected in poultry based on different clades and different provinces of Indonesia. The number of taxa per province and per years are displayed in Supplementary Tables S2 and S3.

The BEAST analysis estimated that the mean nucleotide substitution rate of HA was 0.0042 substitution/site/year (s/s/y) (95% Interval, 0.0038–0.0046) over the course of 13 years. The mean substitution rate of clade 2.1.3.2a was 0.0042 (s/s/y) (95% Interval, 0.0031–0.0054) over an 8-year period, not statistically significant different than that of clade 2.3.2.1c (0.0036 s/s/y; 95% Interval, 0.0026–0.0041) over 4 years (Table 1).

3.2. *Indonesian Viruses in the Phylodynamic of Indonesian Worldwide Avian Influenza H5N1 Virus (AI H5N1v)*

The phylogeny of the worldwide HA including the Indonesian viruses is depicted in Figure 1. Based on analysis of representative sequences, H5N1v clade 2.1 and its subclades were only detected in Indonesia, while clade 2.3 viruses were detected in multiple countries in Asia, Europe, and Africa, including Indonesia, since 2009. Other subclades of clade 2, such as 2.2, 2.4, and 2.5, were circulating in multiple countries, such as China, Egypt, Germany, India, and Japan, but were not detected in Indonesia. These different geographic distributions of viruses also indicate geographic imbalances in virus spread and geographic leaps of multiple viruses from various clades.

The BEAST analysis estimated that the mean nucleotide substitution rate of global HA was 0.0065 substitution/site/year (s/s/y) over the course of 16 years (95% Interval, 0.0061–0.0070) (Table 1).

3.3. *Molecular Dating of HPAIv A(H5N1)*

Results of molecular dating indicated that the common ancestor of HPAIv A(H5N1) detected in Indonesia occurred in May 2001, around 5 to 7 years after the common ancestor of HPAIv A(H5N1) worldwide. The common ancestor of HPAIv A(H5N1) clades 2.1.3.2. and 2.1.3.2a occurred in the first months of 2002 according to this analysis, while the common ancestor of clade 2.3.2.1c occurred in February 2011. Results of the tMRCAs of HA of HPAIv A(H5N1) detected in Indonesian poultry and worldwide, determined by using a relaxed clock, with 95% HPD and posterior values, are displayed in Table 2.

Table 2. tMRCA of HA of HPAIv (H5N1) Indonesian poultry and worldwide H5, determined by using a relaxed clock, with 95% HPD and posterior values.

HA Gene	tMRCA	95% HPD Interval		Posterior
		Begin	End	
		Geographical zone		
Indonesian H5N1 (HA)	27 May 2001	13 September 1999	2 July 2002	1.00
Global H5 (HA)	4 April 1996	27 May 1995	28 December 1996	1.00
		HA clade (Indonesian HA (H5N1))		
Clade 2.1.3.2	8 January 2002	27 May 1997	13 September 2004	1.00
Clade 2.1.3.2a	15 March 2002	2 July 1997	1 January 2006	1.00
Clade 2.3.2.1c	6 February 2011	13 September 2009	13 September 2011	1.00

4. Discussion

4.1. *Temporal Dynamic of Indonesian HPAIv A(H5N1): Time-Measured Phylogenetic Analysis*

In the present study, a time-measured phylogenetic analysis was performed to increase the understanding of the HPAIv A(H5N1) detected in Indonesia from 2003–2016. While phylogenetic analysis of HPAIv A(H5N1) was the focus of a number of studies already [11,22,32], a study including all available Indonesian virus sequences has, to our knowledge, not been performed previously. Posterior analysis of Indonesian HPAIv A(H5N1) 2003–2016 estimated that the HPAIv A(H5N1) clade 2.3.2.1 evolved from the HA clade 2.1.1. In addition, the posterior analyses using BEAST with bModeltest, instead of the maximum likelihood approach, which is used as a criterion in a unified nomenclature system for HPAIv, confirmed the finding of our previous study [11] that HA clade 2.3.2.1c consists of two different clusters [13–15]. The time-measured analysis also showed that after 2012, mainly HPAIv A(H5N1) viruses classified as clade 2.3.2.1c were detected. The observed evolution of HPAIv A(H5N1) viruses, the emergence of new clades, and the emergence of reassortments may have been caused by biosecurity gaps leading to reassortment and limited vaccine efficacy and poor vaccination coverage, although we cannot exclude circulation of these viruses in wild birds due to the very limited surveillance of avian influenza in wild birds in Indonesia [11,33–35].

The substitution rate of avian influenza viruses worldwide has been studied extensively [18,36,37]. A previous study [38] estimated viral RNA substitution rates in the range of 0.01 to 0.001 s/s/y. Additionally, the rapid evolutionary dynamics of avian influenza viruses were estimated by a previous study with a substitution rate range of 0.0018–0.0084 s/s/y [39]. The estimated substitution rate in this study showed the fast substitution rate of Indonesian poultry HPAIv A(H5N1) and HA of worldwide H5, which was in line with previous reports by Duffy et al. (10^{-2} to 10^{-5} subs/site/year) and Chen et al. (1.8 to 8.4 $\times$ 10^{-3} subs/site/year) [38,39], but different from those reported by Ducatez et al. (3.32 $\pm$ 0.05 $\times$ 10^{-3} subs/site/year) [40]. The variation in the substitution rates between the HPAIv A(H5N1) genes can be caused by many factors, such as the differences in viral biologies such as viral genome architecture, replication speeds within-host and viral polymerase enzyme fidelities [41], and environmental selectivity related to the host factors such as species [38], vaccination status [37], contact rate, and age of infection, epidemic, and endemic status in a region during infection [41]. Positive selection pressures related to environmental selectivity have been identified at several antigenic sites of the HA gene in the previous study [22]. Meanwhile, the mean substitution rate of global HA was higher than in Indonesian poultry HPAIv A(H5N1); this observation might, however, be biased by sampling differences.

The phylogenetic analysis estimated that HA clades 2.3.2.1a and 2.3.2.1c shared a common ancestor and were rooted in the clade 2.3.2.1b. The H5N1v clade 2.3.2.1c and 2.3.2.1a diverged from clade 2.3.2.1b in agreement with a previous study [13,15]. A gap in the H5N1v clades in Indonesia is indicated by the lack of report of clade 2.3.2.1b, the clade that has been reported in Vietnam and Hong Kong [15,42]. This clade gap was assumed based on the finding in Indonesia that the HPAIv A(H5N1) clade 2.3.2.1c was rooted in HA clade 2.1.1. Bird migration and/or poultry trade could have driven the transmission and evolution of the H5N1v clade 2.3.2.1a to clade 2.3.2.1c. Additionally, unrecognized clinical signs in poultry and the reluctance of farmers to report the H5N1 outbreaks, particularly in sector 1 farms, might have contributed to the absence of some clades of H5N1v in the data set. This gap shows the need for regular and intensive surveillance to control the evolution of H5N1v, not only in poultry but also in wild birds.

The most recent ancestor of the H5N1 influenza virus in Indonesia has been previously studied [22,32]. The first study [22] estimated the tMRCA of Indonesian H5N1 HPAIV in June 2003 (November 2002 and October 2003) and the second study [32] estimated the tMRCA of reassortant H5N1v in July 2005. This study revealed that the common ancestor of Indonesian poultry HPAIv H5N1 was introduced into Indonesia 5–7 years (2001; 95% Interval: 1999–2002) after the original ancestor of HPAI A(H5Nx) arose worldwide (1996; 95% Interval: 1995–1996). The introduction of HPAIv A(H5N1) 5–7 years after worldwide outbreaks suggested the importance of sustainability of surveillance and control measures in around 5–7 years before the new introduction of new emerging and re-emerging HPAIv into Indonesia, either from outside Indonesia via wild birds or poultry trading of the virus, evolves within themselves in Indonesia.

4.2. Limitations and Benefits of the Study

We acknowledge several limitations in this study. First, the limited data, particularly the number of taxa or samples, may have affected the inferences of evolutionary analysis. Surveillance data and avian influenza virus sequences in wild birds in Indonesia are very limited or absent. All avian influenza sequence data in public genome databases were obtained from domesticated birds. Differences in sampling over time and space may affect the outcomes of the analysis. Therefore, improved surveillance with good competency for clinical and laboratory diagnosis and collection of metadata, as well as the willingness to share the information, is crucial to raise the number of viral genomes in the public database. Surveillance in wild birds is also crucial to reveal the clade gap and study the evolution of the avian influenza virus. Furthermore, additional studies are needed to identify key amino acid changes and evaluate their impact on the viral phenotype, and also on the relationship with the possible role of vaccination programs on the observed evolution of HPAIV A(H5N1).

This study is of importance not only for virus identification but also for studying virus evolution in Indonesia. This study shows that probably only two introductions occurred, after which HPAIv A(H5N1) continued to circulate among poultry in Indonesia. Continuous surveillance of poultry farms in all sectors and live bird markets in Indonesia with global support and collaboration are essential to take adequate measures and prevent further evolution of the virus. In addition, compartmentalization, inspection, and certification [43,44] of poultry farms are also important to control the evolution of HPAIv A(H5N1) in Indonesia. Estimation of temporal characteristics of HPAIv A(H5N1) across Indonesia in association with the viral dynamics is essential in conducting prevention controls such as quarantine, movement restriction, diagnostic tools, surveillance systems, and vaccine development [45,46] for future outbreaks. The discovery of different clades in only a few regions and the fact that some Indonesian HPAIv A(H5N1) clades were not detected in other countries indicates the importance of area- and country-specific preventive measures for HPAI outbreaks [45]. The Indonesian archipelago, with the ocean as a geographical barrier between islands and between continents, can be an advantage for the country and region-specific preventive measures, as well as reconstructions of intensive poultry farming locations and mapping of wild bird captive areas. In parallel, capacity building is of great importance for each country, and an agreed consensus between countries is a necessity in studying the viral phylodynamics, combined with regular genomic surveillance, to prevent future HPAIv pandemics.

5. Conclusions

This study demonstrated that introductions of HPAIv A(H5N1) into Indonesia are infrequent and most of the observed changes in the virus originate from within Indonesia. The lack of detection of H5N1v clade 2.3.2.1b and the limited Indonesian HPAIv A(H5N1) genomic sequences in the database indicate that there is room for improvement in molecular surveillance of HPAIv in Indonesia. Furthermore, the evolutionary dynamics of the Indonesian HPAIv A(H5N1) highlight the need for continuing genomic surveillance and adequate control measures to prevent viral introduction and evolution, within and between farm transmission in different regions.

Supplementary Materials: The following supporting information can be downloaded at: https://www.mdpi.com/article/10.3390/v14102216/s1, Table S1: The GISAID ID of the HA sequences (taxa) for the Phylodynamic analysis in this study; Table S2: The number of Indonesian H5N1 taxa per location; Table S3: The number of Indonesian H5N1 taxa per year.

Author Contributions: D.K.-k.: Conceptualization, data curation, investigation, methodology, formal analysis, project administration, visualization, writing—original draft; R.B.: Supervision, methodology, validation, writing—review, and editing; R.R.: Formal analysis, methodology, validation, writing—review and editing; A.B.: Formal analysis, methodology, validation, writing—review and editing; A.A.: Formal analysis, methodology, validation, writing—review and editing; G.K.: Methodology, formal analysis, validation, funding acquisition, project administration, supervision, writing—review and editing; J.A.S.: Funding acquisition, project administration, resources, supervision, writing—review and editing; D.H.M.: Methodology, formal analysis, validation, supervision, project administration, writing—review and editing. All authors have read and agreed to the published version of the manuscript.

Funding: This study was funded by the Royal Netherlands Academy of Art and Sciences (KNAW) Project No. SPIN3-JRP-61 as a part of the Scientific Programme Indonesia-Netherlands (SPIN).

Data Availability Statement: This study did not report any data.

Acknowledgments: The authors acknowledge all owners of Indonesian and worldwide HA sequences in the GISAID database used in this study. The authors also acknowledge the Animal Health Authority under West Java Province, especially in the Subang, Indramayu, Bandung, Sukabumi, Tasikmalaya, Purwakarta, and Ciamis districts. We thank other collaborators: the Cikole Animal Health Laboratory of the Animal Husbandry Service of West Java Province, District Investigation Center Subang, and Centre of Tropical Animal Health Studies under Bogor Agricultural University (CENTRAS-IPB), Eijkman Institute for Molecular Biology for collecting, screening, and sequencing the samples. I Made Artika is thanked for discussion and sequencing from 2015–2016.

Conflicts of Interest: The authors declare that the research was conducted in the absence of potential conflicts of interest.

Appendix A

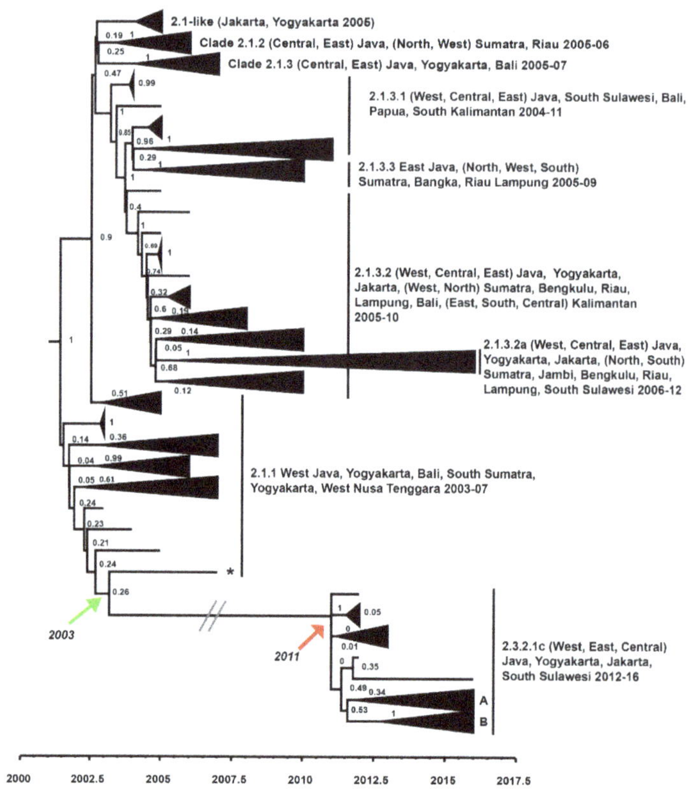

Figure A1. Time-measured phylogeny of HA genes of Indonesian poultry HPAIv A(H5N1) 2003–2016. The estimated root of HA clade 2.3.2.1c was highlighted by asterisk symbol, the tMRCA of HA clade 2.3.2.1c is pointed out by the red arrow, and the tMRCA between HA clade 2.1.1 and 2.3.2.1c is pointed out by the green arrow. The HPAIv A(H5N1) clade 2.3.2.1c diverged into subgroups (A and B). The node labels display the posterior value. The two gray lines between the clades 2.1.1 and 2.3.2.1c represent the presence of multiple viruses between these two clades as presented in Figure 1. The original sequences (GISAID ID) of HA genes of Indonesian poultry HPAIv A(H5N1) 2003–2016 phylogeny is displayed in Supplementary Table S1.

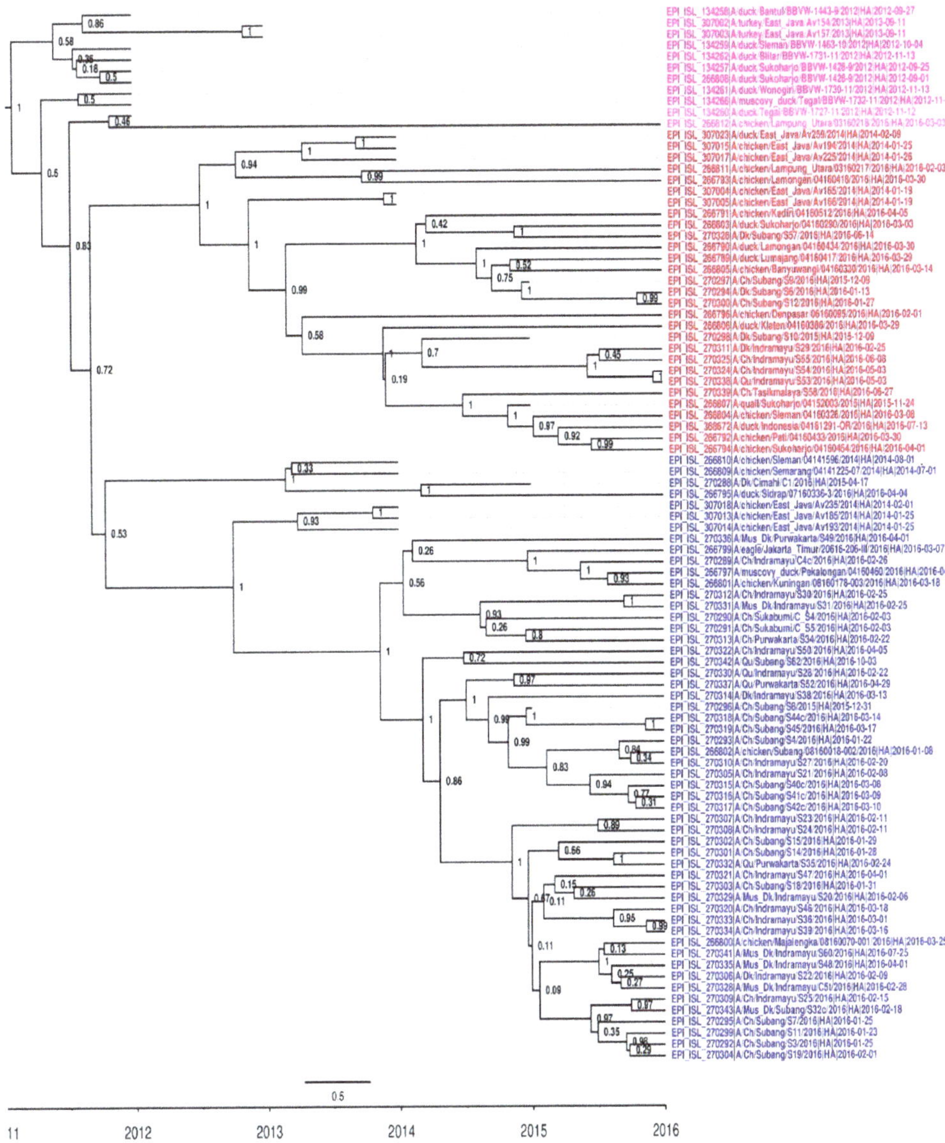

Figure A2. Time-scale phylogeny of Indonesian HA of H5N1v of HA genes of Indonesian poultry HPAIv A(H5N1) clade 2.3.2.1c. The clade 2.3.2.1c subgroup A is highlighted in red and clade 2.3.2.1c subgroup B is highlighted in blue. Another clade, 2.3.2.1c, is highlighted in pink. The node labels display the posterior value.

References

1. WHO. *Cumulative Number of Confirmed Human Cases for Avian Influenza A(H5N1) Reported to WHO, 2003–2022*; WHO: Geneva, Switzerland, 2022.
2. Smith, G.J.; Naipospos, T.; Nguyen, T.; de Jong, M.; Vijaykrishna, D.; Usman, T.; Hassan, S.; Dao, T.; Bui, N.; Leung, Y.; et al. Evolution and adaptation of H5N1 influenza virus in avian and human hosts in Indonesia and Vietnam. *Virology* **2006**, *350*, 258–268. [CrossRef]

3. Wibawa, H.; Karo-Karo, D.; Pribadi, E.S.; Bouma, A.; Bodewes, R.; Vernooij, H.; Sugama, A.; Muljono, D.H.; Koch, G. Exploring contacts facilitating transmission of influenza A(H5N1) virus between poultry farms in West Java, Indonesia: A major role for backyard farms? *Prev. Veter.-Med.* **2018**, *156*, 8–15. [CrossRef]
4. Henning, J.; Wibawa, H.; Morton, J.; Usman, T.B.; Junaidi, A.; Meers, J. Scavenging Ducks and Transmission of Highly Pathogenic Avian Influenza, Java, Indonesia. *Emerg. Infect. Dis.* **2010**, *16*, 1244–1250. [CrossRef]
5. Karo-Karo, D.; Diyantoro; Pribadi, E.S.; Sudirman, F.X.; Kurniasih, S.W.; Sukirman; Indasari, I.; Muljono, D.H.; Koch, G.; Stegeman, J.A. Highly Pathogenic Avian Influenza A(H5N1) Outbreaks in West Java Indonesia 2015–2016: Clinical Manifestation and Associated Risk Factors. *Microorganisms* **2019**, *7*, 327. [CrossRef]
6. Bouwstra, R.J.; Koch, G.; Heutink, R.; Harders, F.; van der Spek, A.; Elbers, A.R.; Bossers, A. Phylogenetic analysis of highly pathogenic avian influenza A(H5N8) virus outbreak strains provides evidence for four separate introductions and one between-poultry farm transmission in the Netherlands, November 2014. *Eurosurveillance* **2015**, *20*, 21174. [CrossRef] [PubMed]
7. Centre for Disease Control and Prevention, C. H5N1 Genetic Changes Inventory: A Tool for Influenza Surveillance and Preparedness. *Natl. Cent. Immun. Respir. Dis. Influenza Div.* **2012**, *1*, 1–8.
8. Suttie, A.; Deng, Y.-M.; Greenhill, A.R.; Dussart, P.; Horwood, P.F.; Karlsson, E.A. Inventory of molecular markers affecting biological characteristics of avian influenza A viruses. *Virus Genes* **2019**, *55*, 739–768. [CrossRef] [PubMed]
9. Beerens, N.; Heutink, R.; Harders, F.; Bossers, A.; Koch, G.; Peeters, B. Emergence and Selection of a Highly Pathogenic Avian Influenza H7N3 Virus. *J. Virol.* **2020**, *94*, e01818-19. [CrossRef]
10. Sitaras, I.; Kalthoff, D.; Beer, M.; Peeters, B.; de Jong, M. Immune Escape Mutants of Highly Pathogenic Avian Influenza H5N1 Selected Using Polyclonal Sera: Identification of Key Amino Acids in the HA Protein. *PLoS ONE* **2014**, *9*, e84628. [CrossRef] [PubMed]
11. Karo-karo, D.; Bodewes, R.; Wibawa, H.; Artika, I.M.; Pribadi, E.S.; Pramono, W.; Sugama, A.; Hendrayani, N.; Indasari, I. Reassortments Among Avian Influenza A(H5N1) Viruses Circulating in Indonesia, 2015–2016. *Emerg. Infect. Dis.-Cent. Dis. Control. Prev. (EID-CDC)* **2019**, *25*, 465. [CrossRef] [PubMed]
12. Beerens, N.; Heutink, R.; Bergervoet, S.A.; Harders, F.; Bossers, A.; Koch, G. Multiple Reassorted Viruses as Cause of Highly Pathogenic Avian Influenza A(H5N8) Virus Epidemic, the Netherlands, 2016. *Emerg. Infect. Dis.* **2017**, *23*, 1974–1981. [CrossRef] [PubMed]
13. Smith, G.J.D.; Donis, R.O. Nomenclature updates resulting from the evolution of avian influenza A(H5) virus clades 2.1.3.2a, 2.2.1, and 2.3.4 during 2013–2014. *Influ. Other Respir. Viruses* **2015**, *9*, 271–276. [CrossRef]
14. WHO/OIE/FAO/H5N1 Evolution Working Group. Toward a Unified Nomenclature System for Highly Pathogenic Avian Influenza Virus (H5N1). *Emerg. Infect. Dis.* **2008**, *14*, e1. [CrossRef] [PubMed]
15. WHO/OIE/FAO. H5N1 Evolution Working Group Revised and updated nomenclature for highly pathogenic avian influenza A (H5N1) viruses. *Influ. Other Respir. Viruses* **2014**, *8*, 384–388. [CrossRef]
16. Dharmayanti, N.L.P.I.; Hartawan, R.; Pudjiatmoko, H.; Wibawa, H.; Hardiman, A.; Al., N.D.E.; Donis, C.T.; Davis, C.T.; Samaan, G. Genetic Characterization of Clade 2.3.2.1 Avian Influenza A(H5N1) Viruses, Indonesia, 2012. *Emerg. Infect. Dis.* **2014**, *20*, 677–680. [CrossRef] [PubMed]
17. FAO-ECTAD. *Emergency Centre for Transboundary Animal Diseases Ectad Annual Report Laporan Tahunan 2014 Indonesia*; FAO-ECTAD: Jakarta, Indonesia, 2015.
18. Kwon, J.-H.; Lee, D.-H.; Criado, M.F.; Killmaster, L.; Ali, Z.; Giasuddin, M.; Samad, M.A.; Karim, R.; Hasan, M.; Brum, E.; et al. Genetic Evolution and Transmission Dynamics of Clade 2.3.2.1a Highly Pathogenic Avian Influenza A/H5N1 Viruses in Bangladesh. *Virus Evol.* **2021**, *7*, veaa096. [CrossRef] [PubMed]
19. WHO/OIE/FAO. H5N1 Evolution Working Group Continued evolution of highly pathogenic avian influenza A (H5N1): Updated nomenclature. *Influ. Other Respir. Viruses* **2011**, *6*, 1–5. [CrossRef]
20. Henning, J.; Henning, K.A.; Morton, J.M.; Long, N.T.; Ha, N.T.; Vu, L.T.; Vu, P.P.; Hoa, D.M.; Meers, J. Highly pathogenic avian influenza (H5N1) in ducks and in-contact chickens in backyard and smallholder commercial duck farms in Viet Nam. *Prev. Veter-Med.* **2011**, *101*, 229–240. [CrossRef]
21. Takano, R.; Nidom, C.A.; Kiso, M.; Muramoto, Y.; Yamada, S.; Sakai-Tagawa, Y.; Macken, C.; Kawaoka, Y. Phylogenetic characterization of H5N1 avian influenza viruses isolated in Indonesia from 2003–2007. *Virology* **2009**, *390*, 13–21. [CrossRef]
22. Lam, T.T.-Y.; Hon, C.-C.; Lemey, P.; Pybus, O.G.; Shi, M.; Tun, H.M.; Li, J.; Jiang, J.; Holmes, E.C.; Leung, F.C.-C. Phylodynamics of H5N1 avian influenza virus in Indonesia. *Mol. Ecol.* **2012**, *21*, 3062–3077. [CrossRef] [PubMed]
23. Zhang, Y.; Xu, Z.; Cao, Y. Host–Virus Interaction: How Host Cells Defend against Influenza A Virus Infection. *Viruses* **2020**, *12*, 376. [CrossRef] [PubMed]
24. Edgar, R.C. MUSCLE: A multiple sequence alignment method with reduced time and space complexity. *BMC Bioinform.* **2004**, *5*, 113. [CrossRef] [PubMed]
25. Kumar, S.; Stecher, G.; Tamura, K. MEGA7: Molecular Evolutionary Genetics Analysis Version 7.0 for Bigger Datasets. *Mol. Biol. Evol.* **2016**, *33*, 1870–1874. [CrossRef] [PubMed]
26. Huang, Y.; Niu, B.; Gao, Y.; Fu, L.; Li, W. CD-HIT Suite: A web server for clustering and comparing biological sequences. *Bioinformatics* **2010**, *26*, 680–682. [CrossRef] [PubMed]
27. Li, W.; Godzik, A. Cd-hit: A fast program for clustering and comparing large sets of protein or nucleotide sequences. *Bioinformatics* **2006**, *22*, 1658–1659. [CrossRef]

28. Li, W.; Jaroszewski, L.; Godzik, A. Clustering of highly homologous sequences to reduce the size of large protein databases. *Bioinformatics* **2001**, *17*, 282–283. [CrossRef]
29. Drummond, A.J.; Bouckaert, R.R. *Bayesian Evolutionary Analysis with BEAST*; Cambridge University Press: Cambridge, MA, USA, 2015.
30. Bouckaert, R.R.; Drummond, A.J. bModelTest: Bayesian phylogenetic site model averaging and model comparison. *BMC Evol. Biol.* **2017**, *17*, 1–11. [CrossRef] [PubMed]
31. Rambaut, A.; Drummond, A.J.; Xie, D.; Baele, G.; Suchard, M.A. Posterior Summarization in Bayesian Phylogenetics Using Tracer 1.7. *Syst. Biol.* **2018**, *67*, 901–904. [CrossRef] [PubMed]
32. Lam, T.T.-Y.; Hon, C.-C.; Pybus, O.G.; Pond, S.L.K.; Wong, R.T.-Y.; Yip, C.-W.; Zeng, F.; Leung, F.C.-C. Evolutionary and Transmission Dynamics of Reassortant H5N1 Influenza Virus in Indonesia. *PLoS Pathog.* **2008**, *4*, e1000130. [CrossRef] [PubMed]
33. Villanueva-Cabezas, J.P.; Coppo, M.J.; Durr, P.A.; McVernon, J. Vaccine efficacy against Indonesian Highly Pathogenic Avian Influenza H5N1: Systematic review and meta-analysis. *Vaccine* **2017**, *35*, 4859–4869. [CrossRef]
34. Poetri, O.N.; Van Boven, M.; Koch, G.; Stegeman, A.; Claassen, I.; Wisaksana, I.W.; Bouma, A. Different cross protection scopes of two avian influenza H5N1 vaccines against infection of layer chickens with a heterologous highly pathogenic virus. *Res. Veter-Sci.* **2017**, *114*, 143–152. [CrossRef]
35. Poetri, O.; Bouma, A.; Claassen, I.; Koch, G.; Soejoedono, R.; Stegeman, A.; van Boven, M. A single vaccination of commercial broilers does not reduce transmission of H5N1 highly pathogenic avian influenza. *Veter.-Res.* **2011**, *42*, 74. [CrossRef]
36. Fourment, M.; Holmes, E.C. Avian influenza virus exhibits distinct evolutionary dynamics in wild birds and poultry. *BMC Evol. Biol.* **2015**, *15*, 120. [CrossRef] [PubMed]
37. Cattoli, G.; Fusaro, A.; Monne, I.; Coven, F.; Joannis, T.; El-Hamid, H.S.A.; Hussein, A.A.; Cornelius, C.; Amarin, N.M.; Mancin, M.; et al. Evidence for differing evolutionary dynamics of A/H5N1 viruses among countries applying or not applying avian influenza vaccination in poultry. *Vaccine* **2011**, *29*, 9368–9375. [CrossRef]
38. Duffy, S.; Shackelton, L.A.; Holmes, E.C. Rates of evolutionary change in viruses: Patterns and. *Nat. Rev. Genet.* **2008**, *9*, 267–276. [CrossRef]
39. Chen, R.; Holmes, E. Avian Influenza Virus Exhibits Rapid Evolutionary Dynamics. *Mol. Biol. Evol.* **2006**, *23*, 2336–2341. [CrossRef]
40. Ducatez, M.; Olinger, C.M.; Owoade, A.A.; Tarnagda, Z.; Tahita, M.C.; Sow, A.; De Landtsheer, S.; Ammerlaan, W.; Ouedraogo, J.B.; Osterhaus, A.D.M.E.; et al. Molecular and antigenic evolution and geographical spread of H5N1 highly pathogenic avian influenza viruses in western Africa. *J. Gen. Virol.* **2007**, *88*, 2297–2306. [CrossRef]
41. Scholle, S.O.; Ypma, R.J.F.; Lloyd, A.L.; Koelle, K. Viral Substitution Rate Variation Can Arise from the Interplay between within-Host and Epidemiological Dynamics. *Am. Nat.* **2013**, *182*, 494–513. [CrossRef]
42. Tung, D.H.; Van Quyen, D.; Nguyen, T.; Xuan, H.T.; Nam, T.N.; Duy, K.D. Molecular characterization of a H5N1 highly pathogenic avian influenza virus clade 2.3.2.1b circulating in Vietnam in 2011. *Veter.-Microbiol.* **2013**, *165*, 341–348. [CrossRef]
43. USDA. *Compartmentalization for Protection against Avian Influenza and Newcastle Disease in Primary Poultry Breeding Companies in the United States of America*; USDA: Washington, DC, USA, 2020.
44. Zepeda, C. Surveillance and compartmentalisation as a tool to control avian influenza. *Dev. Biol.* **2006**, *124*, 163–169.
45. An, M.; Vitale, J.; Han, K.; Ng'Ombe, J.; Ji, I. Effects of Spatial Characteristics on the Spread of the Highly Pathogenic Avian Influenza (HPAI) in Korea. *Int. J. Environ. Res. Public Health* **2021**, *18*, 4081. [CrossRef]
46. Farnsworth, M.L.; Ward, M. Identifying spatio-temporal patterns of transboundary disease spread: Examples using avian influenza H5N1 outbreaks. *Veter.-Res.* **2009**, *40*, 1–14. [CrossRef] [PubMed]

viruses MDPI

Correction

Correction: Mase et al. Genetic Analysis of the Complete S1 Gene in Japanese Infectious Bronchitis Virus Strains. *Viruses* 2022, *14*, 716

Masaji Mase [1,2,3,*], Kanae Hiramatsu [4], Satoko Watanabe [1] and Hiroshi Iseki [1]

1 National Institute of Animal Health, National Agriculture and Food Research Organization, 3-1-5 Kannondai, Tsukuba 305-0856, Japan
2 United Graduate School of Veterinary Sciences, Gifu University, 1-1 Yanagido, Gifu 501-1193, Japan
3 Graduate School of Life and Environmental Sciences, Osaka Prefecture University, Izumisano 598-8531, Japan
4 Oita Livestock Hygiene Service Center of Oita Prefecture, 442 Onozuru, Oita 870-1153, Japan
* Correspondence: masema@affrc.go.jp; Tel.: +81-29-838-7738

Citation: Mase, M.; Hiramatsu, K.; Watanabe, S.; Iseki, H. Correction: Mase et al. Genetic Analysis of the Complete S1 Gene in Japanese Infectious Bronchitis Virus Strains. *Viruses* 2022, *14*, 716. *Viruses* **2022**, *14*, 2098. https://doi.org/10.3390/v14102098

Received: 14 September 2022
Accepted: 20 September 2022
Published: 22 September 2022

Publisher's Note: MDPI stays neutral with regard to jurisdictional claims in published maps and institutional affiliations.

Copyright: © 2022 by the authors. Licensee MDPI, Basel, Switzerland. This article is an open access article distributed under the terms and conditions of the Creative Commons Attribution (CC BY) license (https://creativecommons.org/licenses/by/4.0/).

Error in Table

In the original publication [1], there was a mistake in "**Table 2. List of RT-PCR primers**" as published. **The primers in Table 2 are not for infectious bronchitis virus, and the sequence of avian reovirus amplification is mistakenly listed**. The corrected Table 2 appears below. The authors state that the scientific conclusions are unaffected. This correction was approved by the Academic Editor. The original publication has also been updated.

Table 2. List of RT-PCR primers.

Name		Sequence (5′-3′)	Position [a]	Length (bp)	Reference
Forward	Reverse				
15F		AGGAATGGTAAGTTRCTRGTWAGAG	20343–20367	671	Mase et al., 2004 [11]
	26Rm	GCGCAGTACCRTTRAYAAAATAAGC	21013–20989		Mase et al., 2004 [11]
19F		GCAGTGTTTGTTACGCATTG	20689–20708	1333	in this study
	1R	CATAACTAACATAAGGGCAA	22021–22002		in this study

[a] Position is given for the S1 gene of strain Beadette42 (Acc.No.NC001451).

Reference

1. Mase, M.; Hiramatsu, K.; Watanabe, S.; Iseki, H. Genetic Analysis of the Complete S1 Gene in Japanese Infectious Bronchitis Virus Strains. *Viruses* **2022**, *14*, 716. [CrossRef]

viruses

Review

Pathogenicity of Avian Polyomaviruses and Prospect of Vaccine Development

Chen-Wei Wang [1,2], Yung-Liang Chen [3], Simon J. T. Mao [4], Tzu-Chieh Lin [1,2], Ching-Wen Wu [1,5], Duangsuda Thongchan [1,6], Chi-Young Wang [7,8,*] and Hung-Yi Wu [1,*]

1 Department of Veterinary Medicine, College of Veterinary Medicine, National Pingtung University of Science and Technology, Pingtung 912, Taiwan
2 International Degree Program in Animal Vaccine Technology, National Pingtung University of Science and Technology, Pingtung 912, Taiwan
3 Department of Medical Laboratory Science and Biotechnology, Yuan Pei University of Medical Technology, Yuanpei Street, Hsinchu 300, Taiwan
4 Department of Biological Science and Technology, National Chiao Tung University, Hsinchu 300, Taiwan
5 Department of Tropical Agriculture and International Cooperation, National Pingtung University of Science and Technology, Pingtung 912, Taiwan
6 Faculty of Agriculture and Technology, Rajamangala University of Technology Isan, Surin Campus, Nakhon Ratchasima 30000, Thailand
7 Department of Veterinary Medicine, College of Veterinary Medicine, National Chung Hsing University, Taichung 402, Taiwan
8 The iEGG and Animal Biotechnology Center, National Chung Hsing University, Taichung 402, Taiwan
* Correspondence: cyoungwang@dragon.nchu.edu.tw (C.-Y.W.); wuhy@g4e.npust.edu.tw (H.-Y.W.); Tel.: +886-4-22840369 (ext. 48) (C.-Y.W.); +886-8-7703202 (ext. 5072) (H.-Y.W.)

Abstract: Polyomaviruses are nonenveloped icosahedral viruses with a double-stranded circular DNA containing approximately 5000 bp and 5–6 open reading frames. In contrast to mammalian polyomaviruses (MPVs), avian polyomaviruses (APVs) exhibit high lethality and multipathogenicity, causing severe infections in birds without oncogenicity. APVs are classified into 10 major species: Adélie penguin polyomavirus, budgerigar fledgling disease virus, butcherbird polyomavirus, canary polyomavirus, cormorant polyomavirus, crow polyomavirus, *Erythrura gouldiae* polyomavirus, finch polyomavirus, goose hemorrhagic polyomavirus, and Hungarian finch polyomavirus under the genus *Gammapolyomavirus*. This paper briefly reviews the genomic structure and pathogenicity of the 10 species of APV and some of their differences in terms of virulence from MPVs. Each gene's genomic size, number of amino acid residues encoding each gene, and key biologic functions are discussed. The rationale for APV classification from the *Polyomavirdae* family and phylogenetic analyses among the 10 APVs are also discussed. The clinical symptoms in birds caused by APV infection are summarized. Finally, the strategies for developing an effective vaccine containing essential epitopes for preventing virus infection in birds are discussed. We hope that more effective and safe vaccines with diverse protection will be developed in the future to solve or alleviate the problems of viral infection.

Keywords: avian polyomavirus; *Gammapolyomavirus*; genomic structure; pathogenicity; vaccine

Citation: Wang, C.-W.; Chen, Y.-L.; Mao, S.J.T.; Lin, T.-C.; Wu, C.-W.; Thongchan, D.; Wang, C.-Y.; Wu, H.-Y. Pathogenicity of Avian Polyomaviruses and Prospect of Vaccine Development. *Viruses* **2022**, *14*, 2079. https://doi.org/10.3390/v14092079

Academic Editor: Feng Li

Received: 27 June 2022
Accepted: 13 September 2022
Published: 19 September 2022

Publisher's Note: MDPI stays neutral with regard to jurisdictional claims in published maps and institutional affiliations.

Copyright: © 2022 by the authors. Licensee MDPI, Basel, Switzerland. This article is an open access article distributed under the terms and conditions of the Creative Commons Attribution (CC BY) license (https://creativecommons.org/licenses/by/4.0/).

1. Introduction

Avian polyomaviruses (APVs) have a nonenveloped, icosahedral capsid with a diameter of 40–50 nm and encode a double-stranded circular DNA genome of approximately 5000 bp in length. Because budgerigar fledgling disease virus (BFDV) persistently infects many bird species globally, we used BFDV as a typical example (Figure 1) to describe the common genomic structure of avian polyomaviruses. The genome has early and late coding regions, with a noncoding regulatory control element located between the two regions for DNA replication and gene regulation (Figure 1, top). The early coding region encodes two proteins, large tumor antigen (large T-Ag) and small tumor antigen (small T-Ag), while the late coding region encodes structural proteins VP1, VP2, and VP3. Some polyomaviruses have a putative VP4 or a putative open reading frame (ORF) X that is located between the start and stop codons. In BFDV, the late coding region expresses VP4, and the VP4 ORF contains genes for VP4 and VP4 delta [1,2]. APVs have 5–6 ORFs depending on the species. The number of amino acid residues of putative ORF-X products may have functional similarities with the VP4 of BFDV. Using BFDV as a typical example, the physiologic properties of each gene are listed in Table 1.

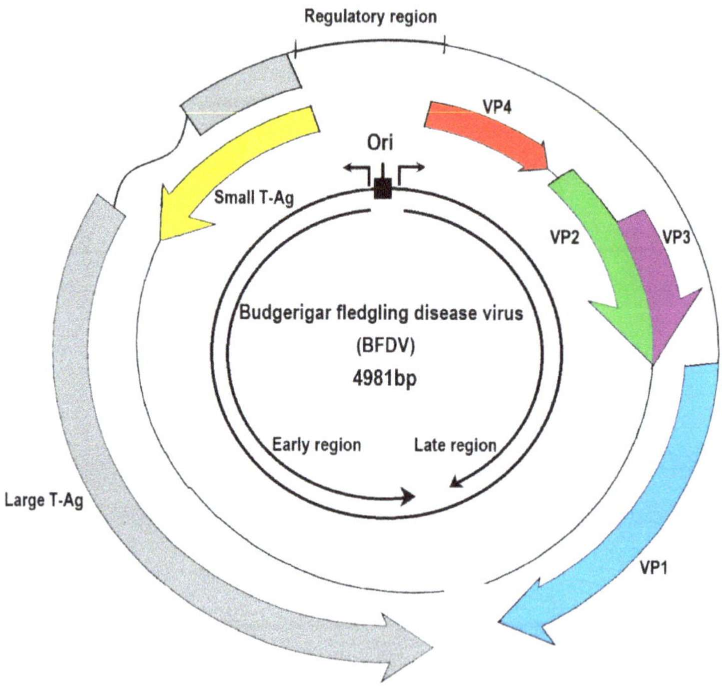

Figure 1. A genomic structure of avian polyomavirus using budgerigar fledgling disease virus as a typical example. In general, all polyomaviruses containing 5–6 open reading frames (ORFs) are composed of an early coding region, which encodes the large T-Ag (colored in gray) and small T-Ag (yellow), and a late region, which encodes structural proteins VP1 (blue), VP2 (green), and VP3 (violet). The VP4 (red) in the genome of the budgerigar fledgling disease virus is not always present in APVs. Some APVs (e.g., butcherbird polyomavirus, cormorant polyomavirus, crow polyomavirus, *Erythrura gouldiae* polyomavirus, finch polyomavirus, goose hemorrhagic polyomavirus, and Hungarian finch polyomavirus) have a putative VP4 or ORF-X protein. The key biological role of each gene is given in Table 1.

Table 1. Name of the genes with respect to their nucleotide size, amino acid residues, and functions of Avian Polyomaviruses.

	Nucleotide Size (bp)	Amino Acid Residues	Functions [Reference]
Gene: gp5 Protein: Large T-Ag	1995 (1995–2166) *	599 (599–660) *	Interacts with the tandem repeat sequences in the noncoding control region to regulate DNA replication and RNA transcription [3,4]

Table 1. *Cont.*

	Nucleotide Size (bp)	Amino Acid Residues	Functions [Reference]
Gene: gp6 Protein: Small T-Ag	483 (483–537)	145 (145–178)	The large T-Ag from the same ATG codon shares an N-terminal domain [5] and is involved in cell transformation and tumorigenicity [6,7]
Gene: gp4 Protein: VP1	1032 (1032–1083)	343 (343–360)	VP1, a capsid protein, binds to the host cell receptor for infection and forms a pentamer for its stability [8,9], which is further reinforced with a single copy of VP2 and VP3.
Gene: gp2 Protein: VP2	1026 (981–1110)	341 (331–369)	
Gene: gp3 Protein: VP3	708 (654–738)	235 (217–245)	
Gene: gp1 Protein: VP4	675 (485–755) except canary polyomavirus and Adélie penguin polyomavirus	176 (112–205) except canary polyomavirus and Adélie penguin polyomavirus	Suppresses immune responses, induces apoptosis, and increases in pathogenicity [1,10,11] and scaffolding function during virion assembly [1].
Gene: gp1 Protein: VP4d	675	112	VP4 delta (deleted a.a. 69–132 in VP4) contains a leucine zipper-like motif [10]. Induces cell apoptosis and affects releases of viral progeny [1].

* BFDV is used as a typical example. Numbers in parenthesis indicates the ranges in other nine APVs.

In 2015, the International Committee on Taxonomy of Viruses conducted a comprehensive revision of polyomavirus taxonomy in order to clearly define virus species and created a nomenclature using the combination of host name and viral genomic sequence. Even though the virus name is used to highlight the host specificity, there are few APVs including BFDV, finch polyomavirus (FPyV), and goose hemorrhagic polyomavirus (GHPyV) that can infect many other birds [12–16].

In 2017, based on gene differences among various viral proteins, the International Committee proposed using the relatively conserved large T-Ag amino acid sequence as a base for the classification of polyomavirus genus and species [12,17]. The family *Polyomaviridae* is classified into six genera. All the APVs mentioned in this review under the genus *Gammapolyomavirus* and their classification with the MPVs are illustrated in Figure 2.

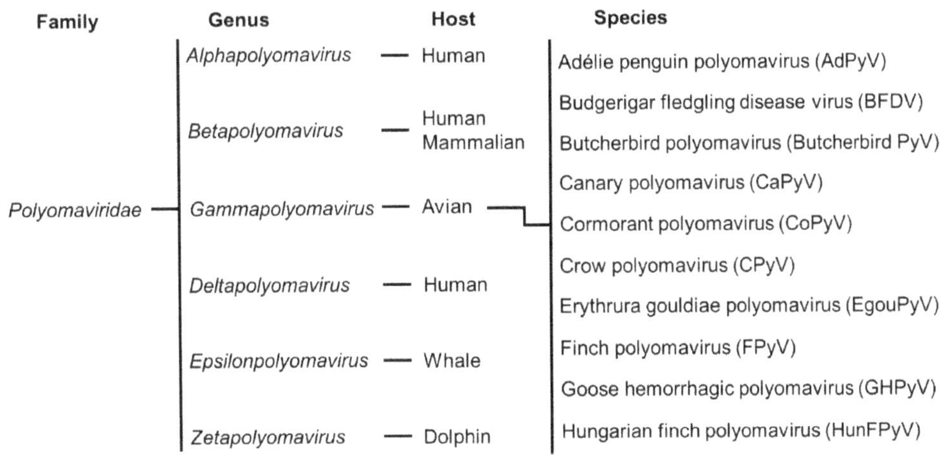

Figure 2. Classification of APVs under the genus *Gammapolyomavirus*. The *Polyomaviridae* family is generally divided into six genera. There are 10 APVs that belong to the genus *Gammapolyomavirus*.

The *Polyomaviridae* family is generally divided into six genera: *Alphapolyomavirus*, *Betapolyomavirus*, *Gammapolyomavirus*, *Deltapolyomavirus*, *Epsilonpolyomavirus*, and *Zetapolyomavirus* [12,18]. Different polyomavirus genera usually infect different types of animals with some host specificity: alpha, beta, and delta polyomaviruses mostly infect humans and other mammals, while gamma, epsilon, and zeta polyomaviruses infect birds, whales, and dolphins, respectively. Further, the study by Burck et al. (2016) revealed that some polyomavirus-like sequences detected by the shot-gun approach were found in several fish species, but they are not classified as polyomaviruses [19].

Under the genus *Gammapolyomaviruses*, there are 10 APVs officially identified as Adélie penguin polyomavirus (AdPyV), BFDV, butcherbird polyomavirus (Butcherbird PyV), canary polyomavirus (CaPyV), cormorant polyomavirus (CoPyV), crow polyomavirus (CPyV), *Erythrura gouldiae* polyomavirus (EgouPyV), FPyV, GHPyV, and Hungarian finch polyomavirus (HunFPyV) [2,17,20–22]. The nucleotide size and the number of amino acid residues corresponding to each gene and its potential biological role using BFDV as a typical example are listed in Table 1.

2. Major Pathogenic Difference between Avian and Mammalian Polyomaviruses

MPVs often cause subclinical or asymptomatic infection in their natural hosts; some symptoms appear only in severely immunocompromised hosts. Interestingly, these viruses can live in their hosts for life [23]. Compared to MPVs, APVs have unique biological characteristics, including a broad host range, substantial pathogenicity, and a preference for tissue tropism [24]. In addition, VP4 is relatively important in APV infection and is thought to be associated with viral genome packaging and cell apoptosis in cell cultures (Table 1). A previous study showed that deleting or inactivating VP4 attenuates BFDV virulence [1,10]. Another reason MPVs, such as JC polyomavirus, are less or not lethal in terms of pathogenicity is that they may induce the expression of interferon (IFN)-α and IFN-β (two components responsible for host innate immunity) in various human primary cells, thereby inhibiting their own replication [25,26]. On the contrary, in an ex vivo study, it showed that APV replicated efficiently in chicken embryo fibroblast cells. It also showed that BFDV proteins (VP4 and VP4 delta) could directly bind to the IFN-β promoter and downregulate the expression of IFN-β, allowing the virus to replicate in host cells. It provides the molecular mechanism by which BFDV can escape the host's innate immunity [27].

3. Clinical Signs and Disease Progression of Avian Polyomavirus Infection

An overview of the 10 APV species with respect to their hosts and the clinical symptoms that they cause is summarized in Table 2.

Table 2. Overview of avian polyomaviruses with their hosts and clinical symptoms developed.

Virus	Host	Clinical Symptoms [Reference]
Adélie penguin polyomavirus (AdPyV)	Adélie penguins (*Pygoscelis adeliae*)	Feather loss [28]
Butcherbird polyomavirus (Butcherbird PyV)	Grey butcherbird (*Cracticus torquatus*)	No apparent signs were observed except for periocular nodule growth. It is unclear whether these symptoms are directly associated with Butcherbird PyV [29]
Budgerigar fledgling disease virus (BFDV)	Parrots, chickens, vultures, falcons, canaries, ostriches, pigeons, ducks, geese, finches, gulls, common ravens, pheasants, Eurasian jays, and starlings	Development of hepatitis, ascites, pericardial effusion, and abdominal distension [30,31]; abnormal feather growth often occurs in adult and chronically infected parrots

Table 2. *Cont.*

Virus	Host	Clinical Symptoms [Reference]
Canary polyomavirus (CaPyV)	Canaries (*Serinus canaria*)	Subcutaneous bleeding; patosplenomegaly; extensive centrilobular degeneration with hemorrhage in the liver; splenic depletion; or polyomavirus-like intranuclear inclusion bodies in the liver and spleen, occasionally found in the epithelium of some renal glomeruli [32]
Cormorant polyomavirus (CoPyV)	Great cormorant (*Phalacrocorax carbo*)	Detection in the liver, but detailed pathological findings are not available [21]
Crow polyomavirus (CPyV)	Western jackdaws (*Corvus monedula)*	Enteritis and death with Salmonella co-infection [2]
Erythrura gouldiae polyomavirus (EgouPyV)	Gouldian finch (*Erythrura gouldiae*)	Detection in the liver, but detailed pathological findings are not available [20]
Finch polyomavirus (FPyV)	European goldfinches, grey-headed bullfinches, and Gouldian finches	Liver necrosis, membranous nephropathy, liver and renal cell nuclei hypertrophy, large amounts of transparent-to-basophilic intranuclear inclusion bodies, with no inclusion bodies in periarteriolar lymphoid sheaths [13,14]
Goose hemorrhagic polyomavirus (GHPyV)	Geese and ducks	Subcutaneous edema, ascites, renal pallor, and swelling; gastrointestinal hemorrhage in some cases [15,33–35]; in acute cases, hemorrhagic necrosis in the small intestine, with endothelial and vascular wall necrosis in the proventriculus, gizzard, and intestines; inflammatory cell infiltration, with mucosal bleeding in gizzard; in subacute and chronic disease cases, gout [35,36], renal and intestinal vascular necrosis, bleeding, and edema [15,33,35] are seen; distal renal tubule necrosis is more severe than proximal tubule necrosis, resulting in urate deposition, calcification, and fibrosis [36,37]; ducks are considered asymptomatic carriers [16,38]
Hungarian finch polyomavirus (HunFPyV)	White-headed munia (*Lonchura maja*)	Liver failure, nephritis, and myocarditis [22]

In 1981, BFDV was reported in budgerigars and identified as the pathogen responsible for causing BFD. The genomic structure of BFDV is depicted in Table 1 [1,2]. The signs of BFDV infection symptoms include hepatitis, ascites, pericardial effusion, and abdominal distension [30,31]. Adult and chronically infected parrots often exhibit abnormal feather growth [39]. The virus is shed in feces for up to 6 months after infection [40]. In addition, budgerigars are suspected to be a bird species that can be persistently infected with this virus and existed for life. However, another study showed that the progeny of serum-positive adult budgerigars housed in a completely clean and BFDV-free environment for 7 months before breeding are clinically normal without the presence of antibodies [40]. BFDV has no strict host specificity and can infect many birds, including chicken, vulture, falcon, canary, ostrich, pigeon, duck, goose, finch, gull, common raven, pheasant, Eurasian jay, and starling [30,41,42]. According to the literature, BFDV infection has been observed globally in parrots in Canada, Australia, Germany, the US, Switzerland, Slovakia, Italy, Japan, Taiwan, China, Thailand, Costa Rica, Turkey, Pakistan, Peru, and Chile [43–52]. Alignment with viral genomic sequences from Asia, North America, Europe, and Australia revealed that BFDV is widespread globally. The lack of host specificity is concerning as it may both transmit and contribute to the virus being widely spread. In addition, the possible routes of BFDV transmission, including bird migration, free trade in poultry, and bird transportation, have been highlighted and discussed by Kou et al. [49].

GHPyV is the pathogen responsible for hemorrhagic nephritis and enteritis in geese, which has a prevalence rate of 4–67% in 3–10-week-old geese [15,34]. A study conducted in 2004 found that neurological symptoms appeared in 1-day-old goslings after GHPyV inoculation, with a 100% mortality rate [35]. The full-length genomic DNA of this virus is approximately 5252–5256 bp [53]. The disease in geese was first reported in Hungary in 1969, and the main symptoms included subcutaneous edema, ascites, renal pallor, and swelling, with occasional gastrointestinal hemorrhage [15]. Histological examinations commonly reveal renal and intestinal vascular necrosis, bleeding, and edema [15]. In acute cases, hemorrhagic necrosis is observed in the small intestine as well as in the endothelial and vascular walls of blood vessels. It is often accompanied by vasculitis with bleeding and inflammatory cell infiltration as well as occasionally by mucosal bleeding in the proventriculus and gizzard. Necrotizing vasculitis is observed in subacute or chronic cases. In term of renal lesions, distal renal tubule necrosis is more severe than proximal tubule necrosis. Uric acid deposition, calcification, and fibrosis occur as the disease progresses are responsible for the gout formation. These symptoms are more severe in subacute and chronic cases than in acute cases [35–37]. In addition, in situ hybridization results revealed that GHPyV is primarily found in the nucleus and cytoplasm of endothelial cells in systemic arteries, veins, and capillaries, but it is absent in cells from other tissues. This also explains why GHPyV infection results in anemia, subcutaneous edema, ascites, or intestinal bleeding. Furthermore, it damages vascular endothelial cells, resulting in vascular dysfunction, thereby increasing vascular permeability and causing edema, hematoma, vessel rupture, and bleeding [37,54]. Although it was previously thought that GHPyV caused lymphoid depletion in the spleen and bursa of Fabricius, the in situ hybridization conducted in 2021 suggested that GHPyV is not present in the lymphocytes of these organs [35,36].

Initially, GHPyV was predominantly found in Europe, with the clinical cases of hemorrhagic nephritis enteritis associated with geese [35]. In 2008, it was found that GHPyV could infect ducks [19]. Currently, it is believed that geese that have survived GHPyV infection can continuously transmit this virus, while ducks are asymptomatic carriers [30,34,36,38,55]. Subsequently, in 2018, GHPyV was identified in ducks across China via polymerase chain reaction; however, it was not identified in geese [56]. In 2021, the first case of this disease was reported in Taiwan from geese outside of Europe [37]. Furthermore, GHPyV has been suspected of being able to infect geese indefinitely [55].

The genomic sequence of FPyV from Eurasian bullfinches isolated in 2006 revealed that it is remarkably different from BFDV and was thus named finch polyomavirus [2]. The FPyV genome is 5278-bp long [2]. Subsequently, identical or similar FPyV sequences were reported in European goldfinches [57], grey-headed bullfinches [58], and Gouldian finches [13,22]. Earlier research had identified polyomavirus infections in various finches; however, there was no way of determining whether these infections were caused by FPyV or BFDV. Examples include canaries, European goldfinches, European greenfinches, and other Fringillidae birds [59–61], as well as Gouldian finches, seed crackers, western bluebills, tricolored munias, cordon-bleu finches, and painted finches [14,47,62]. Infected finches developed typical lesions, such as liver necrosis, membranous nephropathy, and liver and renal cell nuclei hypertrophy, in which a large number of transparent-to-basophilic intranuclear inclusion bodies were found but not in periarteriolar lymphoid sheaths [63].

In 2006, a study on the clinical outcomes of CPyV was conducted [2]. Spleen samples of dead wild western jackdaws collected in northeast Spain in early 2005 were found to carry this virus. The genome of this virus is 5079-bp long [2]. Clinically, it was speculated that the death of jackdaws from CPyV was mainly associated with Salmonella co-infection [2].

CaPyV was discovered in 2010 as a result of a decrease in egg laying and hatching rates of canaries. The offspring of the animal were found dead at 40 days of age, with a 50% mortality rate. The genomic DNA of this virus is 5421-bp, and the most critical difference between CaPyV and other APVs is that CaPyV does not express VP4. VP4 promotes apoptosis in host cells and accelerates the infection rate [1]. It appears to be controversial because CaPyV lacking VP4 can still cause gross lesions, including subcutaneous bleeding and hepatosplenomegaly, in host animals. In the liver and spleen, there was extensive centrilobular degeneration with hemorrhage and intranuclear inclusion bodies. Some inclusion bodies were found in the renal glomerular epithelium [32].

Butcherbird PyV is 5084-bp long. The clinical symptoms of Butcherbird PyV differ from those of typical APVs. The source of the virus was accidentally discovered in wild adult grey butcherbirds with no apparent signs other than periocular nodule growth. It is unclear whether these symptoms in the periocular nodule were related to Butcherbird PyV; the animal recovered after the nodule was resected [29].

AdPyV was discovered in the Cape Crozier colony on Ross Island in Antarctica in 2014, primarily causing feather loss in Adélie penguins [28]. Its genomic DNA is 4988-bp in length [28]. Varsani et al. found that AdPyV has a direct relationship with Adélie penguins in that south polar skuas prey on Adélie penguin eggs and fledglings; therefore, south polar skuas may carry the polyomavirus and spread it in the environment of the South Pole [64].

EgouPyV was discovered in Gouldian finches (*Erythrura gouldiae*) and reported in 2015. The genome of the virus consists of 5172-bp, and its nucleotide sequence has the highest homology (71%) with CPyV but only 41% with Butcherbird PyV. The associated clinical signs have not yet been thoroughly reported [20].

HunFPyV was first discovered in white-headed munia (*Lonchura maja*). The virus genome is 5284-bp long and shares 91% homology with FPyV. In EgouPyV (the other FPyV), the values were less than 65%. Clinically, liver failure, nephritis, and myocarditis are associated with the affected birds [22].

Most recently, in 2022, Feher et al. reported a novel polyomavirus called cormorant polyomavirus (CoPyV) in great cormorants (*Phalacrocorax carbo*) [21]. The DNA is 5133-bp in length and has a genome similar to that of GHPyV (Figure 3). There were no detailed clinical signs reported [21].

(A)

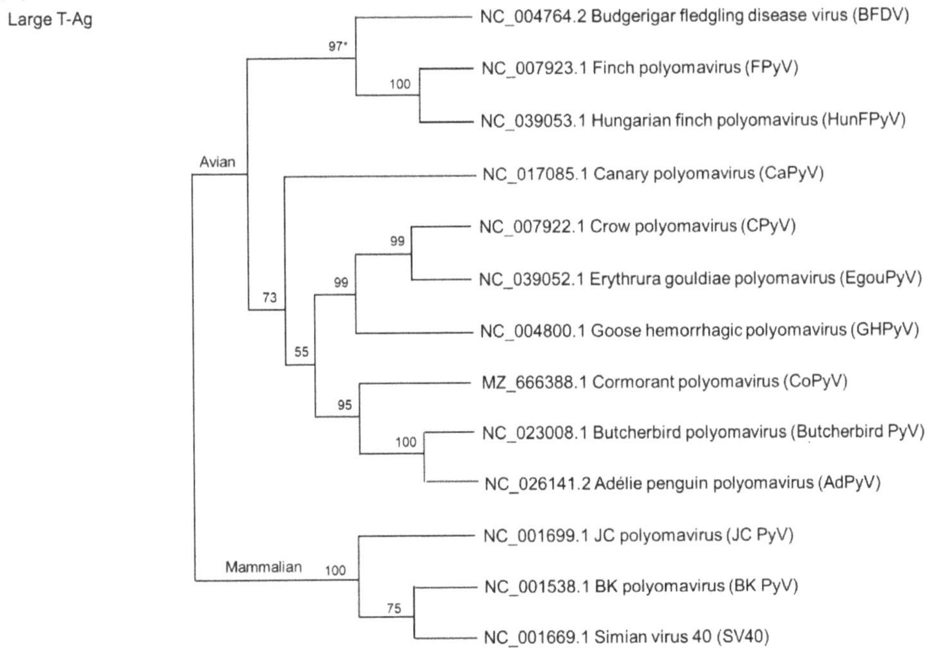

Figure 3. *Cont.*

(B)

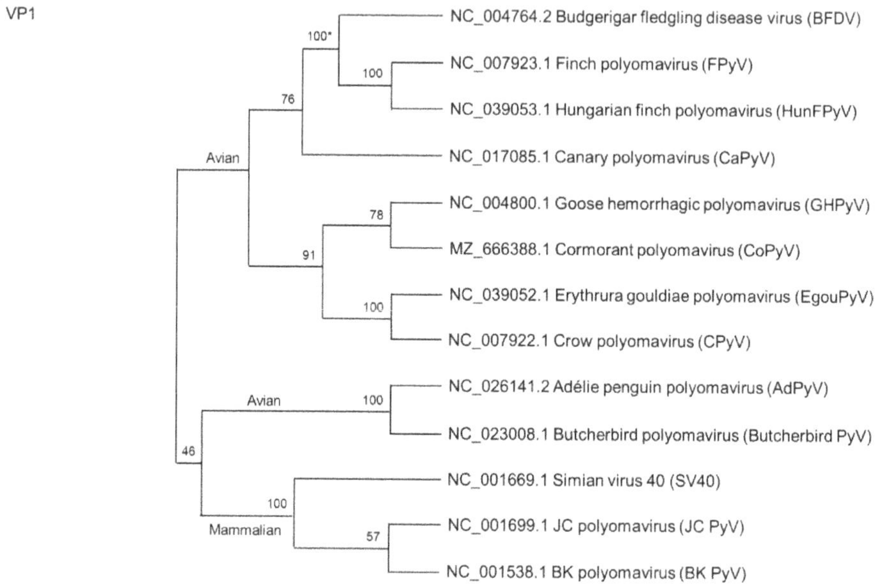

Figure 3. Phylogenetic analyses using amino acid sequences of the large T-Ag (**A**) and VP1 (**B**) from the *Gammapolyomaviruses*. The phylogeny, which uses the large T-Ag, can distinguish between APVs and MPVs. Interestingly, using VP1 for the analysis, Butcherbird PyV and AdPyV show the closest relationship with MPVs on a common branch. Pathogenically, their clinical symptoms are mild and there is no mortality (Table 2) [28,29]. Virus names and GenBank accession numbers are shown on the phylogenetic tree. * The bootstrap value on the branch represents the percentage while using 1000 bootstrap replicates.

4. Some Molecular Characteristics Involved in the Pathogenicity of Avian Polyomaviruses

As shown in Figure 1, the polyomavirus genome is divided into two regions: the early region, which encodes the large T-Ag and small T-Ag, and the late region, which encodes the structural proteins VP1, VP2, VP3, and VP4; a noncoding control element is present between these two regions [1,2]. The noncoding control region contains a bidirectional replication domain with an element for binding virus and host cell transcriptional factors to regulate the expression of early and late genes of the virus [65]. In all BFDV genomic sequences, two sets of the unique palindromic tandem repeats (TCC$(A/T)_6$GGG/A) are present in the noncoding control region, while NCC$(A/T)_6$GGN is present in GHPyV, CPyV, FPyV, and CaPyV [2,66]. Further studies on the palindromic sequence of BFDV reveal that this tandem repeat may bind to the large T-Ag. The nuclear localization sequence in large T-Ags could specifically bind to the replication region in the noncoding control region. In BFDV and FPyV, nuclear localization sequences contain both lysine and arginine residues, whereas in GHPyV, CPyV, and CaPyV, they contain only lysine. Insertion, deletion, and point mutations in the noncoding control region are thought to be one of the main and critical mechanisms by which polyomaviruses change their adaptive behavior, such as host jumping [67]. In short, the main function of large T-Ag in APVs is to interact with tandem repeat sequences in the noncoding control region to regulate DNA replication and RNA transcription for the virus [3,4,68].

BFDV has been proven to infect many bird species. These animals can also be simultaneously co-infected by other APVs [30,41,42]. Related studies found that one MPV species could recombine with other MPV species in the host animal. For example, a

2017 study demonstrated that African bat mammalian polyomavirus might recombine in the bat. This recombination is usually associated with hotspots that are expressed for viral capsid proteins [67]. As mentioned above, the polyomavirus VP1 is involved in host cell recognition [69], and thus recombination could alter the specificity for host transmission. In 2017, Carr et al. showed that recombination events in polyomavirus evolution require a long period of time [67]. As recombination hotspots include the domains of capsid proteins (e.g., VP1), it may increase in changing the host preference and in the probability of species jumps. It may also make the APV infection more complex and concerning.

In MPV, the large T-Ag has functionally conserved domains, such as the DnaJ-, LXCXE, nuclear localization sequence, helicase, and p53-binding domain, which are found in most polyomaviruses [70]. One study reported that these domains can bind to tumor suppressors in the host (p53 and Rb), thereby promoting cell cycle changes to enhance tumorigenesis [71]. In contrast, these mechanisms are rarely observed in APV. APVs induce apoptosis in host cells during infection, and this mechanism is thought to effectively spread progeny viruses, resulting in different virulence [11]. Analysis of the complete sequence of DnaJ and helicase domains of the large T-Ag gene revealed that these domains play important roles in the biochemical function of the large T-Ag and that these domains are relatively stable among different polyomaviruses compared to other domains [72].

The small T-Ag shares an amino-terminal domain with the large T-Ag because both proteins originate from the same ATG codon [5]. Because the gene sequences of large T-Ag and small T-Ag overlap to some extent (Figure 1), the two proteins share some functional similarities. The small T-Ag in MPV also exhibits cell transformation potential in tissue culture and is tumorigenic in experimental animals [6,7]. In 2022, a study on the sequence analyses of all available APV isolates in Taiwan found that the large T-Ag was the most conserved protein, whereas the small T-Ag was the least conserved [66,73].

VP1 is a major capsid protein whose primary function is to bind to sialylated glycan receptors of host cells. In virus-like particles of Simian virus 40, VP1 forms a pentamer and is stabilized by calcium ions and disulfide bonds [74,75]. In the cavity of the pentamer, a single copy of VP2 or VP3 binds to VP1 in a hairpin-like manner [76–79]. VP1 recognizes host cells primarily through its specific determinant, of which glutamic acid residue at 92 is the most important for host cell recognition. It plays an essential role in viral tropism, transmission, and cell targeting [69]. VP1 requires a 12-amino-acid fragment to bind to a calcium ion to promote capsid formation. This sequence in BFDV is 237-DENGVGPLCKGD-248 and differs from that of MPV by only one amino acid residue [80,81]. The study showed that VP4 of BFDV and ORF-X of GHPyV have the highest variation in APVs [66,82]. The VP4 and VP4 delta proteins suppress host immune responses by inducing apoptosis. Variations in these proteins may thus affect the pathogenicity of BFDV [30,66,73]. The VP4 protein is enriched with basic amino acids located in the central region between residues 70 and 77 to mediate DNA binding but not for that of VP4 delta. In addition, VP4 may stabilize itself via dimerization before binding with DNA [22].

5. Phylogenetic Relationship among Avian Polyomaviruses

In 2021, Kaszab et al. conducted a comprehensive study on APV evolution. They calculated the evolutionary rates of APV in substitutions/nucleotide site/year (s/s/y) [82]. The mean evolutionary rate of BFDV is 1.39×10^{-4} s/s/y ($7.18 \times 10^{-5} - 2.10 \times 10^{-4}$ s/s/y), which is 10 times faster than that of GHPyV (mean 3.03×10^{-5} s/s/y; $1.09 \times 10^{-5} - 5.33 \times 10^{-5}$ s/s/y) but similar to those of FPyV (mean 2.63×10^{-4} s/s/y; $1.60 \times 10^{-8} - 6.26 \times 10^{-4}$ s/s/y) and CaPyV (mean 1.41×10^{-4} s/s/y; 5.75×10^{-10}–7.17×10^{-4} s/s/y) [82].

Using large T-Ag and VP1 as a base, we performed the phylogenetic analyses using Molecular Evolutionary Genetics Analysis version 11 (MEGA, https://www.megasoftware.net in assessed date 1 July 2022) in which the Clustal W method was used for amino acid sequence alignment, while the maximum likelihood method was employed with the LG model for large T-Ag and VP1 [83]. The sequence of each GenBank accession number is indicated in Figure 3.

As shown in Figure 3A, we used the large T-Ag amino acid sequences from 10 APVs as a model to analyze their phylogenetic relationship because the sequences are conserved and have been utilized as the key protein for APV classification [12]. As expected, there is a significant distance in their relationship between the avian and mammalian polyomaviruses, which is consistent with previous views on their structural differences [23,24]. As shown in Figure 3B, we used the VP1 protein for the phylogenetic analyses. It showed a unique result. The Butcherbird PyV and AdPyV have a closer relationship with three MPVs than with other APVs. Currently, it is unclear whether the sequence similarity identified in the phylogenetic relationship can be used as a reference to determine some partial similarity of pathogenicity among the virus species. Additional studies with detailed pathogenicity among the species are needed because of the limited studies conducted thus far.

6. Historical Development of Vaccines for Avian Polyomaviruses

In terms of vaccine development, some studies on BFDV have been conducted [84–87]. In 1996, an inactivated vaccine against BFDV was developed using a BFDV isolate [84]. The vaccine was safe, capable of neutralizing viruses, and protected the host animals from infection [84]. In 1998, a study was conducted to further examine the safety, neutralizing capability, and efficacy of the viral vaccine in parrots (non-budgerigar) of different ages and species. The results revealed that the vaccine effectively prevents BFDV from infection [88].

The vaccine containing inactivated BFDV was effective, but there were no subsequent follow-ups to determine the long-term efficacy in terms of host immunity [88]. Because Kaszab et al. showed that the mean evolutionary rate of BFDV is 1.39×10^{-4} s/s/y, with the VP1–VP3 genes evolving at the fastest rates, the protection efficacy of previously developed vaccines would be potentially attenuated with the emergence of new infected strains [82]. In addition, this vaccine was mainly prepared using primary chicken embryo fibroblast in cell culture, which is time-consuming, costly, and requires advanced techniques for production.

Compared to poultry, parrots are more prone to stress and exhibit highly aggressive behavior, making subcutaneous injection difficult. Injections may be feasible for personal pets, but it is extremely difficult to inject a large number of parrots on a farm. Under this situation, it would be better to use alternative inoculation methods (such as intranasal administration, nebulization, and oral feeding via drinking water) with inactivated viral particles. Therefore, attenuation of virulence or deletion of major virulent determinants by creating a live nonlethal virus is needed. For example, fiber-2 was deleted from highly pathogenic fowl adenovirus serotype 4 in 2022 to protect against this fatal viral infection [89]. Another strategy has been tested in which a nonlethal virus was inserted with a lethal viral structure protein. A vaccine against highly lethal avian bornavirus was developed using recombinant virus to protect parrots from Newcastle disease virus [90–92].

An early GHPyV vaccine-related study revealed that this virus cannot proliferate in cell culture. The viral vaccine was then prepared using the recombinant technique to express viral protein in yeast or Sf9 insect cells [85,93]. Effective antibodies prepared against specific epitopes of GHPyV were successfully raised in geese, demonstrating that this vaccine can induce antibodies in birds [85]. In a recent article, GHPyV production has been successfully propagated in cultured primary goose embryo fibroblasts, showing the availability of information regarding the gene expression profile [87]. This should allow for the future preparation for the GHPyV vaccine when needed.

Another study was conducted using GHPyV vaccines prepared from a recombinant virus expressing the VP1 gene under the control of the polyhedron promoter in insect cells. After two vaccinations at days 1 and 18, the survival rate was 100% against a challenge at week 5. [85]. In 2010, Gelfi et al. found that using $Al(OH)_3$ as an adjuvant for vaccine preparation resulted in poor immunogenicity, whereas the vaccine preparation using acrylate polymer carbopol mixed with β-propiolactone-inactivated GHPyV was safe and immunogenic [84]. Thus, the adjuvant also plays a crucial role in the development of an effective vaccine. Interestingly, this vaccine also induces the transfer of maternally derived neutralizing antibodies to goslings, which can still be detected 15 days after hatching, indicating that the vaccine effectively protects goslings from GHPyV infection [86]. Although low-cost Sf9 insect cells were used to express the VP1 protein [85], the overall cost of producing vaccines is still high. Therefore, it is difficult to commercialize these two vaccines for farm animals. The lack of commercially available vaccines may explain why many GHPyV cases continue to occur in Europe [33,34,53,87].

The virus-like particles are commonly used in the development of single-virus vaccines [85,93–97]. Regardless of vaccine cost, the capsid protein VP1 is successfully expressed in eukaryotic systems in sufficient quantities for vaccine preparation [98]. The development of virus-like particles of APV began in 2002, when Sasnauskas et al. used insect and yeast cells to express recombinant proteins of BFDV, thereby encouraging the preparation of the VP1 subunit as a vaccine for GHPyV in 2006 [93,98]. It also revealed that, with the exception of FPyV, the receptor-binding mechanisms of BFDV, GHPyV, and CPyV are similar [94]. We anticipate that effective polyvalent vaccines will be developed once the molecular structure of different APV species is better understood.

It has been further hypothesized that some of the receptors, such as the sialylated glycan receptor, have specifically evolved within polyomaviruses. Because the binding mechanisms for VP1 are similar across APV species, there is some immunochemical cross-reactivity [2,15,99]. This could be used to look for the VP1 protein of APVs as well as other receptor-binding domains for the epitope-based vaccine [100,101]. Polypeptides containing multiple APV epitopes may thus provide comprehensive protection compared to subunit fragments alone. It adds a new dimension to vaccine preparation, making it more flexible and effective. For example, Wang et al. (2017) used a multi-variant epitope ensemble vaccine to raise antibodies against an avian leucosis virus subgroup J [100].

7. Some Prospects with Respect to the Development of Vaccines Utilizing the Epitope of Receptor-Binding Domain

It makes sense to develop a vaccine that generates antibodies capable of directly inactivating or blocking the viral receptor-binding domain involved in the recognition between the virus and host cells. This functionally based epitope would be an excellent target for vaccine development. However, using monoclonal antibodies prepared against human low-density lipoprotein (LDL), Mao et al. first demonstrated that the binding of one specific monoclonal antibody to LDL can drastically enhance the binding of another specific monoclonal antibody via cooperative binding. This enhancement is not additive but synergistic, with a 10-fold increase in binding to LDL [102,103]. It has also been discovered that a single monoclonal antibody can even precipitate LDL by altering its conformation, presumably by exposing the hydrophobic region of the apolipoprotein of LDL [104]. Notably, LDL, one of the largest proteins in human plasma, has a diameter of approximately 30 nm, which is comparable to those of avian polyomaviruses (40–50 nm). The current life-threatening pandemic virus, SARS-CoV-2, is approximately 120 nm in size, which is four times larger than LDL. Because of the constant mutations in SARS-CoV-2 [105], it is clear that a new strategy for developing a new generation of vaccines using additional epitopes other than those epitopes involved in the receptor-binding domain is required. Some considerations are as follows:

1. One monoclonal antibody can enhance the binding of another monoclonal antibody via cooperative binding, as proposed by Mao et al. (1983) [102,103]. This should not be limited to two antigenic domains because each epitope is relatively small, containing only approximately seven amino acid residues [106].
2. The selection of a specific region of a viral protein for vaccine preparation should not only be made against a functional domain, but other structural proteins with "constant" regions or regions with minimal mutations should also be considered as a target to stabilize the antigen–antibody interaction. Other region(s) that will cause a drastic conformational change in the virus structure could also be considered as

targets, such as precipitating antibodies found against LDL [104] that could directly immobilize viral activity.

3. Finally, because the genomes of the 10 major avian polyomaviruses are relatively small, containing only six genes, it is possible to prepare a "cocktail" multivalent vaccine to produce "APV-capturing antibodies" within a single preparation.

8. Conclusions

APVs have a broad host range appearing to be almost distributed worldwide and may be highly pathogenic (Table 2). Although there is currently no evidence that an APV will recombine with another APV species, awareness must be maintained because APVs cause high mortality in birds. Using the VP1 protein sequence for phylogenetic analyses, both Butcherbird PyV and AdPyV of APVs and MPVs have low pathogenicity in terms of developing clinical symptoms and they are closely related on the phylogenetic tree (Table 2). It is unknown whether these two avian species will cause cross-species infections between avians and mammals. Nevertheless, the Newcastle disease virus found in fowls has been confirmed to be transmissible to humans, causing fatal encephalitis in a child [107].

Migration of birds is not the only route through which viral infections can spread globally. The overlap among bird transportation, intensive farming, the existence of wild animals, human routine travel activities, and genome exchange may increase the chances of viral infection and pandemics. Therefore, preventive measures must be implemented to prevent viruses from spreading among susceptible animal hosts [82]. In addition, breeder biosafety assays, pathogen testing, and effective quarantine measures can also be involved in preventing infection by APVs [108].

Vaccines still play an indispensable and reliable role in the prevention of viral infection. In the APV vaccine study, the traditional, subunit, and inactivated vaccines and virus-like particles were developed. However, owing to the recent outbreak of SARS-CoV-2 (CoV-19 infection), the most recently developed mRNA-based vaccine has received much attention and has demonstrated remarkable efficacy in humans to protect against coronavirus infection [105].

Regardless of the techniques already developed for APV vaccine preparation, an mRNA-based vaccine could be another option in APV vaccine preparation because it is convenient and safe to synthesize without causing too much biohazard. The information described in Section 6 is also essential for providing the basis and strategy for future mRNA-based vaccine designs. They can also provide the specific protein motif containing the epitope in the receptor recognition determinants without using tissue culture to prepare recombinant protein [109]. The desired synthetic mRNA may not be restricted to the epitope located within the host recognition site but may also include other epitopes within structural proteins or other functional proteins that can enhance the binding of antibodies via cooperative binding [101,103] or immobilize the virus via precipitation [104]. Although it has not been applied to animals, it provides us with the opportunity to utilize such technology and experience in the future development of bird vaccines containing the key domain with virus recognition [109] and the domains undergoing drastic conformational changes upon binding of specific antibodies from a given host [102–104]. If the overall cost is reasonable, it could also produce a cocktail vaccine by combining with several desired mRNAs corresponding to several avian polyomaviruses. Furthermore, mRNA can be designed to trigger specific T cell or B cell recognition in order to optimize the immune response in a short time [110,111]. We hope that more effective and safe vaccines with diverse protections will be developed in the future to alleviate the problem of viral infection.

Author Contributions: Author Contributions: conceptualization, C.-W.W. (Chen-Wei Wang) and H.-Y.W.; methodology, C.-W.W. (Chen-Wei Wang) and H.-Y.W.; software, T.-C.L.; validation, C.-W.W. (Ching-Wen Wu) and D.T.; investigation, Y.-L.C.; resources, C.-W.W. (Chen-Wei Wang); data curation, C.-W.W. (Chen-Wei Wang); writing—original draft preparation, C.-W.W. (Chen-Wei Wang) and

Y.-L.C.; writing—review and editing, S.J.T.M., C.-W.W. (Chen-Wei Wang), and C.-Y.W.; visualization, T.-C.L. All authors have read and agreed to the published version of the manuscript.

Funding: This work was funded by the grant from the iEGG and Animal Biotechnology Center from the Feature Areas Research Center Program within the framework of the Higher Education Sprout Project by the Ministry of Education in Taiwan. This work including Chen-Wei Wang is supported by the National Science and Technology Council (110-26926-I020-001-MY4) and National Pingtung University of Science and Technology (NPUST) in Taiwan.

Institutional Review Board Statement: Not applicable.

Informed Consent Statement: Not applicable.

Data Availability Statement: All the data are summarized from the references listed.

Conflicts of Interest: The authors declare no conflict of interest.

References

1. Johne, R.; Paul, G.; Enderlein, D.; Stahl, T.; Grund, C.; Muller, H. Avian polyomavirus mutants with deletions in the VP4-encoding region show deficiencies in capsid assembly and virus release, and have reduced infectivity in chicken. *J. Gen. Virol.* **2007**, *88 Pt 3*, 823–830. [CrossRef] [PubMed]
2. Johne, R.; Wittig, W.; Fernandez-de-Luco, D.; Hofle, U.; Muller, H. Characterization of two novel polyomaviruses of birds by using multiply primed rolling-circle amplification of their genomes. *J. Virol.* **2006**, *80*, 3523–3531. [CrossRef] [PubMed]
3. Luo, D.; Muller, H.; Tang, X.B.; Hobom, G. Expression and DNA binding of budgerigar fledgling disease virus large T antigen. *J. Gen. Virol.* **1994**, *75 Pt 6*, 1267–1280. [CrossRef] [PubMed]
4. Rott, O.; Kroger, M.; Muller, H.; Hobom, G. The genome of budgerigar fledgling disease virus, an avian polyomavirus. *Virology* **1988**, *165*, 74–86. Available online: https://www.ncbi.nlm.nih.gov/pubmed/2838972 (accessed on 15 January 2022). [CrossRef]
5. Pipas, J.M. Common and unique features of T antigens encoded by the polyomavirus group. *J. Virol.* **1992**, *66*, 3979–3985. [CrossRef] [PubMed]
6. Butel, J.S.; Lednicky, J.A. Cell and molecular biology of simian virus 40: Implications for human infections and disease. *J. Natl. Cancer Inst.* **1999**, *91*, 119–134. [CrossRef]
7. Gordon, J.; Del Valle, L.; Otte, J.; Khalili, K. Pituitary neoplasia induced by expression of human neurotropic polyomavirus, JCV, early genome in transgenic mice. *Oncogene* **2000**, *19*, 4840–4846. [CrossRef]
8. Chen, X.S.; Stehle, T.; Harrison, S.C. Interaction of polyomavirus internal protein VP2 with the major capsid protein VP1 and implications for participation of VP2 in viral entry. *EMBO J.* **1998**, *17*, 3233–3240. [CrossRef]
9. Li, P.P.; Nakanishi, A.; Tran, M.A.; Salazar, A.M.; Liddington, R.C.; Kasamatsu, H. Role of simian virus 40 Vp1 cysteines in virion infectivity. *J. Virol.* **2000**, *74*, 11388–11393. [CrossRef]
10. Johne, R.; Muller, H. Avian polyomavirus agnoprotein 1a is incorporated into the virus particle as a fourth structural protein, VP4. *J. Gen. Virol.* **2001**, *82 Pt 4*, 909–918. [CrossRef]
11. Johne, R.; Jungmann, A.; Muller, H. Agnoprotein 1a and agnoprotein 1b of avian polyomavirus are apoptotic inducers. *J. Gen. Virol.* **2000**, *81 Pt 5*, 1183–1190. [CrossRef]
12. Calvignac-Spencer, S.; Feltkamp, M.C.; Daugherty, M.D.; Moens, U.; Ramqvist, T.; Johne, R.; Ehlers, B. A taxonomy update for the family Polyomaviridae. *Arch. Virol.* **2016**, *161*, 1739–1750. [CrossRef]
13. Alley, M.R.; Rasiah, I.; Lee, E.A.; Howe, L.; Gartrell, B.D. Avian polyomavirus identified in a nestling Gouldian finch (*Erythrura gouldiae*) in New Zealand. *N. Z. Vet. J.* **2013**, *61*, 359–361. [CrossRef] [PubMed]
14. Rossi, G.; Ceccherelli, R.; Piersigilli, A.; Tarantino, C. Sertoli cell tumor associated with polyomavirus infection in a Gouldian finch (*Erythrura gouldiae*). *Avian Dis.* **2003**, *47*, 240–243. [CrossRef]
15. Guerin, J.L.; Gelfi, J.; Dubois, L.; Vuillaume, A.; Boucraut-Baralon, C.; Pingret, J.L. A novel polyomavirus (goose hemorrhagic polyomavirus) is the agent of hemorrhagic nephritis enteritis of geese. *J. Virol.* **2000**, *74*, 4523–4529. [CrossRef]
16. Pingret, J.L.; Boucraut-Baralon, C.; Guerin, J.L. Goose haemorrhagic polyomavirus infection in ducks. *Vet. Rec.* **2008**, *162*, 164. [CrossRef]
17. Moens, U.; Calvignac-Spencer, S.; Lauber, C.; Ramqvist, T.; Feltkamp, M.C.W.; Daugherty, M.D.; Verschoor, E.J.; Ehlers, B.; Ictv Report, C. ICTV Virus Taxonomy Profile: Polyomaviridae. *J. Gen. Virol.* **2017**, *98*, 1159–1160. [CrossRef] [PubMed]
18. Walker, P.J.; Siddell, S.G.; Lefkowitz, E.J.; Mushegian, A.R.; Adriaenssens, E.M.; Dempsey, D.M.; Dutilh, B.E.; Harrach, B.; Harrison, R.L.; Hendrickson, R.C.; et al. Changes to virus taxonomy and the Statutes ratified by the International Committee on Taxonomy of Viruses (2020). *Arch. Virol.* **2020**, *165*, 2737–2748. [CrossRef] [PubMed]
19. Buck, C.B.; Van Doorslaer, K.; Peretti, A.; Geoghegan, E.M.; Tisza, M.J.; An, P.; Katz, J.P.; Pipas, J.M.; McBride, A.A.; Camus, A.C.; et al. The Ancient Evolutionary History of Polyomaviruses. *PLoS Pathog.* **2016**, *12*, e1005574. [CrossRef]
20. Heenemann, K.; Sieg, M.; Rueckner, A.; Vahlenkamp, T.W. Complete Genome Sequence of a Novel Avian Polyomavirus Isolated from Gouldian Finch. *Genome Announc.* **2015**, *3*, e01001-15. [CrossRef]

21. Feher, E.; Kaszab, E.; Bali, K.; Hoitsy, M.; Sos, E.; Banyai, K. A novel gammapolyomavirus in a great cormorant (*Phalacrocorax carbo*). *Arch. Virol.* **2022**, *167*, 1721–1724. [CrossRef] [PubMed]
22. Marton, S.; Erdelyi, K.; Dan, A.; Banyai, K.; Feher, E. Complete Genome Sequence of a Variant Pyrrhula pyrrhula polyomavirus 1 Strain Isolated from White-Headed Munia (*Lonchura maja*). *Genome Announc.* **2016**, *4*, e01172-16. [CrossRef] [PubMed]
23. Jiang, M.; Abend, J.R.; Johnson, S.F.; Imperiale, M.J. The role of polyomaviruses in human disease. *Virology* **2009**, *384*, 266–273. [CrossRef] [PubMed]
24. Johne, R.; Buck, C.B.; Allander, T.; Atwood, W.J.; Garcea, R.L.; Imperiale, M.J.; Major, E.O.; Ramqvist, T.; Norkin, L.C. Taxonomical developments in the family Polyomaviridae. *Arch. Virol.* **2011**, *156*, 1627–1634. [CrossRef] [PubMed]
25. Assetta, B.; De Cecco, M.; O'Hara, B.; Atwood, W.J. JC Polyomavirus Infection of Primary Human Renal Epithelial Cells Is Controlled by a Type I IFN-Induced Response. *mBio* **2016**, *7*, e00903-16. [CrossRef] [PubMed]
26. Co, J.K.; Verma, S.; Gurjav, U.; Sumibcay, L.; Nerurkar, V.R. Interferon-alpha and-Beta restrict polyomavirus JC replication in primary human fetal glial cells: Implications for progressive multifocal leukoencephalopathy therapy. *J. Infect. Dis.* **2007**, *196*, 712–718. [CrossRef]
27. Ma, J.; Wu, R.; Tian, Y.; Zhang, M.; Wang, W.; Li, Y.; Tian, F.; Cheng, Y.; Yan, Y.; Sun, J. Isolation and characterization of an Aves polyomavirus 1 from diseased budgerigars in China. *Vet. Microbiol.* **2019**, *237*, 108397. [CrossRef]
28. Varsani, A.; Kraberger, S.; Jennings, S.; Porzig, E.L.; Julian, L.; Massaro, M.; Pollard, A.; Ballard, G.; Ainley, D.G. A novel papillomavirus in Adelie penguin (*Pygoscelis adeliae*) faeces sampled at the Cape Crozier colony, Antarctica. *J. Gen. Virol.* **2014**, *95 Pt 6*, 1352–1365. [CrossRef]
29. Bennett, M.D.; Gillett, A. Butcherbird polyomavirus isolated from a grey butcherbird (*Cracticus torquatus*) in Queensland, Australia. *Vet. Microbiol.* **2014**, *168*, 302–311. [CrossRef]
30. Johne, R.; Muller, H. Avian polymavirus in wild birds: Genome analysis of isolates from Falconiformes and Psittaciformes. *Arch. Virol.* **1998**, *143*, 1501–1512. [CrossRef]
31. Krautwald, M.E.; Muller, H.; Kaleta, E.F. Polyomavirus infection in budgerigars (*Melopsittacus undulatus*): Clinical and aetiological studies. *J. Vet. Med. Ser. B* **1989**, *36*, 459–467. [CrossRef] [PubMed]
32. Halami, M.Y.; Dorrestein, G.M.; Couteel, P.; Heckel, G.; Muller, H.; Johne, R. Whole-genome characterization of a novel polyomavirus detected in fatally diseased canary birds. *J. Gen. Virol.* **2010**, *91 Pt 12*, 3016–3022. [CrossRef] [PubMed]
33. Gawel, A.; Wozniakowski, G.; Samorek-Salamonowicz, E.; Kozdrun, W.; Bobrek, K.; Bobusia, K.; Nowak, M. Hemorrhagic nephritis and enteritis in a goose flock in Poland–disease course analysis and characterization of etiologic agent. *Avian Dis.* **2014**, *58*, 518–522. [CrossRef] [PubMed]
34. Garmyn, A.; Verlinden, M.; Bosseler, L.; Adriaensen, C.; Martel, A. Persistent Goose Hemorrhagic Polyomavirus Infection on a Belgian Goose Farm. *Avian Dis.* **2017**, *61*, 536–538. [CrossRef]
35. Lacroux, C.; Andreoletti, O.; Payre, B.; Pingret, J.L.; Dissais, A.; Guerin, J.L. Pathology of spontaneous and experimental infections by Goose haemorrhagic polyomavirus. *Avian Pathol.* **2004**, *33*, 351–358. [CrossRef]
36. Palya, V.; Ivanics, E.; Glavits, R.; Dan, A.; Mato, T.; Zarka, P. Epizootic occurrence of haemorrhagic nephritis enteritis virus infection of geese. *Avian Pathol.* **2004**, *33*, 244–250. [CrossRef]
37. Tu, Y.C.; Li, W.T.; Lee, F.; Huang, C.W.; Chang, J.C.; Hsu, W.C.; Hu, S.C.; Chiou, C.J.; Chen, Y.P. Localization of goose haemorrhagic polyomavirus in naturally infected geese using in situ hybridization. *Avian Pathol.* **2021**, *50*, 41–51. [CrossRef]
38. Corrand, L.; Gelfi, J.; Albaric, O.; Etievant, M.; Pingret, J.L.; Guerin, J.L. Pathological and epidemiological significance of goose haemorrhagic polyomavirus infection in ducks. *Avian Pathol.* **2011**, *40*, 355–360. [CrossRef]
39. Davis, R.B.; Bozeman, L.H.; Gaudry, D.; Fletcher, O.J.; Lukert, P.D.; Dykstra, M.J. A viral disease of fledgling budgerigars. *Avian Dis.* **1981**, *25*, 179–183. Available online: https://www.ncbi.nlm.nih.gov/pubmed/7271654 (accessed on 15 January 2022). [CrossRef]
40. Phalen, D.N.; Wilson, B.G.; Graham, D.L. Production of avian polyomavirus seronegative budgerigars (*Melopsittacus undulatus*) from seropositive adults. *Avian Dis.* **1995**, *39*, 897–899. Available online: https://www.ncbi.nlm.nih.gov/pubmed/8719226 (accessed on 15 January 2022). [CrossRef]
41. Stoll, R.; Luo, D.; Kouwenhoven, B.; Hobom, G.; Muller, H. Molecular and biological characteristics of avian polyomaviruses: Isolates from different species of birds indicate that avian polyomaviruses form a distinct subgenus within the polyomavirus genus. *J. Gen. Virol.* **1993**, *74 Pt 2*, 229–237. [CrossRef] [PubMed]
42. Li, Q.C.; Niu, K.; Sun, H.J.; Xia, Y.J.; Sun, S.J.; Li, J.; Wang, F.; Feng, Y.; Peng, X.W.; Zhu, L.Q.; et al. Complete Genome Sequence of an Avian Polyomavirus Strain First Isolated from a Pigeon in China. *Microbiol. Resour. Ann.* **2019**, *8*, e01490-18. [CrossRef] [PubMed]
43. Bert, E.; Tomassone, L.; Peccati, C.; Navarrete, M.G.; Sola, S.C. Detection of beak and feather disease virus (BFDV) and avian polyomavirus (APV) DNA in psittacine birds in Italy. *J. Vet. Med. Ser. B* **2005**, *52*, 64–68. [CrossRef] [PubMed]
44. Hsu, C.M.; Ko, C.Y.; Tsaia, H.J. Detection and sequence analysis of avian polyomavirus and psittacine beak and feather disease virus from psittacine birds in Taiwan. *Avian Dis.* **2006**, *50*, 348–353. [CrossRef]
45. Pass, D.A. A papova-like virus infection of lovebirds (*Agapornis* sp.). *Aust. Vet. J.* **1985**, *62*, 318–319. [CrossRef]
46. Latimer, K.S.; Niagro, F.D.; Steffens, W.L., 3rd; Ritchie, B.W.; Campagnoli, R.P. Polyomavirus encephalopathy in a Ducorps' cockatoo (*Cacatua ducorpsii*) with psittacine beak and feather disease. *J. Vet. Diagn. Investig.* **1996**, *8*, 291–295. [CrossRef]

47. Sandmeier, P.; Gerlach, H.; Johne, R.; Muller, H. Polyomavirus infections in exotic birds in Switzerland. *Schweiz. Arch. Tierheilkd.* **1999**, *141*, 223–229. Available online: https://www.ncbi.nlm.nih.gov/pubmed/10354740 (accessed on 15 January 2022).
48. Literak, I.; Smid, B.; Dubska, L.; Bryndza, L.; Valicek, L. An outbreak of the polyomavirus infection in budgerigars and cockatiels in Slovakia, including a genome analysis of an avian polyomavirus isolate. *Avian Dis.* **2006**, *50*, 120–123. [CrossRef]
49. Kou, Z.; Zhang, Z.; Chen, S.; Fan, Z.; Tang, S.; Zhao, L.; Li, T. Molecular characterizations of avian polyomavirus isolated from budgerigar in China. *Avian Dis.* **2008**, *52*, 451–454. [CrossRef]
50. Gilardi, K.V.; Lowenstine, L.J.; Gilardi, J.D.; Munn, C.A. A survey for selected viral, chlamydial, and parasitic diseases in wild dusky-headed parakeets (*Aratinga weddellii*) and tui parakeets (*Brotogeris sanctithomae*) in Peru. *J. Wildl. Dis.* **1995**, *31*, 523–528. [CrossRef]
51. Riaz, A.; Yousaf, A.; Moaeen-Ud-Din, M.; Shah, M.A.A.; Zainab, T.; Masood, S.; Akhter, N.; Ali, A. First detection and molecular characterization of avian polyomavirus in young parrots in Pakistan. *Vet. Res. Commun.* **2019**, *43*, 197–202. [CrossRef] [PubMed]
52. Ogawa, H.; Chahota, R.; Hagino, T.; Ohya, K.; Yamaguchi, T.; Fukushi, H. A survey of avian polyomavirus (APV) infection in imported and domestic bred psittacine birds in Japan. *J. Vet. Med. Sci.* **2006**, *68*, 743–745. [CrossRef] [PubMed]
53. Feher, E.; Lengyel, G.; Dan, A.; Farkas, S.L.; Banyai, K. Whole genome sequence of a goose haemorrhagic polyomavirus detected in Hungary. *Acta Microbiol. Immunol. Hung.* **2014**, *61*, 221–227. [CrossRef] [PubMed]
54. Dobos-Kovacs, M.; Horvath, E.; Farsang, A.; Nagy, E.; Kovacs, A.; Szalai, F.; Bernath, S. Haemorrhagic nephritis and enteritis of geese: Pathomorphological investigations and proposed pathogenesis. *Acta Vet. Hung.* **2005**, *53*, 213–223. [CrossRef]
55. Kaszab, E.; Marton, S.; Dan, A.; Farsang, A.; Balint, A.; Banyai, K.; Feher, E. Molecular epidemiology and phylodynamics of goose haemorrhagic polyomavirus. *Transbound. Emerg. Dis.* **2020**, *67*, 2602–2608. [CrossRef]
56. Wan, C.; Chen, C.; Cheng, L.; Liu, R.; Fu, G.; Shi, S.; Chen, H.; Fu, Q.; Huang, Y. Genomic analysis of Sheldrake origin goose hemorrhagic polyomavirus, China. *J. Vet. Sci.* **2018**, *19*, 782–787. [CrossRef]
57. Manarolla, G.; Liandris, E.; Pisoni, G.; Moroni, P.; Piccinini, R.; Rampin, T. Mycobacterium genavense and avian polyomavirus co-infection in a European goldfinch (*Carduelis carduelis*). *Avian Pathol.* **2007**, *36*, 423–426. [CrossRef]
58. Wittig, W.; Hoffmann, K.; Muller, H.; Johne, R. Detection of DNA of the finch polyomavirus in diseases of various types of birds in the order Passeriformes. *Berl. Münch. Tierärztl. Wochenschr.* **2007**, *120*, 113–119. Available online: https://www.ncbi.nlm.nih.gov/pubmed/17416133 (accessed on 15 January 2022).
59. Sironi, G. Concurrent papovavirus-like and atoxoplasma infections in a goldfinch (*Carduelis carduelis*). *Avian Pathol.* **1991**, *20*, 725–729. [CrossRef]
60. Fitzgerald, S.D.; Williams, S.M.; Reed, W.M. Development of a chicken model for studying avian polyomavirus infection. *Avian Dis.* **1996**, *40*, 377–381. Available online: https://www.ncbi.nlm.nih.gov/pubmed/8790889 (accessed on 15 January 2022). [CrossRef]
61. Shivaprasad, H.L.; Kim, T.; Tripathy, D.; Woolcock, P.R.; Uzal, F. Unusual pathology of canary poxvirus infection associated with high mortality in young and adult breeder canaries (*Serinus canaria*). *Avian Pathol.* **2009**, *38*, 311–316. [CrossRef] [PubMed]
62. Garcia, A.; Latimer, K.S.; Niagro, F.D.; Norton, T.M.; Campagnoli, R.P.; Harmon, B.G.; Howerth, E.W.; Ritchie, B.W. Diagnosis of polyomavirus infection in seedcrackers (*Pyrenestes* sp.) and blue bills (*Spermophaga haematina*) using DNA in situ hybridization. *Avian Pathol.* **1994**, *23*, 525–537. [CrossRef] [PubMed]
63. Rinder, M.; Schmitz, A.; Peschel, A.; Moser, K.; Korbel, R. Identification and genetic characterization of polyomaviruses in estrildid and fringillid finches. *Arch. Virol.* **2018**, *163*, 895–909. [CrossRef]
64. Varsani, A.; Porzig, E.L.; Jennings, S.; Kraberger, S.; Farkas, K.; Julian, L.; Massaro, M.; Ballard, G.; Ainley, D.G. Identification of an avian polyomavirus associated with Adelie penguins (*Pygoscelis adeliae*). *J. Gen. Virol.* **2015**, *96 Pt 4*, 851–857. [CrossRef]
65. DeCaprio, J.A.; Garcea, R.L. A cornucopia of human polyomaviruses. *Nat. Rev. Microbiol.* **2013**, *11*, 264–276. [CrossRef] [PubMed]
66. Liu, F.L.; Chang, S.P.; Liu, H.J.; Liu, P.C.; Wang, C.Y. Genomic and phylogenetic analysis of avian polyomaviruses isolated from parrots in Taiwan. *Virus Res.* **2022**, *308*, 198634. [CrossRef]
67. Carr, M.; Gonzalez, G.; Sasaki, M.; Ito, K.; Ishii, A.; Hang'ombe, B.M.; Mweene, A.S.; Orba, Y.; Sawa, H. Discovery of African bat polyomaviruses and infrequent recombination in the large T antigen in the Polyomaviridae. *J. Gen. Virol.* **2017**, *98*, 726–738. [CrossRef]
68. Hu, X.; Cai, D.; Liu, S.; Li, Y.; Chen, L.; Luo, G.; Pu, H.; He, Y.; Liu, X.; Zhao, L.; et al. Molecular Characterization of a Novel Budgerigar Fledgling Disease Virus Strain From Budgerigars in China. *Front. Vet. Sci.* **2021**, *8*, 813397. [CrossRef]
69. Freund, R.; Garcea, R.L.; Sahli, R.; Benjamin, T.L. A single-amino-acid substitution in polyomavirus VP1 correlates with plaque size and hemagglutination behavior. *J. Virol.* **1991**, *65*, 350–355. [CrossRef]
70. Ahuja, D.; Saenz-Robles, M.T.; Pipas, J.M. SV40 large T antigen targets multiple cellular pathways to elicit cellular transformation. *Oncogene* **2005**, *24*, 7729–7745. [CrossRef]
71. Borchert, S.; Czech-Sioli, M.; Neumann, F.; Schmidt, C.; Wimmer, P.; Dobner, T.; Grundhoff, A.; Fischer, N. High-affinity Rb binding, p53 inhibition, subcellular localization, and transformation by wild-type or tumor-derived shortened Merkel cell polyomavirus large T antigens. *J. Virol.* **2014**, *88*, 3144–3160. [CrossRef] [PubMed]
72. Cho, M.; Kim, H.; Son, H.S. Codon usage patterns of LT-Ag genes in polyomaviruses from different host species. *Virol. J.* **2019**, *16*, 137. [CrossRef] [PubMed]
73. Katoh, H.; Ohya, K.; Une, Y.; Yamaguchi, T.; Fukushi, H. Molecular characterization of avian polyomavirus isolated from psittacine birds based on the whole genome sequence analysis. *Vet. Microbiol.* **2009**, *138*, 69–77. [CrossRef] [PubMed]

74. Ben-nun-Shaul, O.; Bronfeld, H.; Reshef, D.; Schueler-Furman, O.; Oppenheim, A. The SV40 capsid is stabilized by a conserved pentapeptide hinge of the major capsid protein VP1. *J. Mol. Biol.* **2009**, *386*, 1382–1391. [CrossRef]
75. Liddington, R.C.; Yan, Y.; Moulai, J.; Sahli, R.; Benjamin, T.L.; Harrison, S.C. Structure of simian virus 40 at 3.8-A resolution. *Nature* **1991**, *354*, 278–284. [CrossRef]
76. Barouch, D.H.; Harrison, S.C. Interactions among the major and minor coat proteins of polyomavirus. *J. Virol.* **1994**, *68*, 3982–3989. [CrossRef]
77. Griffith, J.P.; Griffith, D.L.; Rayment, I.; Murakami, W.T.; Caspar, D.L. Inside polyomavirus at 25-A resolution. *Nature* **1992**, *355*, 652–654. [CrossRef]
78. Hurdiss, D.L.; Morgan, E.L.; Thompson, R.F.; Prescott, E.L.; Panou, M.M.; Macdonald, A.; Ranson, N.A. New Structural Insights into the Genome and Minor Capsid Proteins of BK Polyomavirus using Cryo-Electron Microscopy. *Structure* **2016**, *24*, 528–536. [CrossRef]
79. Sapp, M.; Day, P.M. Structure, attachment and entry of polyoma- and papillomaviruses. *Virology* **2009**, *384*, 400–409. [CrossRef]
80. Rodgers, R.E.; Consigli, R.A. Characterization of a calcium binding domain in the VP1 protein of the avian polyomavirus, budgerigar fledgling disease virus. *Virus Res.* **1996**, *44*, 123–135. [CrossRef]
81. Haynes, J.I., 2nd; Chang, D.; Consigli, R.A. Mutations in the putative calcium-binding domain of polyomavirus VP1 affect capsid assembly. *J. Virol.* **1993**, *67*, 2486–2495. [CrossRef] [PubMed]
82. Kaszab, E.; Marton, S.; Erdelyi, K.; Banyai, K.; Feher, E. Genomic evolution of avian polyomaviruses with a focus on budgerigar fledgling disease virus. *Infect. Genet. Evol.* **2021**, *90*, 104762. [CrossRef] [PubMed]
83. Kumar, S.; Stecher, G.; Li, M.; Knyaz, C.; Tamura, K. MEGA X: Molecular Evolutionary Genetics Analysis across Computing Platforms. *Mol. Biol. Evol.* **2018**, *35*, 1547–1549. [CrossRef]
84. Ritchie, B.W.; Niagro, F.D.; Latimer, K.S.; Pritchard, N.; Campagnoli, R.P.; Lukert, P.D. An inactivated avian polyomavirus vaccine is safe and immunogenic in various Psittaciformes. *Vaccine* **1996**, *14*, 1103–1107. [CrossRef]
85. Mato, T.; Penzes, Z.; Rueda, P.; Vela, C.; Kardi, V.; Zolnai, A.; Misak, F.; Palya, V. Recombinant subunit vaccine elicits protection against goose haemorrhagic nephritis and enteritis. *Avian Pathol.* **2009**, *38*, 233–237. [CrossRef]
86. Gelfi, J.; Pappalardo, M.; Claverys, C.; Peralta, B.; Guerin, J.L. Safety and efficacy of an inactivated Carbopol-adjuvanted goose haemorrhagic polyomavirus vaccine for domestic geese. *Avian Pathol.* **2010**, *39*, 111–116. [CrossRef] [PubMed]
87. Kaszab, E.; Szabadi, L.; Kepner, A.; Bajnoczi, P.; Lengyel, G.; Banyai, K.; Feher, E. Viral gene expression profile of goose haemorrhagic polyomavirus in susceptible primary cells. *Avian Pathol.* **2021**, *50*, 447–452. [CrossRef]
88. Ritchie, B.W.; Vaughn, S.B.; Leger, J.S.; Rich, G.A.; Rupiper, D.J.; Forgey, G.; Greenacre, C.B.; Latimer, K.S.; Pesti, D.; Campagnoli, R.; et al. Use of an inactivated virus vaccine to control polyomavirus outbreaks in nine flocks of psittacine birds. *J. Am. Vet. Med. Assoc.* **1998**, *212*, 685–690. Available online: https://www.ncbi.nlm.nih.gov/pubmed/9524641 (accessed on 15 January 2022).
89. Xie, Q.; Wang, W.; Kan, Q.; Mu, Y.; Zhang, W.; Chen, J.; Li, L.; Fu, H.; Li, T.; Wan, Z.; et al. FAdV-4 without Fiber-2 Is a Highly Attenuated and Protective Vaccine Candidate. *Microbiol. Spectr.* **2022**, *10*, e0143621. [CrossRef]
90. Olbert, M.; Romer-Oberdorfer, A.; Herden, C.; Malberg, S.; Runge, S.; Staeheli, P.; Rubbenstroth, D. Viral vector vaccines expressing nucleoprotein and phosphoprotein genes of avian bornaviruses ameliorate homologous challenge infections in cockatiels and common canaries. *Sci. Rep.* **2016**, *6*, 36840. [CrossRef]
91. Runge, S.; Olbert, M.; Herden, C.; Malberg, S.; Romer-Oberdorfer, A.; Staeheli, P.; Rubbenstroth, D. Viral vector vaccines protect cockatiels from inflammatory lesions after heterologous parrot bornavirus 2 challenge infection. *Vaccine* **2017**, *35*, 557–563. [CrossRef] [PubMed]
92. Rall, I.; Amann, R.; Malberg, S.; Herden, C.; Rubbenstroth, D. Recombinant Modified Vaccinia Virus Ankara (MVA) Vaccines Efficiently Protect Cockatiels Against Parrot Bornavirus Infection and Proventricular Dilatation Disease. *Viruses* **2019**, *11*, 1130. [CrossRef] [PubMed]
93. Zielonka, A.; Gedvilaite, A.; Ulrich, R.; Luschow, D.; Sasnauskas, K.; Muller, H.; Johne, R. Generation of virus-like particles consisting of the major capsid protein VP1 of goose hemorrhagic polyomavirus and their application in serological tests. *Virus Res.* **2006**, *120*, 128–137. [CrossRef] [PubMed]
94. Zielonka, A.; Gedvilaite, A.; Reetz, J.; Rosler, U.; Muller, H.; Johne, R. Serological cross-reactions between four polyomaviruses of birds using virus-like particles expressed in yeast. *J. Gen. Virol.* **2012**, *93 Pt 12*, 2658–2667. [CrossRef] [PubMed]
95. Grgacic, E.V.; Anderson, D.A. Virus-like particles: Passport to immune recognition. *Methods* **2006**, *40*, 60–65. [CrossRef]
96. Ludwig, C.; Wagner, R. Virus-like particles-universal molecular toolboxes. *Curr. Opin. Biotechnol.* **2007**, *18*, 537–545. [CrossRef]
97. Roldao, A.; Mellado, M.C.; Castilho, L.R.; Carrondo, M.J.; Alves, P.M. Virus-like particles in vaccine development. *Expert Rev. Vaccines* **2010**, *9*, 1149–1176. [CrossRef]
98. Sasnauskas, K.; Bulavaite, A.; Hale, A.; Jin, L.; Knowles, W.A.; Gedvilaite, A.; Dargeviciute, A.; Bartkeviciute, D.; Zvirbliene, A.; Staniulis, J.; et al. Generation of recombinant virus-like particles of human and non-human polyomaviruses in yeast Saccharomyces cerevisiae. *Intervirology* **2002**, *45*, 308–317. [CrossRef]
99. Muller, H.; Nitschke, R. A polyoma-like virus associated with an acute disease of fledgling budgerigars (*Melopsittacus undulatus*). *Med. Microbiol. Immunol.* **1986**, *175*, 1–13. [CrossRef]
100. Wang, X.; Zhou, D.; Wang, G.; Huang, L.; Zheng, Q.; Li, C.; Cheng, Z. A novel multi-variant epitope ensemble vaccine against avian leukosis virus subgroup J. *Vaccine* **2017**, *35 Pt B*, 6685–6690. [CrossRef]

101. Qin, Y.; Tu, K.; Teng, Q.; Feng, D.; Zhao, Y.; Zhang, G. Identification of Novel T-Cell Epitopes on Infectious Bronchitis Virus N Protein and Development of a Multi-epitope Vaccine. *J. Virol.* **2021**, *95*, e0066721. [CrossRef] [PubMed]
102. Mao, S.J.; Patton, J.G.; Badimon, J.J.; Kottke, B.A.; Alley, M.C.; Cardin, A.D. Monoclonal antibodies to human plasma low-density lipoproteins. I. Enhanced binding of 125I-labeled low-density lipoproteins by combined use of two monoclonal antibodies. *Clin. Chem.* **1983**, *29*, 1890–1897. Available online: https://www.ncbi.nlm.nih.gov/pubmed/6627627 (accessed on 15 January 2022). [CrossRef]
103. Patton, J.G.; Badimon, J.J.; Mao, S.J. Monoclonal antibodies to human plasma low-density lipoproteins. II. Evaluation for use in radioimmunoassay for apolipoprotein B in patients with coronary artery disease. *Clin. Chem.* **1983**, *29*, 1898–1903. Available online: https://www.ncbi.nlm.nih.gov/pubmed/6627628 (accessed on 15 January 2022). [CrossRef] [PubMed]
104. Marcovina, S.; Kottke, B.A.; Mao, S.J. Monoclonal antibodies can precipitate low-density lipoprotein. II. Radioimmunoassays with single and combined monoclonal antibodies for determining apolipoprotein B in serum of patients with coronary artery disease. *Clin. Chem.* **1985**, *31*, 1659–1663. Available online: https://www.ncbi.nlm.nih.gov/pubmed/3930092 (accessed on 25 April 2022). [CrossRef] [PubMed]
105. Forchette, L.; Sebastian, W.; Liu, T. A Comprehensive Review of COVID-19 Virology, Vaccines, Variants, and Therapeutics. *Curr. Med. Sci.* **2021**, *41*, 1037–1051. [CrossRef]
106. Song, C.Y.; Chen, W.L.; Yang, M.C.; Huang, J.P.; Mao, S.J. Epitope mapping of a monoclonal antibody specific to bovine dry milk: Involvement of residues 66-76 of strand D in thermal denatured beta-lactoglobulin. *J. Biol. Chem.* **2005**, *280*, 3574–3582. [CrossRef]
107. Winter, S.; Lechapt, E.; Gricourt, G.; N'Debi, M.; Boddaert, N.; Moshous, D.; Blauwblomme, T.; Kossorotoff, M.; Fouyssac, F.; Chareyre, J.; et al. Fatal encephalitis caused by Newcastle disease virus in a child. *Acta Neuropathol.* **2021**, *142*, 605–608. [CrossRef]
108. Wang, Q.; Vlasova, A.N.; Kenney, S.P.; Saif, L.J. Emerging and re-emerging coronaviruses in pigs. *Curr. Opin. Virol.* **2019**, *34*, 39–49. [CrossRef]
109. Adiguzel, M.C.; Timurkan, M.O.; Cengiz, S. Investigation and Sequence Analysis of Avian Polyomavirus and Psittacine Beak and Feather Disease Virus from Companion Birds in Eastern Turkey. *J. Vet. Res.* **2020**, *64*, 495–501. [CrossRef]
110. Vitiello, A.; Ferrara, F. Brief review of the mRNA vaccines COVID-19. *Inflammopharmacology* **2021**, *29*, 645–649. [CrossRef]
111. Szabo, G.T.; Mahiny, A.J.; Vlatkovic, I. COVID-19 mRNA vaccines: Platforms and current developments. *Mol. Ther.* **2022**, *30*, 1850–1868. [CrossRef] [PubMed]

viruses

MDPI

Article

Dynamics of the Emerging Genogroup of Infectious Bursal Disease Virus Infection in Broiler Farms in South Korea: A Nationwide Study

Tuyet Ngan Thai [1,†], Dae-Sung Yoo [2,†], Il Jang [3,†], Yong-Kuk Kwon [1] and Hye-Ryoung Kim [1,*]

1 Avian Disease Division, Animal and Plant Quarantine Agency, Gimcheon 39660, Gyeongsangbuk-do, Korea; ttn267@korea.kr (T.N.T.); kwonyk66@korea.kr (Y.-K.K.)

2 Veterinary Epidemiology Division, Animal and Plant Quarantine Agency, Gimcheon 39660, Gyeongsangbuk-do, Korea; shanuar@korea.kr

3 Veterinary Drugs and Biologics Division, Animal and Plant Quarantine Agency, Gimcheon 39660, Gyeongsangbuk-do, Korea; moa305ho@korea.kr

* Correspondence: dvmkim77@korea.kr

† These authors contributed equally to this work.

Abstract: Infectious bursal disease (IBD), caused by IBD virus (IBDV), threatens the health of the poultry industry. Recently, a subtype of genogroup (G) 2 IBDV named G2d has brought a new threat to the poultry industry. To determine the current status of IBDV prevalence in South Korea, active IBDV surveillance on 167 randomly selected broiler farms in South Korea from August 2020 to July 2021 was conducted. The bursas of Fabricius from five chickens from each farm were independently pooled and screened for IBDV using virus-specific RT-PCR. As a result, 86 farms were found to be infected with the G2d variant, 13 farms with G2b, and 2 farms with G3. Current prevalence estimation of IBDV infection in South Korea was determined as 17.8% at the animal level using pooled sampling methods. G2d IBDV was predominant compared to other genogroups, with a potentially high-risk G2d infection area in southwestern South Korea. The impact of IBDV infection on poultry productivity or *Escherichia coli* infection susceptibility was also confirmed. A comparative pathogenicity test indicated that G2d IBDV caused severe and persistent damage to infected chickens compared with G2b. This study highlights the importance of implementation of regular surveillance programs and poses challenges for the comprehensive prevention of IBDV infections.

Keywords: infectious bursal disease virus; antigenic variant; broilers; nation surveillance; random sampling

Citation: Thai, T.N.; Yoo, D.-S.; Jang, I.; Kwon, Y.-K.; Kim, H.-R. Dynamics of the Emerging Genogroup of Infectious Bursal Disease Virus Infection in Broiler Farms in South Korea: A Nationwide Study. *Viruses* **2022**, *14*, 1604. https://doi.org/10.3390/v14081604

Academic Editor: Chi-Young Wang

Received: 30 June 2022
Accepted: 20 July 2022
Published: 22 July 2022

Publisher's Note: MDPI stays neutral with regard to jurisdictional claims in published maps and institutional affiliations.

Copyright: © 2022 by the authors. Licensee MDPI, Basel, Switzerland. This article is an open access article distributed under the terms and conditions of the Creative Commons Attribution (CC BY) license (https://creativecommons.org/licenses/by/4.0/).

1. Introduction

Infectious bursal disease (IBD) is a highly contagious viral disease that is responsible for significant losses in the poultry industry worldwide [1]. The disease is caused by infectious bursal disease virus (IBDV) which attacks and destroys B-lymphocytes in the bursa of Fabricius (BF), leading to not only the surge of mortality in young chicks but also the increase in immunodeficient flocks vulnerable to secondary pathogen infection [2]. The IBDV genome consists of two segments, A and B, and the hypervariable (hv) VP2 region (amino acids 206–350) in segment A has been widely used as the primary determinant of the genetic evolution and antigenic variation of IBDV [3,4]. IBDV was first reported in Gumboro, Delaware in 1957 [5]. Since then, two serotypes (I and II) have been identified, of which only serotype I IBDV causes immunosuppressive disease in chickens.

Based on their antigenicity and pathogenicity, serotype I strains were originally categorized into four phenotypes: classic virulent (cv), antigenic variant (av), highly virulent (vv), and attenuated (at) [6]. However, owing to rapid genetic mutation in the hvVP2 region, a new classification of IBDVs with seven genogroups (G) has been proposed [7,8]. According to this improved scheme for IBDV genotype classification, the cv/atIBDV, avIBDV, and vvIBBV strains were identified as G1, G2, and G3, respectively. G2 was mostly prevalent in poultry farms in North America and Europe and further geographically distributed into three sub-lineages: G2a, G2b, and G2c [8,9]. Other genogroups G4 to G7 were later isolated and classified according to their geographic distribution or distinct molecular characterization from the original ones, indicating continuous mutation and recombination of the IBDV genome [8,10].

The first IBD outbreak in South Korea was reported in 1980, and since 1992, the prevalence of G3 IBDV has caused severe damage to chicken farms [11–13]. To prevent damage caused by G3 IBDV, the intermediate/intermediate plus vaccine is widely used in Korea. Nevertheless, a number of outbreaks of a novel type of G2 IBDV have been reported in several farms in South Korea [14]. This novel variant of IBDV was first found in China and was later classified as sub-lineage G2d, whose molecular characterization was different from that of the early variant IBDVs originally reported in America [9]. G2d IBDV has become widespread in China, mostly driven by evading the immune protection conferred by available vaccines [15–17]. More recently, G2d IBDV epidemics have also been reported in other Asian countries, including Japan [18] and Malaysia [19]. Under these circumstances, the lack of effective antigen-matched vaccines has contributed to the increasing prevalence of this new sub-genogroup in poultry holdings.

Under the potential circumstances that G2d IBDV may also dominate in South Korea, it is urgent to understand the prevalence rate of IBDV and evaluate its potential adverse impact on poultry production. Therefore, this study aimed to investigate the prevalence of IBDV in broiler farms across the nation from 2020 to 2021 and to identify high-risk areas for infection. Furthermore, the impact of IBDV infection on either productivity or secondary infection susceptibility and comparative pathogenicity between the two subgroups of G2 IBDV were also examined. This study provides a scientific basis for the development of an intervention strategy optimized for new variants of IBD outbreaks.

2. Materials and Methods

2.1. IBDV Surveillance

Nationwide surveillance of IBD in poultry farms across the nation was performed from August 2020 to July 2021. Two hundred meat-type chicken farms were randomly selected, taking samples from five chickens at slaughterhouses with ages ranging from 24 to 38 days that originated from each farm in nine municipalities. The number of farms for IBDV surveillance at slaughterhouses was determined based on the proportion of chickens slaughtered per abattoir in 2019 [20]. Details on the calculation of the designed sample size and the actual sample size can be found in Table S1.

During the surveillance period, 33 meat-chicken farms were excluded from this study for the following reasons. First, 15 meat-chicken farms were excluded because 6 slaughterhouses were not in operation at the time of surveillance. Second, a sample from one farm could not be used because of poor storage conditions. Furthermore, among meat-type chicken farms, 17 farms raising minor chicken types such as White semi or Korean Native Chicken were excluded from the surveillance due to differences in age and weight at slaughter, and only commercial broiler farms such as Arbor Acre, Cobb, and Ross were investigated. As a result, a total of 167 farms were included in this study.

2.2. Animal Husbandry Status

The geographical location and number of birds at slaughter of the study farms were obtained from the livestock production safety management system. The production index (*PI*) at the study farm, which corresponds to productivity, was assessed using the following equation [21]. The *PI* for individual farms was obtained from companies operating slaughterhouses. This indicator was compiled from 108 broiler farms, except for 59 farms, where the company did not know or refused to provide information. The *PI* is defined as follows:

$$Production\ index\ (PI) = \frac{mean\ weight\ at\ slaughter(kg) \times (number\ of\ live\ birds\ \div number\ of\ one-day\ chicks)}{days\ at\ slaughter \times (feed\ intake\ \div total\ body\ weight)} \times 100$$

2.3. Laboratory Diagnosis

For the detection of IBDV, BFs of five chickens from each farm were collected during necropsy examination and pooled as one sample. BF samples were homogenized in 10% (*w*/*v*) phosphate buffered saline (PBS) supplemented with 50 µg/mL gentamicin. The homogenate was then centrifuged at 3000× *g* for 10 min at 4 °C, and the supernatant was collected for subsequent IBDV detection and analysis. IBDV in these samples was detected using a virus-specific RT-PCR assay as described previously [8]. Briefly, viral RNA was extracted from bursa homogenates using TANBead nucleic acid extraction kits (Taiwan Advanced Nanotech Inc., Taoyuan City, Taiwan). RT-PCR was then conducted using Maxime RT-PCR Premix (iNtRON Biotechnology, Gyeonggi, Korea) under the following thermocycling conditions: 48 °C for 30 min; 94 °C for 5 min; 35 cycles of denaturation at 94 °C for 40 s, 57 °C for 40 s, and extension at 72 °C for 1 min; and final extension at 72 °C for 5 min. Primer05/01/95s 743-F (5′-GCCCAGAGTCTACACCAT-3′) and 1331-R (5′-ATGGCTCCTGGGTCAAATCG-3′) were used to amplify the 579-bp fragment of the hypervariable region of the VP2 gene (hvVP2). Amplified PCR products were electrophoresed on 1.5% agarose gels to detect an amplification of hvVP2 fragments.

Positive RT-PCR products were purified using a QIAquick PCR Purification Kit (QIAGEN, Hilden, Germany) and sequenced by Sanger sequencing (Bionics, Seoul, Korea). A 545-bp fragment of hvVP2 was assembled and aligned with reference IBDV strains using the CLC Workbench version 6.7.2 (CLC bio, Aarhus, Denmark). Sequences with 100% nucleotide similarities with those of IBDV live vaccines commercially used in South Korea were excluded from the prevalence study, and these farms were subsequently considered as negative for the virus infection. A phylogenetic tree was constructed using the neighbor-joining method and 1000 bootstrap replications using MEGA X software [22] and then displayed with iTOL (Interactive Tree Of Life) [23]. The partial VP2 gene sequences used in this study were deposited in the GenBank database under accession numbers ON715016-ON715115.

To diagnose colibacillosis caused by pathogenic *Escherichia coli* (*E. coli*), typical macroscopic lesions of colibacillosis were observed in organs such as the liver, air sac, and pericardium during necropsy. If such lesions were observed, pathogen isolation and identification were performed subsequently. Samples of the organs were cultured on a blood agar plate and MacConkey agar plate per sample, and the plates were incubated for 24 h at 37 °C. Bacterial species identification was performed using the VITEK-2 system (VITEK2 GN-card; bioMérieux, Marcy-l'Étoile, France) according to the manufacturer's instructions [24].

2.4. Prevalence Estimation of IBDV

As pooled testing commenced with collection of BF samples from the 167 farms, we followed the protocol of animal prevalence estimation using pooled sampling methods, where 100% sensitivity and specificity for the test was assumed because the golden standard of virus detection by RT-PCR assay was applied to diagnose the infection. One to five BF samples were collected randomly from each broiler farm. Given these sample sizes, we assumed that the dilution did not affect the sensitivity of the test. Additionally, we assumed that all pooled samples were independent of each other. Under this assumption, the outcome of testing IBDV infection at farm i was denoted by y_i, which was coded as one for positive and zero for negative from sample size s_i, and followed a pooled Bernoulli distribution as follows [25]:

$$\prod_i Bern\ (y_i|s_i,\ p) = (1-p)^{\sum_i s_i(1-y_i)} \prod_i (1-(1-p)^{s_i})^{y_i}$$

where p is the animal prevalence, which is related to the probability of a positive result from the pooled sample, $1 - (1 - p)s_i$. Using these likelihood functions, we inferred animal prevalence using Bayesian inference with the Hamiltonian Markov-chain Monte Carlo method. The PoolTestR package version 0.1.2. was used based on the R software version 4.2.1. (Vienna, Austria, R core Team 2020).

2.5. *Identification of Risk Area of IBDV Occurrence in Broiler Farms*

Since the novel antigenic variant IBDV infection (i.e., G2d) has recently been reported, the geographical distribution of outbreaks could be limited to small areas, which is indicative of the initial phase of infectious disease outbreaks. Therefore, we conducted a spatial analysis to identify whether IBDV occurrence in broiler farms was clustered in a specific area. Three broiler farms were excluded from the cluster analysis because their geographical information was not available. For this analysis, the elliptic scan statistics were employed to calculate Bernoulli likelihood ratio (coded as one if the farm had a certain genogroup of IBDV infection (i.e., G2b, G2d, G3, or mixed infection of G2b and G2d) of interest, otherwise zero) for searching candidate clusters, limiting the maximum size to 50% of the total population of study at risk as default setting. The significance for the detection of clusters was determined by the p-value, which was obtained from a Monte Carlo simulation of 999 iterations based on the Bernoulli likelihood ratio function. The SaTScan software version 10.0.2 was used for spatial cluster analysis [26].

2.6. *Adverse Effect of IBDV Infection on Broiler Production*

IBDV is known to mainly attack B lymphocytes in the BF, leading to suppression of immunological function in chickens, increased vulnerability to secondary infection [27], and consequently declines in productivity, such as low feed conversion. Therefore, we conducted Bayesian multivariate mixed-effects linear regression to identify the association between IBDV infection and PI at the study farm.

Furthermore, Bayesian multivariate mixed-effect logistic regression was applied to examine the contribution of IBDV infection to *E. coli* infection. *E. coli* infection was diagnosed using samples collected from the same farms for IBDV testing.

$$E.coli\ infection_{ij} \sim B(\pi_{ij}),\ E(E.coli\ infection_{ij}) = \pi_{ij}$$

$$logit(\pi_{ij}) = \alpha + \beta_1 * G2d_{ij} + \beta_2 * G2b_{ij} + \beta_3 * G3_{ij} + \beta_4 * mixed\ infection_{ij} + random\ effect_j$$

$$random\ effect_j \sim N\left(\mu_j, \sigma_j^2\right)$$

The mixed regression model assigned the farm location with respect to the spatial cluster to the random effect to account for different prevalence levels. In turn, *E. coli* infection ij corresponds to the outcome of *E. coli* infection at broiler farm i located inside or outside spatial cluster j for IBDV infection. The goodness of fit for the Bayesian logistic regression model was assessed using the area under the curve (AUC) of the receiver operating characteristic (ROC) curve for IBDV infection. R2jags package version 0.6–1 with R software was used for logistic regression using Markov-chain Monte Carlo (MCMC) sampling with 3 chains, 50,000 iterations, 4000 burn-in, and thin of 10. The Rubin-Gelman index (<1.05) was used to check the convergence of the MCMC chains [28].

2.7. *Comparative Pathogenicity Study of IBDV Strains Belonging to Different G2 Sub-Genogroups*

IBDV isolates used in this pathogenicity study, G2b strain 19D38 and G2d strain 20D39, were isolated from a passive surveillance performed in 2019 and 2020, respectively [14]. Isolation and titration of IBDV strains were performed in 10-day-old specific pathogen-free (SPF) chicken embryonated eggs via the chorioallantoic membrane (CAM) route, as previously described [29]. Viruses were harvested from CAMs and examined to exclude contamination by other avian pathogens using molecular assays. Determining the median embryo effective dose (EID_{50}) was performed by virus titration following the method described by Reed and Muench [30].

All experimental animal procedures were approved and supervised by the Institutional Animal Care and Use Committee of the Animal and Plant Quarantine Agency of Korea. Sixty-three one-day-old SPF chickens were purchased from a local company (Namduk SPF, Gyeonggi, Korea) and divided into three groups (n = 21 birds/group) housed in separate isolators provided with feed and water ad libitum. At 21 days of age, chickens in group 1 and 2 were challenged through oculo-nasal route with a dose of 10^5 EID_{50}/0.1mL of IBDV G2b strain 19D38 and G2d strain 20D39, respectively. Chickens in the control group were used as negative controls and mock-inoculated with phosphate-buffered saline solution. At 1, 3, 5, 7, and 14 days post-challenge (dpc), three birds from each group were euthanized, weighed, and examined for macroscopic lesions post-mortem. BFs were weighed to calculate the bursa:body weight (B/BW) ratio and bursal body weight index (BBIX). The mean BBIX value was calculated along with the standard deviation [BBIX = (B/BW ratio in the infected group)/(B/BW ratio in the control group)]. BF with a BBIX value less than 0.7 was considered as atrophy [31]. Viral loads in BFs and viral shedding in cloacal (CL) swabs were examined by fluorescent quantitative real-time RT-PCR (qRT-PCR) using IBDV-specific primers and probes. The same titrated stock of challenge virus was used to establish the standard curve for viral load calculation, and data were indicated in EID_{50}/0.1 mL. BF samples collected at 1, 3, 5, and 7 dpc were stored in 10% neutral formalin for histopathological examination. Briefly, paraffin-embedded tissues were routinely processed, and 4-μm-thick sections were prepared and stained with hematoxylin and eosin. Microscopic lesions in BFs were scored in order of severity from 0 for normal BFs to +4 for severely infected birds, according to a previous study [32]. The mean lesion scores (MLSs) were calculated for each group. At 7 and 14 dpc, blood was collected from all chickens, and their sera were tested for anti-IBDV antibodies using the commercial ELISA IBD-XR Ab test kit (IDEXX Laboratories, Maine, USA) according to the manufacturer's instructions.

All statistical analyses were performed using GraphPad Prism version 8.4.3 (GraphPad Software, San Diego, CA, USA). The statistical significance of the differences among the different groups was determined using two-way analysis of variance. Differences were considered significant at $p < 0.05$ (*), $p < 0.01$ (**) and $p < 0.0001$ (****).

3. Results

3.1. IBDV Detection from Surveillance and Its Genogroup

Out of 167 broiler farms in this study, 103 farms were confirmed to have infection with field IBDV, which accounted for 61.67% of tested farms. There were 33 farms not infected with IBDV in which BF samples were negative for IBDV-specific RT-PCR assay. Samples collected from 31 farms were detected with commercial live vaccine strains, which were also considered as not infected with field IBDV. According to the phylogenetic analysis based on hypervariable regions of the VP2 gene, 86 farms were infected with the G2d variant, 13 farms were infected with the G2b variant, 2 farms were infected with G3, and 1 farm named 21RI008 was found to be concurrently infected with both G2b and G2d variants (Figure 1).

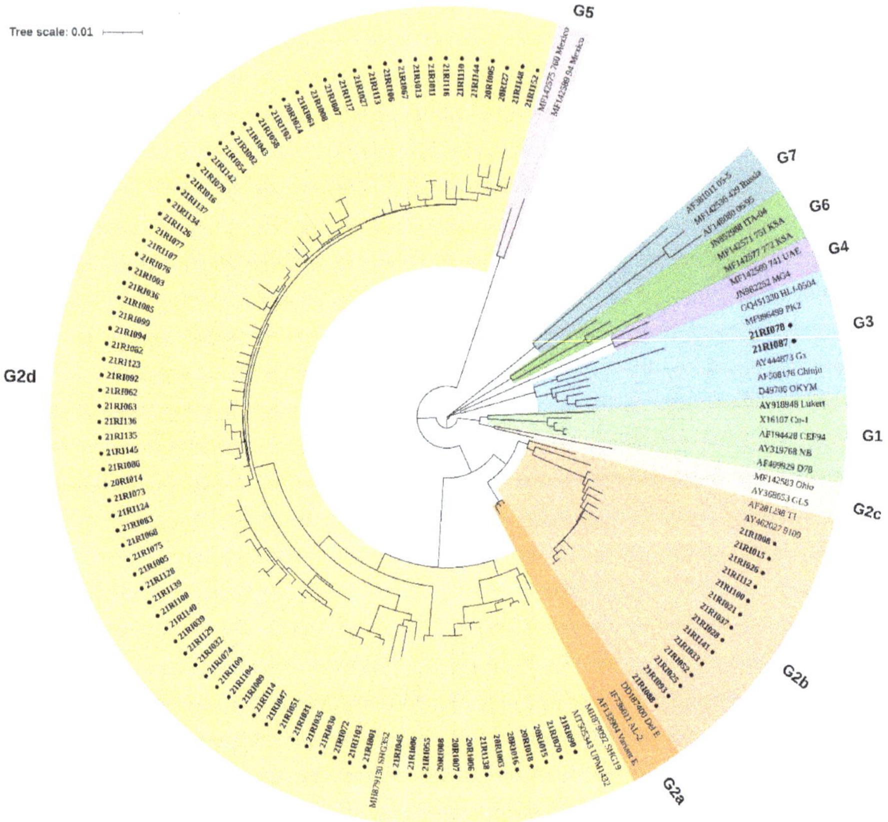

Figure 1. Phylogenetic analysis of the nucleotide sequences of the hypervariable region of the VP2 gene. Broiler farms infected with IBDVs found in this study were indicated as bold with a solid circle mark.

3.2. Prevalence Estimation of IBDV Infection

Table 1 provides the prevalence estimates of IBDV infection in broilers in South Korea, based on randomly sampled pooled tests. At the animal level, approximately 17.8% of broilers in South Korea were infected with various genogroups of IBDV. Specifically, there was evident discrepancy in G2-specific prevalence where the G2d viruses (mean = 0.140, 95% credible intervals [95% CrI] = 0.113 − 0.168) predominantly infected broilers than other genogroups of IBDV.

Table 1. The prevalence estimation of infectious bursal disease of broilers in South Korea.

Category	Mean	95% Credible Intervals	
		Lower	Upper
IBDV			
G2d infection	0.140	0.113	0.168
G2b infection	0.017	0.087	0.026
All IBDV infection [a]	0.178	0.148	0.209

[a] All IBDV infection consisted of G2d, G2b, G3 and mixed infection (G2b and G2d).

3.3. Identification of Risk Area of IBDV Occurrence in Broiler Farms

As shown in Figure 2, one spatial cluster for G2d IBDV outbreaks in broiler farms located in the southwestern part of the country was identified. Inside the cluster, 77.4% of total G2d variant incidence was confirmed, resulting in relative risk of 2.02. In contrast, other genogroups of IBDV did not display a geographically clustered outbreak in broiler farms.

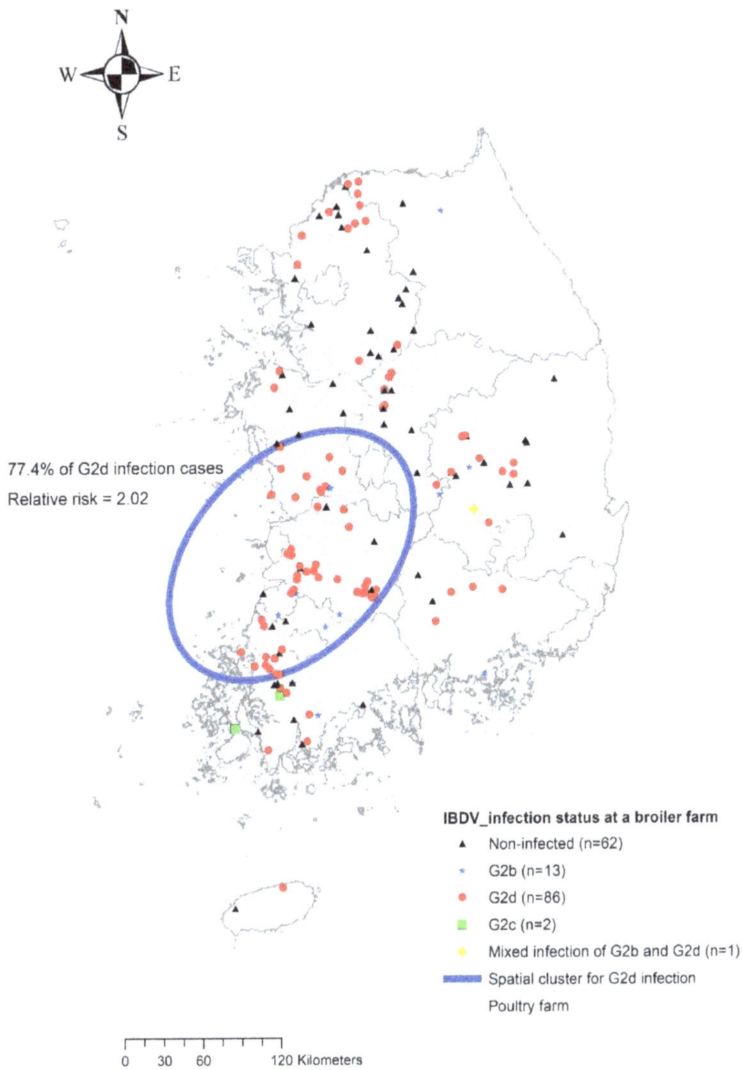

Figure 2. Geographical distribution of IBDV-infected premises and the spatial cluster for G2d variant of IBDV infection denoted by a blue circle.

3.4. Association of Productivity of Broiler and Secondary Infection to E. coli

Table 2 summarizes the regression coefficient estimates of IBDV infection for the productivity index estimated using Bayesian multiple mixed-effect linear regression. According to the results, the broiler farms infected with the G2d IBDV showed a decline in their PI, but not a significant drop compared to non-infected farms. G3 viral infection decreased the PI of broiler farms by the highest number, but no significance was reported. In addition, G2b virus-infected broiler farms showed a lower decline in PI than other types of IBDV infections.

Table 2. Association of IBDV infection with productivity index at broiler farms (n = 108) by Bayesian linear mixed effect regression.

Variables	Coefficient Estimates [a]		
	Mean	95% Credible Interval	
		Lower	Upper
G2d infection	−8.651	−28.243	10.941
G2b infection	−5.340	−38.189	27.510
G3 infection	−30.550	−93.171	32.067
No. of chicken slaughtered	0.081	−0.212	0.374

[a] The coefficient estimates of G2d, G2b and G2c infection for productivity index (PI) represents the change of PI value when the corresponding genogroup of infection occurred against the non-infection.

The association between IBDV infection with *E. coli* infection is shown in Table 3. The broiler farms with G2d IBDV infection reported 2.64-times higher risk of *E. coli* infection (95% Crl = 1.16–6.24) compared to IBDV non-infected premises. However, the other genogroups (i.e., G2b or G3) of IBDV infection did not have a significant impact on *E. coli* infection, with a broader CrI containing an odds ratio of 1.

Table 3. Odds ratio of infectious bursal disease infection for the presence of *Escherichia coli* in commercial broiler farms (n = 167) by Bayesian mixed-effect logistic regression.

IBDV Infection (Reference = Not Infection)	Odds Ratio		
	Mean	95% Credible Interval	
		Lower	Upper
G2d infection	2.64	1.16	6.24
G2b infection	3.05	0.74	11.90
G3 infection	5.09	0.13	20.60
Mixed infection (G2b and G2d)	0.02	0.00	17.20

3.5. Comparative Pathogenicity Study of IBDV Strains Belonging to Different G2 Subgroups

Considering the high prevalence rate of G2 IBDV in South Korea and the possible adverse effects of IBDV infection, an in vivo pathogenicity trial was performed in SPF chickens. During the study, neither clinical signs nor mortality were observed in any of the experimental groups. Atrophy of the bursa was observed from 3 dpc in all virus-challenged groups and lasted until the end of the experiment. No post-mortem lesions were observed in the control birds at any time point. BBIX values for virus-challenged groups were <0.7 from 3 dpc and continued to decrease to 0.25 and 0.27 at 14 dpc in the 19D38 and 20D39 virus-challenged groups, respectively, confirming the results of the post-mortem examination (Figure 3).

Histopathologically, typical lesions of BFs infected with IBDV, including loss of epithelium, lymphocyte depletion, and lymphoid follicle diffusion, were found in the G2 IBDV-infected BFs, although no lesions were found in the BFs of the control group (Figure 4). In particular, according to the mean lesion scoring, the G2d 20D39 strain caused more severe microscopic lesions in the BFs than the G2b 19D38 strain at 5 and 7 dpc (Figure 5).

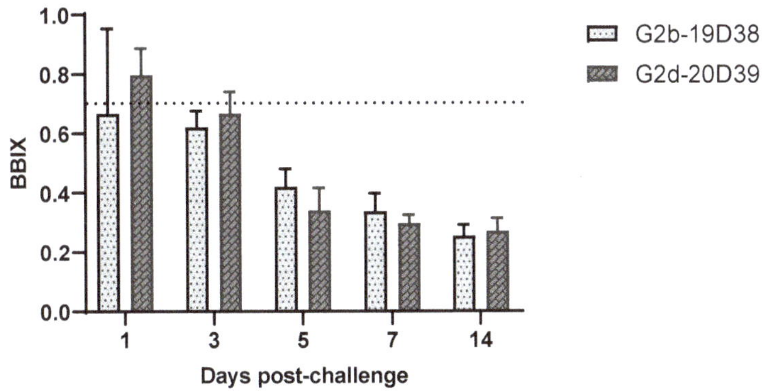

Figure 3. The bursa:body weight index (BBIX) of chickens challenged with IBDV strains 19D38 and 20D39. The dotted horizontal line indicates threshold of bursal atrophy. Error bars indicate standard deviation.

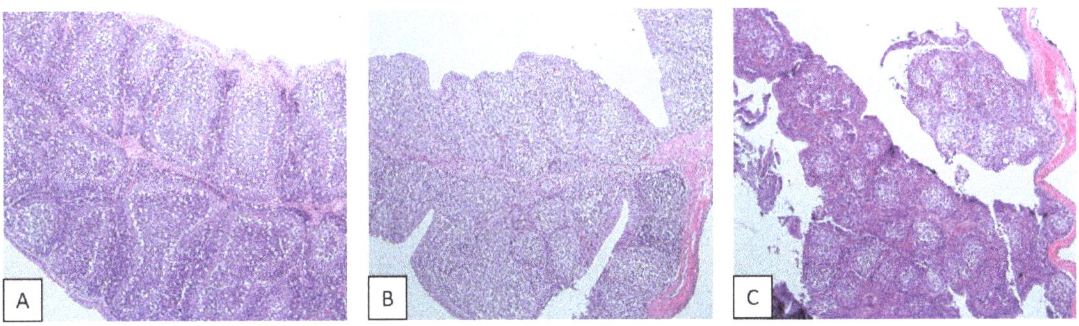

Figure 4. Histopathology of the bursas of Fabricius of specific pathogen-free chickens at 7 days post-challenge (× 100 magnification). (**A**) Control birds: normal lymphoid follicles of the bursa. (**B**) 19D38 virus-infected birds: moderate atrophy in bursal lymphoid follicle is diffuse. (**C**) 20D39 virus-infected birds: severe atrophy, bursal lymphoid follicle diffusion and exfoliation of epithelium.

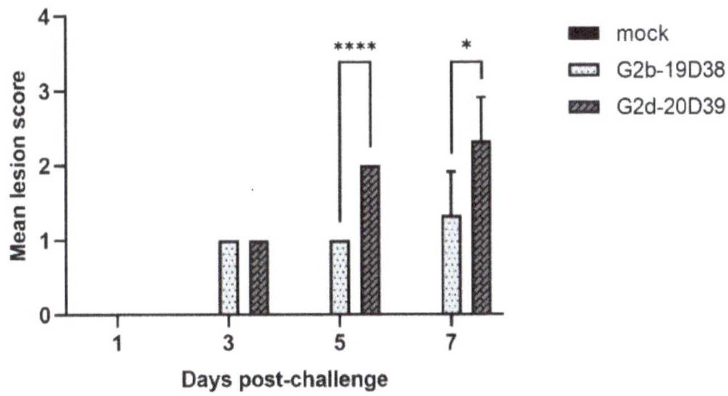

Figure 5. Mean lesion scores in bursas of Fabricius of chickens challenged with variant IBDV strains. Error bars indicate standard deviation. *, $p < 0.05$; ****, $p < 0.001$.

The viral loads in the BFs and CL swabs were determined using qRT-PCR analysis. Viral RNA was not detected in the control group. The virus started to replicate in BFs at 1 dpc in both virus-challenged groups, and the viral loads peaked at 3 dpc, then gradually declined over the course of the experiment. There was no significant difference in viral replication in BFs infected with the different IBDV G2 subgroups. Viral shedding also peaked at 3 dpc for the virus infection groups, and viruses were detected in cloacal swabs until 7 dpc (Figure 6).

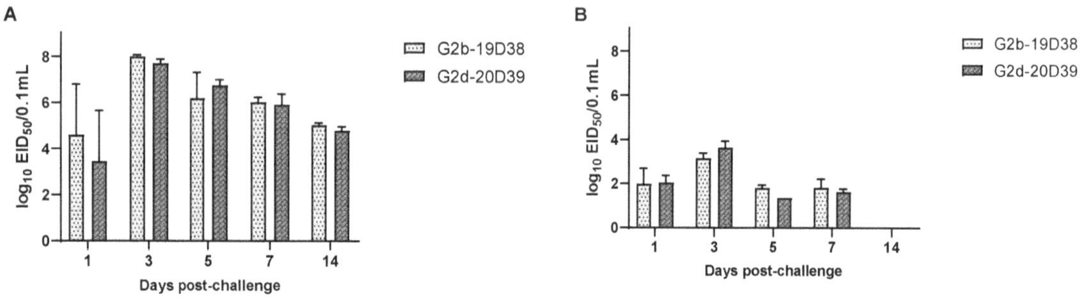

Figure 6. Viral loads of antigenic variant strains in (**A**) bursas of Fabricius and (**B**) cloacal swabs from chickens as determined by qRT-PCR analysis. Error bars indicate standard deviation.

Antibody responses in chickens infected with IBDV strains are shown in Figure 7. No IBDV-specific antibody was obtained for the control group at all time points. A positive antibody response to both variant IBDV strains was observed from 7 dpc onwards, and an increasing antibody titer was observed at 14 dpc. Remarkably, chickens infected with G2d strain 20D39 had significantly higher antibody titers than those infected with G2b strain 19D38 at 14dpc ($p < 0.05$).

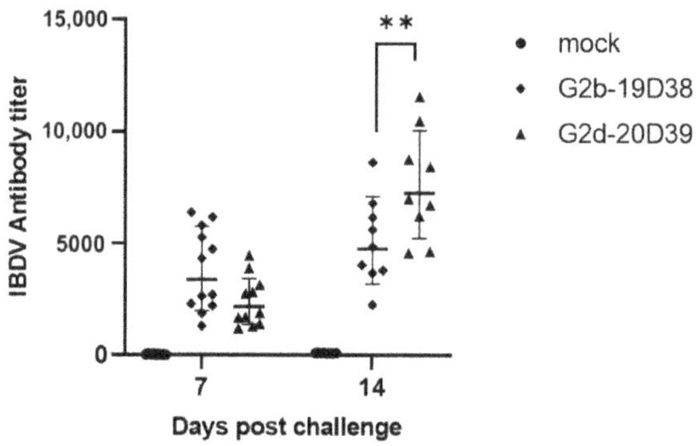

Figure 7. Serum antibody titers to IBDV of the control group and challenged groups. Each data point represents individual antibody titers, and the horizontal bar indicates geometric mean titers. **, $p < 0.01$.

4. Discussion

In an effort to keep track of the current IBDV status in South Korea, we attempted to conduct active surveillance on 167 commercial broiler farms randomly selected across the country. This diagnostic approach could be considered reliable for monitoring the current prevalence of the disease, since the detection of IBDV is not traditionally focused on birds with clinical signs of IBD or any other diseases. Remarkably, 61.67% of tested farms were confirmed to have IBDV infections. Attempts to determine prevalence estimates at the animal level, based on random sampling pooled test, revealed that 17.8% of broilers in South Korea had been infected with IBDV. Moreover, three genogroups, G2b, G2d, and G3, were found to be present in poultry farms across the nation. Notably, G2b continues to circulate in South Korea since two G2b strains were isolated in 2019 [12]. Specifically, there was an evident discrepancy in genogroup-specific prevalence, where the G2d virus predominantly infected broilers compared to other IBDV genogroups. These findings reflect the characteristics of the current IBDV epidemic in South Korea, where there has been an outbreak of IBDV infection in domestic broiler farms with the G2d virus becoming the dominant genogroup.

IBDV G2d infection in broiler farms, the most prevalent genogroup of infection, showed a geographically clustered pattern. This clustered area is known to have the highest density of chicken farms in South Korea and has experienced several highly infectious disease outbreaks, such as highly pathogenic avian influenza [33]. A similar finding was reported in China, where G2d IBDV was also prevalent in the major poultry-raising regions of this country [17]. This implies that the novel sub-genogroup of IBDV is likely to harbor a vulnerable area and facilitate inter-farm transmission at close distances during the course of the epidemic. Our findings highlight the urgent need to formulate an intensive biosecurity standard for highly vulnerable areas to prevent the spread of infectious diseases.

Concerted efforts to control domestic G3 IBDV outbreaks for nearly 30 years, mainly including widespread vaccination and improved poultry feeding management, might have contributed to effectively controlling the G3 epidemic in South Korea. Although vaccination information on the farm could not be fully obtained in this surveillance due to a lack of or wrong information from farm owners, this situation was confirmed by our surveillance results with only 2 out of 167 randomly examined broiler farms across the country found to be infected with this genogroup. However, a recent study showed that current commercial vaccines against G3 IBDV were unable to provide complete protection to chickens against G2 variant viruses [34]. This finding correlated with the fact that G2 variants were isolated from poultry flocks immunized with commercial intermediate vaccines [14,19,35,36]. Several attempts to developing vaccine candidates against G2d variants has been done [37,38]. Moreover, some commercial recombinant vaccines have been reported to induce protection against a wide range of IBDV strains, including G2 variants [39]. The fact that genogroup 2 IBDV, especially G2d, is currently becoming a dominant genogroup in South Korea poses an urgent need for a proper update on the current vaccination program.

Antigenic variant IBDV infection-induced immunodeficiency in chickens has been well reported to make the flock susceptible to secondary infections [6,40]. In this study, we found that broiler farms with G2d virus infection reported a higher risk of *E. coli* infection compared to IBDV non-infected farms. This finding was consistent with a recent study that found that high mortality observed in 30 day-old chickens in low-performance broiler farms infected with G2d IBDV in Japan could be due to secondary infections such as *E. coli*, probably one week after IBDV infection, which was at the time of immunosuppression in these broilers [18].

Due to high resistance to most disinfectants and environmental factors, IBDV continues to be present in poultry houses and tends to reappear in subsequent chicken flocks. Consequently, acute or chronic IBD in the flock reduces the efficiency of broiler production and net profitability. A 5-year study of the incidence and economic impact of various IBDVs on broiler production in Canada found that IBDV prevalence was probably attributed to substantial economic losses of approximately 3.9 million kg per year to the poultry industry [41]. Although the data were not significant, our findings confirmed the possible adverse impact of IBDV infection on poultry productivity, with broiler farms infected with the G2d variant showing a decline in their productivity index compared to non-infected farms. Some data in our study did not show a significant difference. This may have been due to the exclusion of more than 30 farms during the survey (Figure 1). Furthermore, G2 IBDV infection has been considered a subclinical infection; therefore, it may be difficult to see the remarkable adverse association of G2 IBDV with productivity with short-term surveillance.

Since the G2d variant was first identified in China and has been widespread throughout Asian countries, several studies on its pathogenicity have been performed [16,19,42,43]. In this study, we compared the pathogenicity of two IBDV strains belonging to different IBDV G2 sub-genogroups: IBDV G2b strain 19D38 and G2d strain 20D39. Neither strain caused any obvious gross clinical symptoms or mortality, which correlated with previous studies considering variant IBDV infection as a subclinical infection. Moreover, the two isolates showed relatively similar effects on bursal atrophy, viral shedding, and replication efficacy in the bursa. However, at a histological level, G2d isolate 20D39 caused more severe lesions in BFs than G2b isolate 19D38. A protective antibody response occurred from 7dpc for both virus-challenged groups despite the potentially immunosuppressive effects on the bursa. Antibody titers against IBDV in chickens in the G2b-challenged group were slightly higher than those of G2d group at 7dpc, whereas at 14dpc the antibody response was significantly reversed. A similar pattern of antibody response was observed in a recent comparative pathogenicity study of a variant of very virulent IBDV [43]. The results may suggest that there was a weaker humoral immune response to G2b strain than the G2d strain at the acute phase, but a reverse incidence was observed at the chronic phase. Furthermore, in a study evaluating the antibody response against IBDV vaccines with different level of pathogenicity, an intermediate plus vaccine induced higher antibody response than an intermediate vaccine [44]. These findings may indicate that higher pathogenicity levels of IBDV G2d than G2b may correlate with the antibody response against those sub-lineages. To our knowledge, this is the first report to evaluate the pathogenicity of two sub-lineages of G2 IBDV.

5. Conclusions

We found that IBDV was highly prevalent in South Korea, with G2d the dominant genogroup, particularly in dense areas with a high density of broiler farms, and had an adverse impact on productivity efficiency as well as susceptibility to secondary infection. Pathogenicity studies revealed that both G2 sub-genogroups induced bursal atrophy, leading to an immunosuppressive effect, with G2d causing more severe histopathological lesions. Furthermore, many studies have provided adequate evidence that current commercial G1-derived vaccines against G3 IBDV do not effectively protect poultry against variants of IBDV, underscoring the urgency for an updated vaccine program. Without pharmaceutical interventions available for the emergence of novel viruses, this study suggests the development of a control strategy that includes intensive risk management in clustered areas and underscores the importance of the implementation of regular surveillance programs. Therefore, it is necessary to develop systematic intervention strategies to minimize nationwide disease transmission.

Supplementary Materials: The following supporting information can be downloaded at: https://www.mdpi.com/article/10.3390/v14081604/s1, Table S1. Sampling scheme and sample size calculation of broiler chicken farms for the prevalence estimation of infectious bursal disease virus infection in South Korea.

Author Contributions: Conceptualization, I.J. and H.-R.K.; methodology, T.N.T., D.-S.Y. and I.J.; investigation, T.N.T. and I.J.; formal analysis, D.-S.Y.; visualization, T.N.T. and D.-S.Y.; writing—original draft preparation, T.N.T., D.-S.Y. and I.J.; writing—review and editing, T.N.T., D.-S.Y., I.J. and H.-R.K.; supervision, Y.-K.K. and H.-R.K. All authors have read and agreed to the published version of the manuscript.

Funding: This work was supported by the Animal and Plant Quarantine Agency, Republic of Korea, under Grant APQA B-1543084-2020-22-0201.

Institutional Review Board Statement: The study was conducted, approved and supervised by the Institutional Animal Care and Use Committee of the Animal and Plant Quarantine Agency of Korea (permission number 2021-564, December 2021).

Informed Consent Statement: Not applicable.

Data Availability Statement: The hvVP2 sequences of IBDV strains in this study are available in the GenBank database under accession numbers: ON715016-ON715115.

Acknowledgments: We are grateful for the support from regional offices of the Animal and Plant Quarantine Agency across the country and farm owners for providing adequate information and samples to perform this study.

Conflicts of Interest: The authors declare that they have no competing interests.

References

1. Berg, T.P. Acute infectious bursal disease in poultry: A review. *Avian Pathol.* **2000**, *29*, 175–194. [CrossRef] [PubMed]
2. Kibenge, F.S.; Dhillon, A.S.; Russell, R.G. Biochemistry and immunology of infectious bursal disease virus. *J. Gen. Virol.* **1988**, *69 Pt 8*, 1757–1775. [CrossRef] [PubMed]
3. Bayliss, C.; Spies, U.; Shaw, K.; Peters, R.; Papageorgiou, A.; Müller, H.; Boursnell, M. A comparison of the sequences of segment A of four infectious bursal disease virus strains and identification of a variable region in VP2. *J. Gen. Virol.* **1990**, *71 Pt 6*, 1303–1312. [CrossRef] [PubMed]
4. Letzel, T.; Coulibaly, F.; Rey, F.A.; Delmas, B.; Jagt, E.; van Loon, A.A.; Mundt, E. Molecular and structural bases for the antigenicity of VP2 of infectious bursal disease virus. *J. Virol.* **2007**, *81*, 12827–12835. [CrossRef]
5. Cosgrove, A.S. An Apparently New Disease of Chickens: Avian Nephrosis. *Avian Dis.* **1962**, *6*, 385–389. [CrossRef]
6. Muller, H.; Islam, M.R.; Raue, R. Research on infectious bursal disease—The past, the present and the future. *Vet. Microbiol.* **2003**, *97*, 153–165. [CrossRef]
7. Jackwood, D.J.; Schat, K.A.; Michel, L.O.; de Wit, S. A proposed nomenclature for infectious bursal disease virus isolates. *Avian Pathol.* **2018**, *47*, 576–584. [CrossRef]
8. Michel, L.O.; Jackwood, D.J. Classification of infectious bursal disease virus into genogroups. *Arch. Virol.* **2017**, *162*, 3661–3670. [CrossRef]
9. Jackwood, D.J.; Cookson, K.C.; Sommer-Wagner, S.E.; Le Galludec, H.; de Wit, J.J. Molecular characteristics of infectious bursal disease viruses from asymptomatic broiler flocks in Europe. *Avian Dis.* **2006**, *50*, 532–536. [CrossRef]
10. Hernández, M.; Tomás, G.; Marandino, A.; Iraola, G.; Maya, L.; Mattion, N.; Hernández, D.; Villegas, P.; Banda, A.; Panzera, Y.; et al. Genetic characterization of South American infectious bursal disease virus reveals the existence of a distinct worldwide-spread genetic lineage. *Avian Pathol.* **2015**, *44*, 212–221. [CrossRef]
11. Jeon, W.J.; Choi, K.S.; Lee, D.W.; Lee, E.K.; Cha, S.H.; Cho, S.H.; Kwon, J.H.; Yoon, Y.S.; Kim, S.J.; Kim, J.H.; et al. Molecular epizootiology of infectious bursal disease (IBD) in Korea. *Virus Genes* **2009**, *39*, 342–351. [CrossRef]
12. Song, C.S.; Lee, Y.J.; Kim, J.H.; Sung, H.W.; Lee, C.W.; Izumiya, Y.; Miyazawa, T.; Jang, H.K.; Mikami, T. Epidemiological classification of infectious bronchitis virus isolated in Korea between 1986 and 1997. *Avian Pathol.* **1998**, *27*, 409–416. [CrossRef]
13. Kim, H.R.; Kwon, Y.K.; Bae, Y.C.; Oem, J.K.; Lee, O.S. Genetic characteristics of virion protein 2 genes of infectious bursal disease viruses isolated from commercial chickens with clinical disease in South Korea. *Poult. Sci.* **2010**, *89*, 1642–1646. [CrossRef]
14. Thai, T.N.; Jang, I.; Kim, H.A.; Kim, H.S.; Kwon, Y.K.; Kim, H.R. Characterization of antigenic variant infectious bursal disease virus strains identified in South Korea. *Avian Pathol.* **2021**, *50*, 174–181. [CrossRef]
15. Fan, L.; Wu, T.; Hussain, A.; Gao, Y.; Zeng, X.; Wang, Y.; Gao, L.; Li, K.; Wang, Y.; Liu, C.; et al. Novel variant strains of infectious bursal disease virus isolated in China. *Vet. Microbiol.* **2019**, *230*, 212–220. [CrossRef]
16. Xu, A.; Pei, Y.; Zhang, K.; Xue, J.; Ruan, S.; Zhang, G. Phylogenetic analyses and pathogenicity of a variant infectious bursal disease virus strain isolated in China. *Virus Res.* **2020**, *276*, 197833. [CrossRef]
17. Jiang, N.; Wang, Y.; Zhang, W.; Niu, X.; Huang, M.; Gao, Y.; Liu, A.; Gao, L.; Li, K.; Pan, Q.; et al. Genotyping and Molecular Characterization of Infectious Bursal Disease Virus Identified in Important Poultry-Raising Areas of China During 2019 and 2020. *Front. Vet. Sci.* **2021**, *8*, 759861. [CrossRef]
18. Myint, O.; Suwanruengsri, M.; Araki, K.; Izzati, U.Z.; Pornthummawat, A.; Nueangphuet, P.; Fuke, N.; Hirai, T.; Jackwood, D.J.; Yamaguchi, R. Bursa atrophy at 28 days old caused by variant infectious bursal disease virus has a negative economic impact on broiler farms in Japan. *Avian Pathol.* **2021**, *50*, 6–17. [CrossRef]
19. Aliyu, H.B.; Hair-Bejo, M.; Omar, A.R.; Ideris, A. Genetic Diversity of Recent Infectious Bursal Disease Viruses Isolated From Vaccinated Poultry Flocks in Malaysia. *Front. Vet. Sci.* **2021**, *8*, 643976. [CrossRef]
20. Yoo, D.S.; Lee, K.N.; Chun, B.C.; Lee, H.S.; Park, H.; Kim, J.K. Preventive effect of on-farm biosecurity practices against highly pathogenic avian influenza (HPAI) H5N6 infection on commercial layer farms in the Republic of Korea during the 2016–17 epidemic: A case-control study. *Prev. Vet. Med.* **2022**, *199*, 105556. [CrossRef]
21. Sasaki, Y.; Uemura, R.; Sekiguchi, S.; Takahashi, T.; Fujii, Y.; Sueyoshi, M. An analysis of factors affecting production performance in broiler flocks on Japanese commercial farms. *Br. Poult. Sci.* **2014**, *55*, 737–744. [CrossRef]
22. Kumar, S.; Stecher, G.; Li, M.; Knyaz, C.; Tamura, K. MEGA X: Molecular Evolutionary Genetics Analysis across Computing Platforms. *Mol. Biol. Evol.* **2018**, *35*, 1547–1549. [CrossRef]
23. Letunic, I.; Bork, P. Interactive Tree Of Life (iTOL) v5: An online tool for phylogenetic tree display and annotation. *Nucleic Acids Res.* **2021**, *49*, W293–W296. [CrossRef]
24. Lutful Kabir, S.M. Avian colibacillosis and salmonellosis: A closer look at epidemiology, pathogenesis, diagnosis, control and public health concerns. *Int. J. Environ. Res. Public Health* **2010**, *7*, 89–114. [CrossRef]

25. McLure, A.; O'Neill, B.; Mayfield, H.; Lau, C.; McPherson, B. PoolTestR: An R package for estimating prevalence and regression modelling for molecular xenomonitoring and other applications with pooled samples. *Environ. Model. Softw.* **2021**, *145*, 105158. [CrossRef]
26. Kulldorff, M. A spatial scan statistic. *Commun. Stat. Theory Methods* **1997**, *26*, 1481–1496. [CrossRef]
27. Dey, S.; Pathak, D.C.; Ramamurthy, N.; Maity, H.K.; Chellappa, M.M. Infectious bursal disease virus in chickens: Prevalence, impact, and management strategies. *Vet. Med.* **2019**, *10*, 85–97. [CrossRef]
28. Su, Y.-S.; Yajima, M.; Su, M.Y.-S. SystemRequirements: JAGS. Package 'R2jags'. R Package, version 0.03-08. 2015. Available online: http://CRAN.R-project.org/package=R2jags (accessed on 20 November 2021).
29. Hitchner, S.B. Infectivity of infectious bursal disease virus for embryonating eggs. *Poult. Sci.* **1970**, *49*, 511–516. [CrossRef]
30. Reed, L.J.; Muench, H. A Simple Method of Estimating Fifty Per Cent Endpoint. *Am. J. Epidemiol.* **1938**, *27*, 493–497. [CrossRef]
31. Lucio, B.; Hitchner, S.B. Infectious Bursal Disease Emulsified Vaccine: Effect upon Neutralizing-Antibody Levels in the Dam and Subsequent Protection of the Progeny. *Avian Dis.* **1979**, *23*, 466–478. [CrossRef]
32. Henry, C.W.; Brewer, R.N.; Edgar, S.A.; Gray, B.W. Studies on infectious bursal disease in chickens. 2. Scoring microscopic lesions in the bursa of fabricius, thymus, spleen, and kidney in gnotobiotic and battery reared White Leghorns experimentally infected with infectious bursal disease virus. *Poult. Sci.* **1980**, *59*, 1006–1017. [CrossRef] [PubMed]
33. Yoo, D.S.; Lee, K.; Beatriz, M.L.; Chun, B.C.; Belkhiria, J.; Lee, K.N. Spatiotemporal risk assessment for avian influenza outbreak based on the dynamics of habitat suitability for wild birds. *Transbound. Emerg. Dis.* **2021**, *69*, e953–e967. [CrossRef] [PubMed]
34. Fan, L.; Wu, T.; Wang, Y.; Hussain, A.; Jiang, N.; Gao, L.; Li, K.; Gao, Y.; Liu, C.; Cui, H.; et al. Novel variants of infectious bursal disease virus can severely damage the bursa of fabricius of immunized chickens. *Vet. Microbiol.* **2020**, *240*, 108507. [CrossRef] [PubMed]
35. Yamazaki, K.; Ohta, H.; Kawai, T.; Yamaguchi, T.; Obi, T.; Takase, K. Characterization of variant infectious bursal disease virus from a broiler farm in Japan using immunized sentinel chickens. *J. Vet. Med. Sci.* **2017**, *79*, 175–183. [CrossRef]
36. Higgins, J.; Wallner-Pendleton, E.; Michel, L.; Jackwood, D. An unusual case of haemorrhagic bursa in a pullet flock associated with a genogroup 2 infectious bursal disease virus. *Avian Pathol.* **2021**, *50*, 269–276. [CrossRef]
37. Li, G.; Kuang, H.; Guo, H.; Cai, L.; Chu, D.; Wang, X.; Hu, J.; Rong, J. Development of a recombinant VP2 vaccine for the prevention of novel variant strains of infectious bursal disease virus. *Avian Pathol.* **2020**, *49*, 557–571. [CrossRef]
38. Wang, Y.; Jiang, N.; Fan, L.; Gao, L.; Li, K.; Gao, Y.; Niu, X.; Zhang, W.; Cui, H.; Liu, A.; et al. Development of a Viral-Like Particle Candidate Vaccine Against Novel Variant Infectious Bursal Disease Virus. *Vaccines* **2021**, *9*, 142. [CrossRef]
39. Perozo, F.; Villegas, A.P.; Fernandez, R.; Cruz, J.; Pritchard, N. Efficacy of single dose recombinant herpesvirus of turkey infectious bursal disease virus (IBDV) vaccination against a variant IBDV strain. *Avian Dis.* **2009**, *53*, 624–628. [CrossRef]
40. Kurukulsuriya, S.; Ahmed, K.A.; Ojkic, D.; Gunawardana, T.; Gupta, A.; Goonewardene, K.; Karunaratne, R.; Popowich, S.; Willson, P.; Tikoo, S.K.; et al. Circulating strains of variant infectious bursal disease virus may pose a challenge for antibiotic-free chicken farming in Canada. *Res. Vet. Sci.* **2016**, *108*, 54–59. [CrossRef]
41. Zachar, T.; Popowich, S.; Goodhope, B.; Knezacek, T.; Ojkic, D.; Willson, P.; Ahmed, K.A.; Gomis, S. A 5-year study of the incidence and economic impact of variant infectious bursal disease viruses on broiler production in Saskatchewan, Canada. *Can. J. Vet. Res.* **2016**, *80*, 255–261.
42. Lian, J.; Wang, Z.; Xu, Z.; Pang, Y.; Leng, M.; Tang, S.; Zhang, X.; Qin, J.; Chen, F.; Lin, W. Pathogenicity and molecular characterization of infectious bursal disease virus in China. *Poult. Sci.* **2022**, *101*, 101502. [CrossRef]
43. Aliyu, H.B.; Hamisu, T.M.; Hair Bejo, M.; Omar, A.R.; Ideris, A. Comparative pathogenicity of Malaysian variant and very virulent infectious bursal disease viruses in chickens. *Avian Pathol.* **2022**, *51*, 76–86. [CrossRef]
44. Jakka, P.; Reddy, Y.K.; Kirubaharan, J.J.; Chandran, N.D. Evaluation of immune responses by live infectious bursal disease vaccines to avoid vaccination failures. *Eur. J. Microbiol. Immunol.* **2014**, *4*, 123–127. [CrossRef]

viruses MDPI

Article

Detection of *Gyrovirus galga 1* in Cryopreserved Organs from Two Commercial Broiler Flocks in Japan

Masaji Mase [1,2,3,*], Yu Yamamoto [1], Hiroshi Iseki [1], Taichiro Tanikawa [1] and Aoi Kurokawa [1]

[1] National Institute of Animal Health, 3-1-5 Kannondai, Tsukuba 305-0856, Japan; yyu@affrc.go.jp (Y.Y.); hiseki@affrc.go.jp (H.I.); ttanikawa@affrc.go.jp (T.T.); kurokawaa998@affrc.go.jp (A.K.)
[2] United Graduate School of Veterinary Sciences, Gifu University, 1-1 Yanagido, Gifu 501-1193, Japan
[3] Graduate School of Life and Environmental Sciences, Osaka Prefecture University, Izumisano 598-8531, Japan
* Correspondence: masema@affrc.go.jp; Tel.: +81-29-838-7738

Abstract: *Gyrovirus galga 1* (GyVg1, previously recognized as avian gyrovirus 2), which was first reported in chicken in 2011, is a new member of the genus Gyrovirus. The presence of GyVg1 has also been confirmed in different regions of Europe, South America, Africa, and Asia, indicating its global distribution. However, because there are no reports of examining the distribution of GyVg1 in animals in Japan, the epidemiology of this virus is unknown. In this study, we attempted to retrospectively detect GyVg1 in cryopreserved chicken materials derived from different two commercial broiler flocks in 1997. The GyVg1 genome was detected in organ materials derived from both flocks by PCR. GyVg1 detected in both flocks was classified into four genetic groups by analyzing the nucleotide sequences of the detected PCR products. These results suggest that diverse GyVg1 strains were present in commercial chicken flocks as early as 1997 in Japan.

Keywords: *Gyrovirus galga 1*; Japan; PCR

Citation: Mase, M.; Yamamoto, Y.; Iseki, H.; Tanikawa, T.; Kurokawa, A. Detection of *Gyrovirus galga 1* in Cryopreserved Organs from Two Commercial Broiler Flocks in Japan. *Viruses* **2022**, *14*, 1590. https://doi.org/10.3390/v14071590

Academic Editor: Chi-Young Wang

Received: 24 June 2022
Accepted: 20 July 2022
Published: 21 July 2022

Publisher's Note: MDPI stays neutral with regard to jurisdictional claims in published maps and institutional affiliations.

Copyright: © 2022 by the authors. Licensee MDPI, Basel, Switzerland. This article is an open access article distributed under the terms and conditions of the Creative Commons Attribution (CC BY) license (https://creativecommons.org/licenses/by/4.0/).

1. Introduction

Gyrovirus galga 1 (GyVg1, previously recognized as avian gyrovirus 2) is a new member of the genus Gyrovirus with similarity to chicken infectious anemia virus (CAV). GyVg1 was first detected in the sera of chicken exhibiting clinical signs such as apathy and loss of weight in 2011 in Brazil [1]. The GyVg1 genome is approximately 2.3 kb in length and consists of circular single-stranded DNA, and it mainly comprises three overlapping open-reading frames encoding VP1, VP2, and VP3 [1,2].

Abolnik et al. stated that GyVg1 infections in chickens can result in brain damage, mental decline, and weight loss [3]. Although other specific symptoms of GyVg1 infection have not been observed, autopsy-based studies reported clinical manifestations such as hemorrhage, edema, glandular gastric erosion, and facial and head swelling in infected chickens [3].

Although chicken is considered the natural host of GyVg1 [4], experimental infection in chickens using isolated has have not been conducted because the isolation of virus using cultured cells has not been described. Thus, the pathogenicity to chicken of GyVg1 are still unknown.

GyVg1 has also been detected in different regions of Europe [5], South America [5], Africa [6], and Asia [4,7], indicating its global distribution. In addition, Sauvage et al. [8] identified human gyrovirus, which shares 96% nucleotide identity with GyVg1, from the skin swab of a healthy individual, indicating that GyVg1 might also infect humans. Subsequently, GyVg1 was detected in human blood samples [9,10], human feces [11,12], ferret feces [13], snakes [14], ticks [15], and dogs [16], suggesting that the virus can infect a broad range of animals.

In Japan, since the distribution of GyVg1 in animals including chicken has not been investigated at all to date, the epidemiology of this virus in animals is unknown. In this study, we attempted to retrospectively detect GyVg1 in various organs of chicken obtained from two different commercial broiler farms in 1997. To our knowledge, this study is the first report to show the presence of GyVg1 in Japanese chickens. The GyVg1 genome was detected in the examined samples, and thus, we further conducted detailed genetic analysis.

2. Materials and Methods

2.1. Cryopreserved Chicken Organ Samples

At two commercial broiler flocks (Farms M and N) in Japan in 1997, sampling was periodically performed four times for routine viral surveillance (Tables 1 and 2). Significant clinical signs were not observed in these flocks. Major organs, such as the trachea, lungs, liver, spleen, kidneys, pancreas, and rectum, were collected from five birds, and 10–20% tissue homogenates were prepared from these organs. These homogenates were stored at −80 °C at our laboratory.

Table 1. Detection of GyVg1 in various organs in chicken from farm M.

Age (Days)	Positive Birds/Examined Birds	Chick no.	Trachea	Lung	Liver	Spleen	Kidney	Pancreas	Rectum	Genetic Groups
18	0/5	1–5	-	-	-	-	-	-	-	
		Detection rate	0/5	0/5	0/5	0/5	0/5	0/5	0/5	
		1	-	-	-	-	-	-	-	
		2	-	-	-	-	-	-	-	
27	2/5	3	-	-	+	+	-	-	-	G1
		4	+	+	+	+	+	+	+	G1
		5	-	-	-	-	-	-	-	
		Detection rate	1/5	1/5	2/5	2/5	1/5	1/5	1/5	
		1	-	-	+	+	-	-	-	G1
		2	+	+	+ *	+	+	+	+	G1
35	5/5	3	-	+	+	+	+	+	+	G1
		4	-	+	+	+	+	+	+	G1
		5	-	+	+	+	+	-	-	G1
		Detection rate	1/5	4/5	5/5	5/5	4/5	3/5	3/5	
		1	+	+	+	+	+	+	+	G1
		2	+	+	+	+	+	+	+	G1
48	5/5	3	-	-	+	+	+	-	-	G1
		4	-	+	+	+	+	-	-	G1
		5	+	+	+	+	+	+	+	G1
		Detection rate	3/5	4/5	5/5	5/5	5/5	3/5	3/5	
Total Detection rate	12/20		5/20	9/20	12/20	12/20	10/20	7/20	7/20	

* used as representative strain.

Table 2. Detection of GyVg1 in various organs in chicken from farm N.

Age (Days)	Positive Birds/Examined Birds	Chick no.	Trachea	Lung	Liver	Spleen	Kidney	Pancreas	Rectum	Genetic Groups
14	0/5	1–5	-	-	-	-	-	-	-	
23	0/5	1–5	-	-	-	-	-	-	-	
		Detection rate	0/10	0/10	0/10	0/10	0/10	0/10	0/10	
		1	-	-	-	-	-	-	-	
		2	-	-	-	-	-	-	-	
31	2/5	3	-	-	-	-	-	-	-	
		4	-	-	+ *	+	-	-	-	G2
		5	-	-	-	+ *	-	-	-	G3
		Detection rate	0/5	0/5	1/5	2/5	0/5	0/5	0/5	
		1	+	+	+	+ *	+	+	+	G4
		2	+	-	+	-	+	+	+	G4
44	5/5	3	+	-	+	+	+	-	-	G4
		4	+	-	+	-	+	-	-	G2
		5	+	-	+	+	+	+	+	G2
			5/5	1/5	5/5	3/5	5/5	3/5	3/5	
Total Detection rate	7/20		5/20	1/20	6/20	5/20	5/20	3/20	3/20	

* used as representative strain.

2.2. Detection of GyVg1 by PCR

The homogenates were then centrifuged at 3000× *g* for 10 min. Total DNA was extracted from each organ homogenate using a QIAamp DNA Micro Kit (Qiagen Inc., Valencia, CA, USA) following the manufacturer's instructions. The DNA was resuspended in nuclease-free water and used as the template for PCR.

PCR to detect GyVg1 was conducted using GyVg1-specific primers (F, 5′-CGTGTCCGCCAGCAGAAACGAC-3′; R, 5′-GGTAGAAGCCAAAGCGTCCACGA-3′; nucleotides 656–1001). These primers were previously designed from the highly coserved region of the GyVg1 genome, and they amplify a 346-bp fragment and target a genomic region that encodes part of the VP2 and VP3 genes [7]. The PCR mixture contained 5 µL of 10× PCR buffer, 0.5 µL of each primer (20 pM), 1.25 U of Ex Taq DNA polymerase (Takara Bio Co., Ltd., Shiga, Japan), 4 µL of 10 mM dNTP, 1 µL of the sample DNAs, and sterilized Milli-Q water to a total volume of 50 µL. The PCR protocol consisted of 35 cycles of 94 °C for 30 s, 60 °C for 30 s, and 72 °C for 30 s. The PCR products were analyzed by 1.5% agarose gel electrophoresis. The obtained PCR products were purified with Montage (Merck, Darmstadt, Germany) according to the manufacturer's instructions. The PCR products were sequenced directly in both directions using Sanger sequencing and confirmed to be derived from GyVg1 by BLASTn searches.

2.3. Near-Complete Nucleotide Sequence for Representative GyVg1 Strains

The near-complete genome sequences of GyVg1 for representative detected strains were determined using primer sequences based on previous studies [4,8]. The sequenced fragments were assembled using ATGC-Mac ver.5 (GENETYX CORPORATION, Tokyo, Japan). The obtained sequences were deposited in GenBank (accession number: LC716405-8).

2.4. Genetic and Phylogenetic Analyses

The obtained nucleotide sequences were analyzed using GENETYX-Mac software (GENETYX CORPORATION, Tokyo, Japan) and compared with known GyVg1 sequences deposited in GenBank.

We conducted phylogenetic analysis of the nucleotide sequences of GyVg1 with available sequences from GenBank using the ClustalX sequence alignment program, and a phylogenetic tree was constructed by the neighbor-joining method using MEGA7 software [17]. All tools were run with default parameters unless otherwise specified.

3. Results

3.1. Detection of GyVg1 from Cryopreserved Chicken Organ Samples

The results of GyVg1 detection in the organs examined in this study are presented in Tables 1 and 2. Figure 1 presents an example of agarose gel electrophoresis of PCR products. The obtained sequences revealed that all PCR products were derived from GyVg1 genomes.

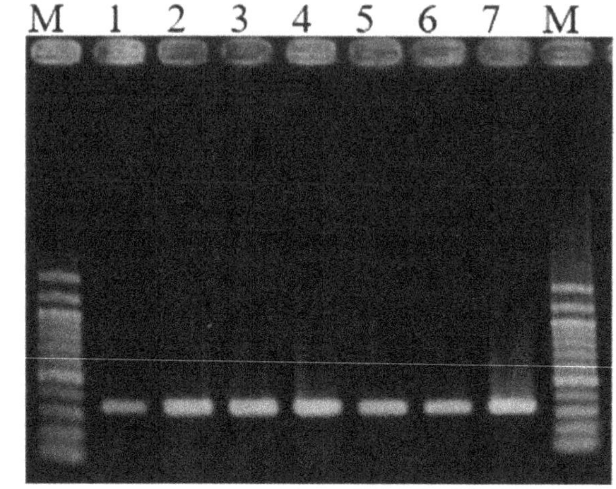

Figure 1. An example of PCR products analyzed by agarose gel electrophoresis. PCR was performed using samples obtained from chicken no. 1 (44 days old) on farm N. M, 100-bp ladder marker; 1, trachea; 2, lung; 3, liver; 4, spleen; 5, kidney; 6, pancreas; 7, rectum.

GyVg1 was detected in chicken older than 27 days at both farms (Tables 1 and 2). The virus was detected in 12 chicken on farm M and 7 chicken on farm N. The organ with the highest detection rate was the liver (18 samples/19 positive birds), followed by the spleen (17 samples/19 positive birds) (Tables 1 and 2). However, in some samples, GyVg1 was detected in all major organs examined, such as chicken No. 1 and 5 (48 days old) on M farm and chicken no. 1 (44 days old) on N farm. The detection rates of GyVg1 were lower in the trachea and lungs than in the liver and spleen.

3.2. *Sequence Analysis of GyVg1 Detected in Organ Samples by PCR*

By determining the nucleotide sequences of the PCR products derived from each organ, GyVg1 samples detected on farm M were identical (Table 1). Conversely, GyVg1 samples detected on farm N could be divided into three genetic groups (Table 2). Interestingly, two different groups of GyVg1 were detected in 31-day-old chicken, and two other groups were detected in 44-day-old chicken (Table 2). Based on phylogenetic analysis, four different groups were identified and termed G1–4 (Figure 2).

All GyVg1 samples detected in chicken on farm M were type G1. Conversely, regarding samples obtained from 31-day-old chicken on N farm, the GyVg1 sample detected in chicken no. 4 was type G2, and the sample detected in chicken no. 5 was type G3. Furthermore, among 44-day-old chicken, GyVg1 samples detected in chicken Nos. 1–3 were type G4, and samples detected in chicken Nos. 4–5 were type G2 (Table 2).

GyVg1 detected on farm M (G1) displayed 100% nucleotide sequence identity with G13 and HLJ1506-2 in BLAST searches [2,13] (Figure 2, Table 3). GyVg1 detected in the liver of chicken no. 4 (31 days old) on farm N (G2) exhibited 100% nucleotide sequence identity with strains JL1511 and HLJ1508, which were isolated from chickens in China [2]. GyVg1 detected in spleen of chicken no. 5 (31 days old) on farm N (G3) displayed 99.7% nucleotide sequence identity with strains GZ1601 and HLJ1506-1 [2]. GyVg1 detected in the spleen of chicken no. 1 (44 days old) on farm N (G4) exhibited 100% nucleotide sequence identity with strain G17 [13], which was isolated from a ferret in Hungary. Thus, the four genetically different strains of GyVg1 detected on two different farms examined in this study displayed high genetic similarity to foreign strains. Additionally, multiple genotypes existed in the same farm, and multiple genotypes were present in chickens of the same age.

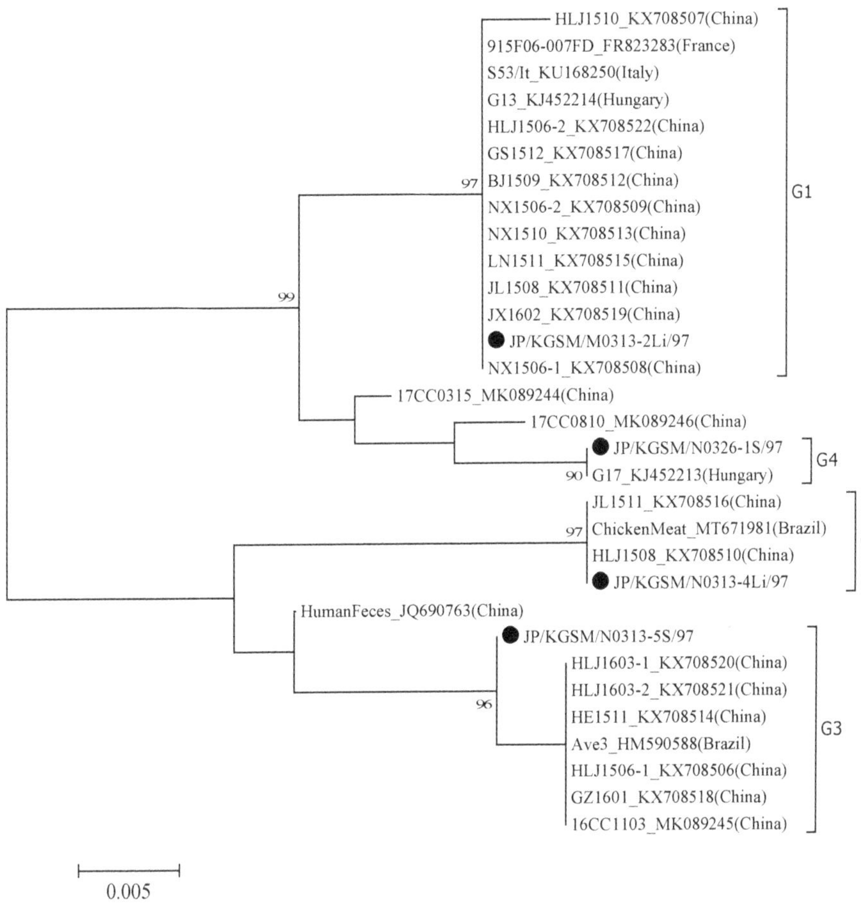

Figure 2. Phylogenetic trees based on the partial gene of GyVg1 strain G13 (GenBank accession no. KJ452214). Nucleotides 678–978 were subjected to phylogenetic analysis. Subsequently, a tree was generated using the neighbor-joining method in MEGA 7 [17] with 1000 bootstrap replicates. All tools were run with default parameters unless otherwise specified. Then, horizontal distances were proportionally set to the minimum number of nucleotide differences required to join nodes and sequences. The Japanese GyVg1 strains are indicated by black circles.

Table 3. Genetic comparison of nucleotide sequences of PCR products detected in first screening PCR by GenBank searches.

Representative Strain Name	Chicken	GyVg1 Groups in This Study	Viruses with the Highest Homology	Percent Homology
JP/KGSM/M0313-2Li/97	Farm M, 35 days old, no.2 chicken liver	G1	G13(KJ452214), HLJ1506-2 (KX708522)	100
JP/KGSM/N0313-4Li/97	Farm N, 31 days old, no.4 chicken liver	G2	JL1511 (KX708516), HLJ1508 (KX708510)	100
JP/KGSM/N0313-5S/97	Farm N, 31 days old, no.5 chicken spleen	G3	GZ1601 (KX708518), HLJ1506-1 (KX708506)	99.7
JP/KGSM/N0326-1S/97	Farm N, 44 days old, no.1 chicken spleen	G4	G17 (KJ452213)	100

3.3. Near-Complete Genome Analysis of Representative GyVg1 Strains

To obtain an in-depth understanding of the genetic characteristics of GyVg1 in Japan, nearly complete genome sequences of selected samples including the VP1, VP2, and VP3 coding regions were determined excluding the GC-rich region of approximately 40 bp in the 3′-untranslated region (UTR). It has been reported that GyVg1 can be classified into different clusters based on the amino acid sequences of VP1, VP2 and VP3 proteins [2].

Via comparisons with GyVg1 nucleotide sequences available in GenBank by BLASTn searches, GyVg1 detected on farm M (G1, representative strain JP/KGSM/M0313-2Li/97) exhibited 99.8% nucleotide sequence identity with strain G13 [13] (Figure 3).

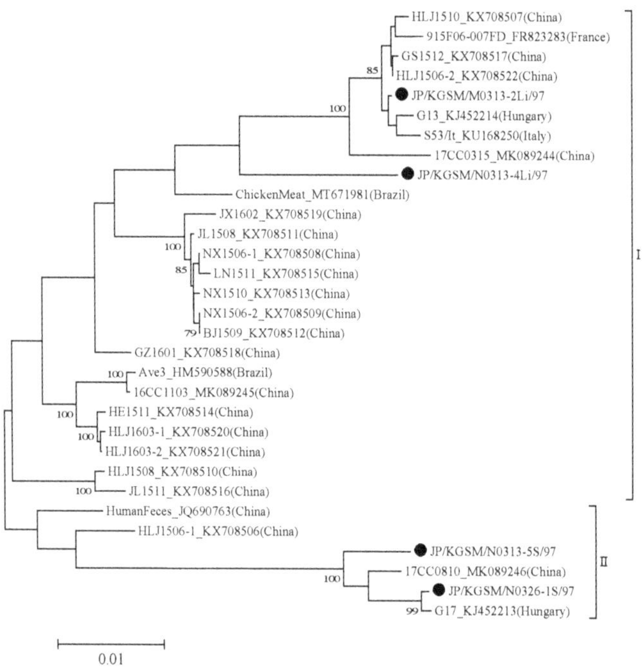

Figure 3. Phylogenetic trees based on the near-complete genome of GyVg1 strain G13 (GenBank accession no. KJ452214). Nucleotides 1–2335 were subjected to phylogenetic analysis. Subsequently, a tree was generated using the neighbor-joining method in MEGA 7 [17] with 1000 bootstrap replicates. All tools were run with default parameters unless otherwise specified. Then, horizontal distances were proportionally set to the minimum number of nucleotide differences required to join nodes and sequences. The Japanese GyVg1 strains are indicated by black circles. The genotypes were defined as described by Yao et al. [2].

Conversely, GyVg1 detected in the liver of chicken no. 4 on farm N (G2, representative strain JP/KGSM/N0313-4Li/97) exhibited 98.5% nucleotide sequence identity with a strain from chicken meat (GenBank accession no. MT671981). GyVg1 detected in the spleen of chicken no. 5 on farm N (G3, representative strain JP/KGSM/N0313-5S/97) exhibited 98.9% nucleotide sequence identity with strain G17 [13]. GyVg1 detected in the spleen of chicken no. 1 on farm N (G4, representative strain JP/KGSM/N0326-1S/97) displayed 99.8% nucleotide sequence identity with strain G17 [13].

3.4. VP1 Amino Acid Sequences of Representative GyVg1 Strains

VP1 is the major viral structural protein of CAV, and it contains an abundant neutralization antigenic epitope and plays a key role in virus growth and transmission [18]. All

VP1 amino acids sequences currently reported are 460 amino acids in length [2], including those found in Japan. At present, the reported GyVg1 strains have been classified into three genetic clusters (I–III) by phylogenetic analysis using the VP1 amino acid sequence [2], but the Japanese strains detected in this study were divided into two genetic clusters (I and II) (Table 4, Figure 4A). Strains JP/KGSM/M0313-2Li/97 and JP/KGSM/N0313-4Li/97 classified into cluster I exhibited 100% identity with the HLJ1506-2 strain detected in China. In addition, JP/KGSM/N0313-5S/97 and JP/KGSM/N0326-1S/97 classified into Cluster II exhibited 100% identity with the G17 strain detected in Hungary [13].

Table 4. Summary of classification based on different genetic regions.

Representative Strain Name	Cluster				
	Regions Detected by First Screening PCR	Near Complete Genome	VP1	VP2	VP3
JP/KGSM/M0313-2Li/97	G1	I	I	I	I
JP/KGSM/N0313-4Li/97	G2	I	I	III	III
JP/KGSM/N0313-5S/97	G3	II	II	IV	V
JP/KGSM/N0326-1S/97	G4	II	II	II	II

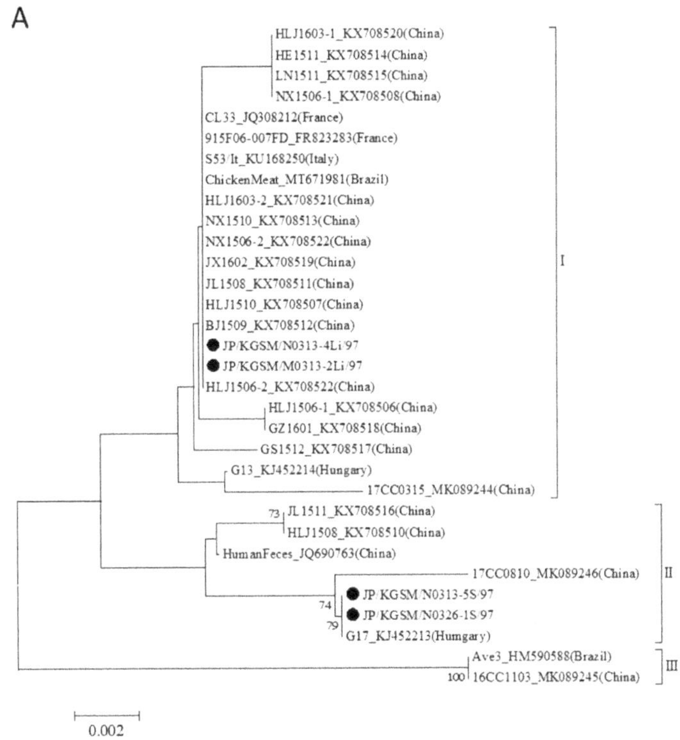

Figure 4. *Cont.*

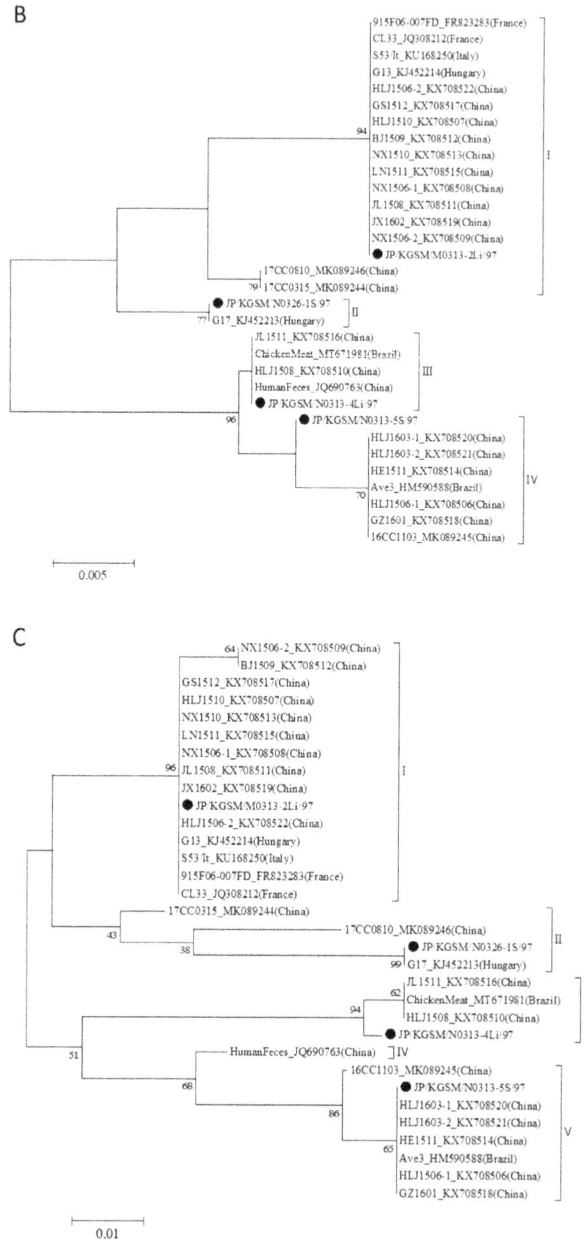

Figure 4. Phylogenetic analysis of the amino acid sequences of VP1 (**A**), VP2 (**B**), and VP3 (**C**) of GyVg1. The predicted amino acid of complete sequences of each protein was subjected to phylogenetic analysis. Subsequently, a tree was generated using the neighbor-joining method in MEGA 7 with 1000 bootstrap replicates. All tools were run with default parameters unless otherwise specified. Then, horizontal distances were proportionally set to the minimum number of nucleotide differences required to join nodes and sequences. The Japanese GyVg1 strains are indicated by black circles. The genotypes were defined as described by Yao et al. [2].

3.5. VP2 Amino Acid Sequences of Representative GyVg1 Strains

Most reported VP2 amino acid sequences are 231 amino acids in length, and those found in Japan, excluding one strain (JP/KGSM/N0326-1S/97, G4), were also this length. This strain is identical to the G17 strain, which has a serine insertion at amino acid 162 as reported previously. The VP2 sequences were well conserved within their major phylogenetic clusters [2]. The currently reported GyVg1 strains were classified into four genetic clusters (I–IV) by phylogenetic analysis using the VP2 amino acid sequence, and interestingly, the Japanese strains detected in this study were also divided into four clusters (Table 4, Figure 4B). The JP/KGSM/N0326-1S/97 strain was classified into Cluster II similarly as the G17 strain. The sequence of phosphatase motif WLRQCARSHDEICTCGRWRSH (amino acids 95–115) was also conserved in Japanese GyVg1 strains, which indicated the importance of VP2 for GyVg1 [2].

3.6. VP3 Amino Acid Sequences of Representative GyVg1 Strains

The VP3 protein in CAV is also known as apoptin, which was demonstrated to induce apoptosis of hematopoietic cells, causing the observed anemia in infected chicken [19]. It was revealed that GyVg1 VP3 protein can also induce apoptosis of tumor cells [20]. Regarding the predicted protein sequences of VP3 of GyVg1, three strains, excluding the JP/KGSM/N0326-1S/97 strain, were 124 amino acids in length. The JP/KGSM/N0326-1S/97 strain had an arginine insertion at amino acid 122, similarly as the G17 strain. Currently reported GyVg1 strains were classified into five genetic clusters (I–V) by phylogenetic analysis using the VP3 amino acid sequence, but interestingly, the Japanese strains detected in this study were divided into four clusters (Table 4, Figure 4C). The aforementioned JP/KGSM/N0326-1S/97 strain was classified into Cluster II similarly as the G17 strain.

3.7. Analysis of Direct Repeats (DRs) of Representative GyVg1 Strains

A previous study described novel motif patterns in the DR region in the 5′-UTR [2]. The 5′-UTR of GyVg1 is by definition located between the canonical polyadenylation site (AATAAA) and the transcription site. Although the function of the noncoding region is not yet known, it may have the same promoter-enhancer function as its CAV counterpart, which includes four or five 21-base DRs and an indispensable 12-bp insert. According to these criteria [2], we observed three different types of DR sequences in Japanese GyVg1 strains. G1 was classified as pattern E, G2 was classified as pattern D, and G3 and G4 were classified as pattern B.

4. Discussion

The existence of GyVg1 was first confirmed in 2008 in Brazil [1], but the present study revealed that it was already present in commercial broiler flocks in the 1990s in Japan.

In this report, we detected GyVg1 in materials collected periodically at two different farms. The results illustrated that chickens older than 27 days were infected by the virus. GyVg1 was detected in many organs in infected chickens, most commonly in the liver and spleen. This is consistent with findings that GyVg1 in chickens in China is most commonly isolated from the liver and spleen, suggesting that these organs are suitable for detecting this virus.

GyVg1 was not detected in younger chickens, but it becomes detectable as chickens age. Regarding CAV infection in commercial broiler flocks, at 35 days of age, the virus was detectable by PCR in 16 of 20 chickens, and by 42 days of age, the virus was detectable in 18 of 20 chickens [21]. This suggests that the mode of infections of GyVg1 in commercial broiler flocks is similar to that of CAV. Importantly, the existence of multiple genotypes of GyVg1 in the flocks, even at the same farm and in chickens of the same age, indicates that viral surveillance should be conducted using individual samples as opposed to pooled samples.

GyVg1 can be classified into multiple genotypes by phylogenetic analysis based on the amino acid sequences of VP1, VP2, and VP3, but the Japanese GyVg1 strains detected in this study were also classified into multiple genotypes. This indicates that multiple genotypes were already present on Japanese farms in the 20th century. Additionally, the amino acid sequences of VP1, VP2, and VP3 remain well-conserved within their major phylogenetic clusters [2].

To date, the epidemiology, host range, transmission route, and pathogenesis of GyVg1 remain poorly understood. In addition, the pathogenicity of GyVg1 against chicken is unclear. The only report suggesting the pathogenicity of this virus is a case reported in South Africa [3]. Low-pathogenic NDV and GyVg1 were detected in this case, but their involvement, including interactions, was not clarified. The possibility of a synergistic pathogenic effect between avirulent NDV and GyVg1 requires further investigation. Interestingly, as GyVg1 infection in ticks was reported recently, it may also be possible that the virus spreads among flocks via ticks [15]. Furthermore, Varela et al. reported the detection of GyVg1 in poultry vaccines, indicating the potential role of contaminated vaccines in the spread of GyVg1 [22].

Most GyVg1 strains are detected mainly in chickens, but they have been also detected in various hosts such as ferrets, snakes, and dogs, suggesting possible cross-species transmission [7,11–14,16]. Additionally, GyVg1 has been detected in the blood of healthy humans, transplant recipients and HIV-positive patients [9,10]. Cases of infection apparently caused by the consumption of infected chicken have been reported in humans and snakes.

Therefore, it is important to understand the epidemiology of GyVg1 in chicken; however, little is known about the viral infection or antibody production in chicken. As the genomic information of GyVg1 has been determined, antibody testing using proteins expressed using recombinant gene techniques would be useful for clarifying the infiltration status in animals. However, the isolation of the virus from clinical samples is absolutely necessary to evaluate its pathogenicity in animals. CAV is known to grow in some cell lines such as MDCC-MSB1 cells [23]. To evaluate the pathogenicity of a virus in animals, it is necessary to establish an effective culture system for GyVg1.

In conclusion, this is the first report to show the presence of GyVg1 in Japanese chickens. GyVg1 was detected in chicken materials derived from two different commercial broiler flocks in 1997. These results suggest that GyVg1 was already present in commercial flocks prior to the first report of GyVg1 infection in 2011 in Brazil. Furthermore, multiple genotypes existed in the same farm, and multiple genotypes were present in chickens of the same age. Unfortunately, the current status of GyVg1 infection in chickens in Japan is unknown. The findings obtained in this study should be useful for further epidemiological research of GyVg1 in commercial flocks.

Author Contributions: M.M.: designed the study and performed experiments; M.M., Y.Y., H.I., T.T. and A.K.: analyzed the data and wrote the manuscript. All authors have read and agreed to the published version of the manuscript.

Funding: This research received no external funding.

Acknowledgments: The authors would like to thank the Shinichiro Hashimoto for his cooperation in collecting the chicken materials.

Conflicts of Interest: The authors declare no conflict of interest.

References

1. Rijsewijk, F.A.M.; dos Santos, H.F.; Teixeira, T.F.; Cibulski, S.P.; Varela, A.P.M.; Dezen, D.; Franco, A.C.; Roehe, P.M. Discovery of a genome of a distant relative of chicken anemia virus reveals a new member of the genus Gyrovirus. *Arch. Virol.* **2011**, *156*, 1097–1100. [CrossRef] [PubMed]
2. Yao, S.; Gao, X.; Tuo, T.; Han, C.; Gao, Y.; Qi, X.; Zhang, Y.; Liu, C.; Gao, H.; Wang, Y.; et al. Novel characteristics of the avian gyrovirus 2 genome. *Sci. Rep.* **2017**, *7*, 41068. [CrossRef] [PubMed]
3. Abolnik, C.; Wandrag, D.B.R. Avian gyrovirus 2 and avirulent newcastle disease virus coinfection in a chicken flock with neurologic symptoms and high mortalities. *Avian Dis.* **2014**, *58*, 90–94. [CrossRef] [PubMed]

4. Yao, S.; Tuo, T.; Gao, X.; Han, C.; Li, Y.; Gao, Y.; Zhang, Y.; Liu, C.; Qi, X.; Gao, H.; et al. Avian gyrovirus 2 in poultry, China, 2015–2016. *Emerg. Microbes Infect.* **2016**, *5*, e112. [CrossRef]
5. dos Santos, H.F.; Knak, M.B.; de Castro, F.L.; Slongo, J.; Ritterbusch, G.A.; Klein, T.A.P.; Esteves, P.A.; Silva, A.D.; Trevisol, I.M.; Claassen, E.A.W.; et al. Variants of the recently discovered avian gyrovirus 2 are detected in Southern Brazil and The Netherlands. *Vet. Microbiol.* **2012**, *155*, 230–236. [CrossRef]
6. Smuts, H.E.M. Novel gyroviruses, including chicken anaemia virus, in clinical and chicken samples from South Africa. *Adv. Virol.* **2014**, *2014*, 321284. [CrossRef]
7. Ye, J.; Tian, X.; Xie, Q.; Zhang, Y.; Sheng, Y.; Zhang, Z.; Wang, C.; Zhu, H.; Wang, Y.; Shao, H.; et al. Avian gyrovirus 2 DNA in fowl from live poultry markets and in healthy humans, China. *Emerg. Infect. Dis.* **2015**, *21*, 1486–1488. [CrossRef]
8. Sauvage, V.; Cheval, J.; Foulongne, V.; Gouilh, M.A.; Pariente, K.; Manuguerra, J.C.; Richardson, J.; Dereure, O.; Lecuit, M.; Burguiere, A.; et al. Identification of the First Human Gyrovirus, a Virus Related to Chicken Anemia Virus. *J. Virol.* **2011**, *85*, 7948–7950. [CrossRef]
9. Maggi, F.; Macera, L.; Focosi, D.; Vatteroni, M.L.; Boggi, U.; Antonelli, G.; Eloit, M.; Pistello, M. Human gyrovirus DNA in human blood, Italy. *Emerg. Infect. Dis.* **2012**, *18*, 956–959. [CrossRef]
10. Biagini, P.; Bédarida, S.; Touinssi, M.; Galicher, V.; de Micco, P. Human gyrovirus in healthy blood donors, France. *Emerg. Infect. Dis.* **2013**, *19*, 1014–1015. [CrossRef]
11. Chu, D.K.W.; Poon, L.L.M.; Chiu, S.S.S.; Chan, K.H.; Ng, E.M.; Bauer, I.; Cheung, T.K.; Ng, I.H.Y.; Guan, Y.; Wang, D.; et al. Characterization of a novel gyrovirus in human stool and chicken meat. *J. Clin. Virol.* **2012**, *55*, 209–213. [CrossRef] [PubMed]
12. Phan, T.G.; da Costa, A.C.; Zhang, W.; Pothier, P.; Ambert-Balay, K.; Deng, X.; Delwart, E. A new gyrovirus in human feces. *Virus Genes* **2015**, *51*, 132–135. [CrossRef] [PubMed]
13. Fehér, E.; Pazár, P.; Kovács, E.; Farkas, S.L.; Lengyel, G.; Jakab, F.; Martella, V.; Bányai, K. Molecular detection and characterization of human gyroviruses identified in the ferret fecal virome. *Arch. Virol.* **2014**, *159*, 3401–3406. [CrossRef] [PubMed]
14. Wu, Q.; Xu, X.; Chen, Q.; Ji, J.; Kan, Y.; Yao, L.; Xie, Q. Genetic analysis of avian gyrovirus 2 variant-related Gyrovirus detected in farmed king ratsnake (Elaphe carinata): The first report from China. *Pathogens* **2019**, *8*, 185. [CrossRef] [PubMed]
15. Yang, Z.; Zhang, J.; Yang, S.; Wang, X.; Shen, Q.; Sun, G.; Wang, H.; Zhang, W. Virome analysis of ticks in a forest region of Liaoning, China: Characterization of a novel hepe-like virus sequence. *Virol. J.* **2021**, *18*, 163. [CrossRef]
16. Liu, Y.; Lv, Q.; Li, Y.; Yu, Z.; Huang, H.; Lan, T.; Wang, W.; Cao, L.; Shi, Y.; Sun, W.; et al. Cross-species transmission potential of chicken anemia virus and avian gyrovirus 2. *Infect. Genet. Evol.* **2022**, *99*, 105249. [CrossRef]
17. Kumar, S.; Stecher, G.; Tamura, K. MEGA7: Molecular Evolutionary Genetics Analysis Version 7.0 for Bigger Datasets. *Mol. Biol. Evol.* **2016**, *33*, 1870–1874. [CrossRef]
18. Renshaw, R.W.; Soiné, C.; Weinkle, T.; O'Connell, P.H.; Ohashi, K.; Watson, S.; Lucio, B.; Harrington, S.; Schat, K.A. A hypervariable region in VP1 of chicken infectious anemia virus mediates rate of spread and cell tropism in tissue culture. *J. Virol.* **1996**, *70*, 8872–8878. [CrossRef]
19. Feng, C.; Liang, Y.; Teodoro, J.G. The role of apoptin in chicken anemia virus replication. *Pathogens* **2020**, *9*, 294. [CrossRef]
20. Bullenkamp, J.; Cole, D.; Malik, F.; Alkhatabi, H.; Kulasekararaj, A.; Odell, E.W.; Farzaneh, F.; Gäken, J.; Tavassoli, M. Human Gyrovirus Apoptin shows a similar subcellular distribution pattern and apoptosis induction as the chicken anaemia virus derived VP3/Apoptin. *Cell Death Dis.* **2012**, *3*, e296. [CrossRef]
21. Sommer, F.; Cardona, C. Chicken Anemia Virus in Broilers: Dynamics of the Infection in Two Commercial Broiler Flocks. *Avian Dis.* **2003**, *47*, 1466–1473. [CrossRef] [PubMed]
22. Varela, A.P.M.; dos Santos, H.F.; Cibulski, S.P.; Scheffer, C.M.; Schmidt, C.; Sales Lima, F.E.; Silva, A.D.A.; Esteves, P.A.; Franco, A.C.; Roehe, P.M. Chicken anemia virus and avian gyrovirus 2 as contaminants in poultry vaccines. *Biologicals* **2014**, *42*, 346–350. [CrossRef] [PubMed]
23. Yuasa, N. Propagation and infectivity titration of the Gifu-1 strain of chicken anemia agent in a cell line (MDCC-MSB1) derived from Marek's disease lymphoma. *Natl. Inst. Anim. Health Q.* **1983**, *23*, 13–20.

viruses

MDPI

Review

Pigeon Circovirus over Three Decades of Research: Bibliometrics, Scoping Review, and Perspectives

Benji Brayan Ilagan Silva [1], Michael Louie R. Urzo [2,3], Jaymee R. Encabo [2], Alea Maurice Simbulan [2], Allen Jerard D. Lunaria [2], Susan A. Sedano [4], Keng-Chih Hsu [5], Chia-Chi Chen [5,6], Yu-Chang Tyan [7,8,9,10,11,12,13,14,*] and Kuo-Pin Chuang [1,8,14,15,16,*]

1 International Degree Program in Animal Vaccine Technology, International College, National Pingtung University of Science and Technology, Pingtung 912, Taiwan; g10985004@mail.npust.edu.tw
2 Microbiology Division, Institute of Biological Sciences, College of Arts and Sciences, University of the Philippines Los Baños, Los Baños 4031, Laguna, Philippines; mrurzo@up.edu.ph (M.L.R.U.); jrencabo@up.edu.ph (J.R.E.); asimbulan@up.edu.ph (A.M.S.); adlunaria@up.edu.ph (A.J.D.L.)
3 Graduate School, University of the Philippines Los Baños, Los Baños 4031, Laguna, Philippines
4 Veterinary Vaccines Laboratory, National Institute of Molecular Biology and Biotechnology, University of the Philippines Los Baños, Los Baños 4031, Laguna, Philippines; sasedano@up.edu.ph
5 Language Center, National Pingtung University of Science and Technology, Pingtung 912, Taiwan; awilliamhsu@yahoo.com.tw (K.-C.H.); ss77ss77.tw@gmail.com (C.-C.C.)
6 You Guan Yi Biotechnology Company, Kaohsiung 807, Taiwan
7 Department of Medical Imaging and Radiological Sciences, Kaohsiung Medical University, Kaohsiung 807, Taiwan
8 School of Medicine, Kaohsiung Medical University, Kaohsiung 807, Taiwan
9 Graduate Institute of Medicine, College of Medicine, Kaohsiung Medical University, Kaohsiung 807, Taiwan
10 Institute of Medical Science and Technology, National Sun Yat-Sen University, Kaohsiung 804, Taiwan
11 Department of Medical Research, Kaohsiung Medical University Hospital, Kaohsiung 807, Taiwan
12 Center for Cancer Research, Kaohsiung Medical University, Kaohsiung 807, Taiwan
13 Research Center for Environmental Medicine, Kaohsiung Medical University, Kaohsiung 807, Taiwan
14 Graduate Institute of Animal Vaccine Technology, College of Veterinary Medicine, National Pingtung University of Science and Technology, Pingtung 912, Taiwan
15 School of Dentistry, Kaohsiung Medical University, Kaohsiung 807, Taiwan
16 Companion Animal Research Center, National Pingtung University of Science and Technology, Pingtung 912, Taiwan
* Correspondence: yctyan@kmu.edu.tw (Y.-C.T.); kpchuang@g4e.npust.edu.tw (K.-P.C.)

Citation: Silva, B.B.I.; Urzo, M.L.R.; Encabo, J.R.; Simbulan, A.M.; Lunaria, A.J.D.; Sedano, S.A.; Hsu, K.-C.; Chen, C.-C.; Tyan, Y.-C.; Chuang, K.-P. Pigeon Circovirus over Three Decades of Research: Bibliometrics, Scoping Review, and Perspectives. *Viruses* **2022**, *14*, 1498. https://doi.org/10.3390/v14071498

Academic Editor: Chi-Young Wang

Received: 17 June 2022
Accepted: 4 July 2022
Published: 8 July 2022

Publisher's Note: MDPI stays neutral with regard to jurisdictional claims in published maps and institutional affiliations.

Copyright: © 2022 by the authors. Licensee MDPI, Basel, Switzerland. This article is an open access article distributed under the terms and conditions of the Creative Commons Attribution (CC BY) license (https://creativecommons.org/licenses/by/4.0/).

Abstract: The pigeon circovirus (PiCV), first described in the literature in the early 1990s, is considered one of the most important infectious agents affecting pigeon health. Thirty years after its discovery, the current review has employed bibliometric strategies to map the entire accessible PiCV-related research corpus with the aim of understanding its present research landscape, particularly in consideration of its historical context. Subsequently, developments, current knowledge, and important updates were provided. Additionally, this review also provides a textual analysis examining the relationship between PiCV and the young pigeon disease syndrome (YPDS), as described and propagated in the literature. Our examination revealed that usages of the term 'YPDS' in the literature are characterizations that are diverse in range, and neither standard nor equivalent. Guided by our understanding of the PiCV research corpus, a conceptualization of PiCV diseases was also presented in this review. Proposed definitions and diagnostic criteria for PiCV subclinical infection (PiCV-SI) and PiCV systemic disease (PiCV-SD) were also provided. Lastly, knowledge gaps and open research questions relevant to future PiCV-related studies were identified and discussed.

Keywords: bibliometrics; circovirus; pigeon circovirus; young pigeon disease syndrome

1. Introduction

Published in 1992 in the "In my experience ... " section of the *Journal of the Association of Avian Veterinarians* is a five-sentence report marking the earliest record on circovirus infections in pigeons. The report describes an infection that is morphologically diagnosable by examination of the presence of inclusion bodies in necropsy tissues, commonly or sometime only found in the bursa of Fabricius of the birds. It was noted that the virus is "apparently distinct from the psittacine virus" [1]. Of importance here is that neither the psittacine virus mentioned, which was later renamed as the beak and feather disease virus (BFDV), nor the family *Circoviridae* were officially recognized yet [2].

While no earlier records were recoverable, the report mentioned above also seemingly reflects what is then an on-going circulation of information among practitioners regarding the existence of and diagnostic methods for the detection of this virus. According to subsequent reports, this virus was first recognized in the late 1980s in Canada and Australia [3]. It is similarly from this report that the current review started tracing the progress in pigeon circovirus (PiCV) knowledge in the 30 years of research and studies that followed up to this day.

Aided by bibliometrics, the current review sought to map the PiCV research landscape as it expanded for the past three decades. Bibliometrics, a multidisciplinary approach crossing fields of mathematics, statistics, and philology to analyze knowledge carriers quantitatively, is commonly employed to assist in organizing records from a field of study or a specific research topic to identify historical trends and locate research hotspots and frontiers. Bibliometric tools are employed to perform a variety of analyses producing measures related to authorship, sources of records, collaborations, citations, and keywords association and co-occurrence, among others. Previously, reviews employing bibliometrics and original bibliometric studies on several viruses, including coronaviruses, oncolytic viruses, human immunodeficiency virus, norovirus, Japanese encephalitis virus, African swine fever virus, and other veterinary and human pathogens, have been published in various scientific journals [4–13].

Primarily divisible into three general goals, this review aims to (1) present a bibliometric assessment of the entire accessible PiCV-related research corpus, and in tandem with bibliometrics; (2) present a synthesis of the historical developments, current knowledge, and important updates; and (3) provide a critical analysis to identify new perspectives, knowledge gaps, and open research questions.

2. PiCV-Related Knowledge Production—Bibliometrics

Preliminary searches for this review were conducted on the Dimensions, PubMed, and Scopus databases. The search terms included were 'pigeon circovirus', 'columbid circovirus', 'young pigeon disease syndrome', and a combination of these three terms joined by the operator 'OR'. The searches were conducted in February 2022, and all studies were included for evaluation regardless of publication time or form. Additional records were retrieved by manual searches in Google Scholar and ResearchGate. Upon retrieval of the potentially relevant articles from the databases, a unified list was generated following the Dimensions database format, and records were deduplicated. Preliminary assessments of the records based on the title and abstract were performed by four reviewers working in two independent groups. Differences were resolved by discussion after the assessment of the recoverable information of each of the records.

Articles were selected if they were at least able to demonstrate detection of pigeon circovirus (either as the main virus of interest, or as a virus concurrently investigated with other pathogens concerning a disease or condition), or if they were able to at least show an appreciable discussion of aspects directly related to PiCV research, for original and review papers, respectively. Bibliometric analyses were performed, and visualizations were created using VOSViewer version 1.6.18 [14].

2.1. Research Corpus: Records, Growth and Sources

The search for pigeon circovirus-related records conducted in February 2022 yielded 1648 potentially relevant records. After the creation of a unified database of all these records following the format of the Dimensions database, 1049 records were removed as duplicates, while an additional 79 entries were deemed as erroneous database entries from the sources. Therefore, 520 records were screened for eligibility based on the minimal criteria set above, wherein 389 records were determined as out-of-scope, and two others were eliminated as they only contain correction notes of relevant records that were already part of the eligible entries. A diagram of the flow of records through the steps of this review is presented in Figure 1.

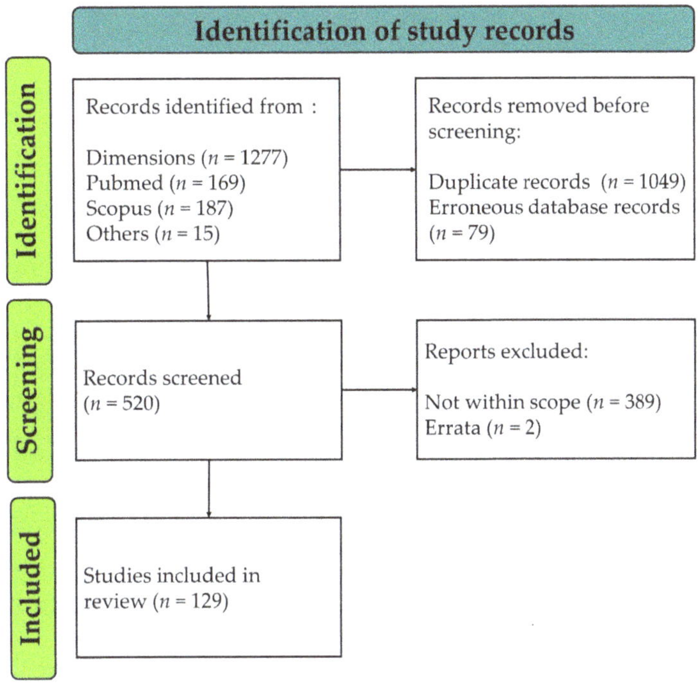

Figure 1. PiCV-related records retrieved from different sources.

Of these 129 included records, 99 (76.7%) were original articles including research publications in journals, short/brief/preliminary communications, a government agency report, a conference paper, a dissertation manuscript, and two pre-prints, while the remaining 30 (23.26%) were review papers and book chapters. A total of 117 records were retrieved in full text, while the remaining 12 were mostly either older records or in languages other than English or Chinese. Regardless, other pertinent information (abstracts, publication type, publication year, source title, authors, authors' institutional affiliations, author's county of affiliation, etc.) on all the 129 records was retrieved as much as we could for subsequent bibliometric analyses.

From the earliest record that we were able to retrieve [1], PiCV-related publications consistently grew in a linear manner in the following three decades, as shown in Figure 2a provided below.

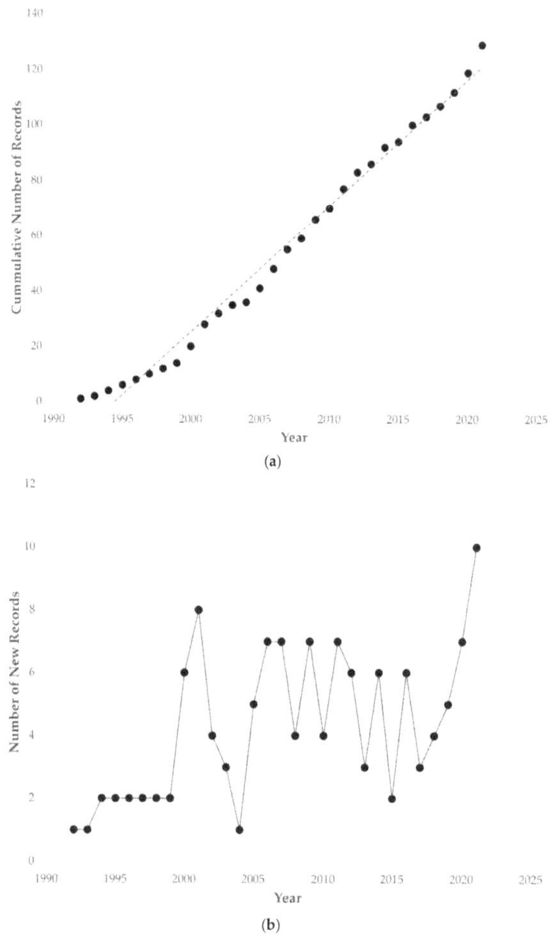

Figure 2. PiCV-related records retrieved from different sources shown as (**a**) cumulative number of records over the years, and as (**b**) new records per year.

While the first decade of PiCV research saw just a couple of new records per year, remarkable is the first peak in the number of new records (Figure 2b) at the turn of the millennium that saw the first use of polymerase chain reaction for the detection and cloning, and the subsequent sequencing, alignment, and phylogenetic analyses of the PiCV genome [15–19], as well as the first report of pigeon circovirus detection from Africa [20] and Asia [21]. It must be noted that the record of PiCV detection from the African continent [20] was not recovered by our search strategy but was however found during the examination of the texts of the included records. This thereby highlights one of the necessities for this review because this record has also been missed in some of the most recent reviews [22–24].

A few succeeding years following the availability of the first whole genome sequence of PiCV seemingly looks like a lull in the history of PiCV research. However, examination of the records produced during the years 2001 to 2004 revealed a period wherein the apparent focus of research works seemed to be the utilization of the sequence information for the development and evaluation of new and more specific molecular methods for the detection of this circovirus infection [15–19,25–29]. The adaptation of these developed methods manifested in the studies that soon followed that demonstrated other potential host of the virus, its transmission, epidemiology, and organ/cellular tropism [30–36].

One hundred fourteen (114) of the 129 included records for this review were made available through publications in peer-reviewed academic journals, and further examination revealed that almost a third (31/114; 27.2%) of all these peer-reviewed publications were published in only three journals. The 11 journals with the greatest number of PiCV-related publications also account for 49.6% (64/129) of all included records even though there were 51 journals identified, suggesting possible authorial preference for highly specific and specialized journals, like *Avian Pathology*, *Avian Diseases*, and Journal of *Avian Medicine and Surgery*, over more general veterinary- or virology-related journals (Table 1).

Table 1. Sources with the greatest number of records.

Sources	Count
Avian Pathology	15
Avian Diseases	8
Veterinary Record	8
Journal of Avian Medicine and Surgery	5
Archives of Virology	5
Medycyna Weterynaryjna	5
Journal of Veterinary Diagnostic Investigation	4
Veterinary Microbiology	4
Journal of General Virology	4
Virus Research	3
PLoS ONE	3

2.2. *International and Institutional Collaborations*

Co-authorship analysis by country showed that there are 39 countries listed as the authors' affiliations. Most productive among which are the United States with 27 records retrieved and included for this review and collaborating with 20 other countries, Northern Ireland with 25 records and 8 international collaborations, and Germany with 17 records and 3 collaborating countries (Figure 3a). These countries are followed by Belgium, Poland, China, Taiwan, Australia, and Italy.

On the other hand, several records were also observed to have been produced, or co-produced from countries that have not had outside collaborations (Figure 3b). Among these countries are Netherlands [37], Japan [21,38], Iraq [39], Czech Republic [29,32], and Brazil [40].

Meanwhile, ranking institutions/organizations by productivity, the Department of Agriculture and Rural Development for Northern Ireland, University of Liege, University of Warmia and Mazury in Olsztyn, Queen's University of Belfast, University of California, Davis, National Pingtung University of Science and Technology, Leipzig University, the Agri-food and Biosciences Institute in Stormont, and the Robert Koch Institute in Berlin produced the most number of records. Additionally, both the Department of Agriculture and Rural Development for Northern Ireland, and University of Liege are both linked with 10 institutions, while University of Warmia and Mazury in Olsztyn has links with six (Figure 3b). However, when ranked by total link strength, Blood Systems Research Institute and University of California, San Francisco were the two top institutions, each with a total of 30 points. Total link strength is defined as "the total strength of the co-authorship links of a given [organization/institution] with other [organizations/institutions] (p. 5)", whereas links are defined as "number of co-authorship links of a given [organization/institution] with other [organization/institution] [41]. These observations highlight the central role of the western countries and institutions both in terms of number of record production (in extension, knowledge production), and international institutional collaborations in the field of PiCV research, compared to countries and institutions from other regions.

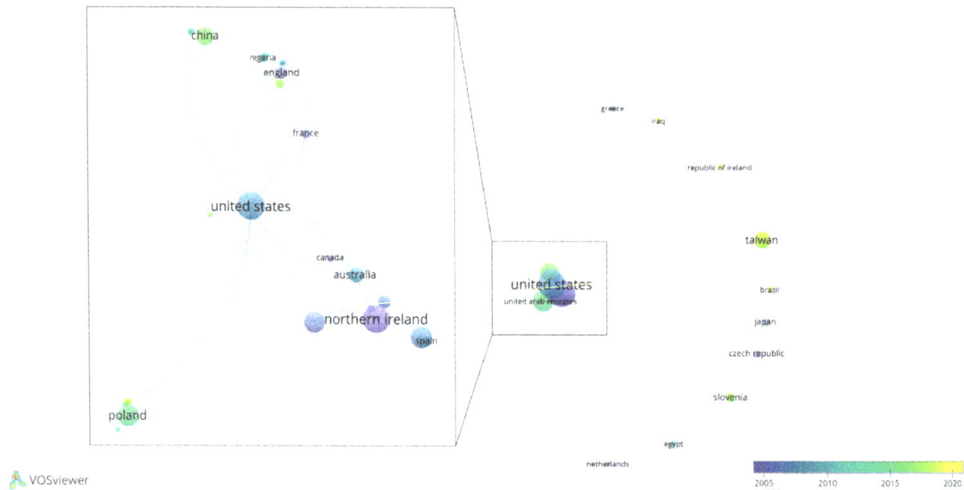

(**a**)

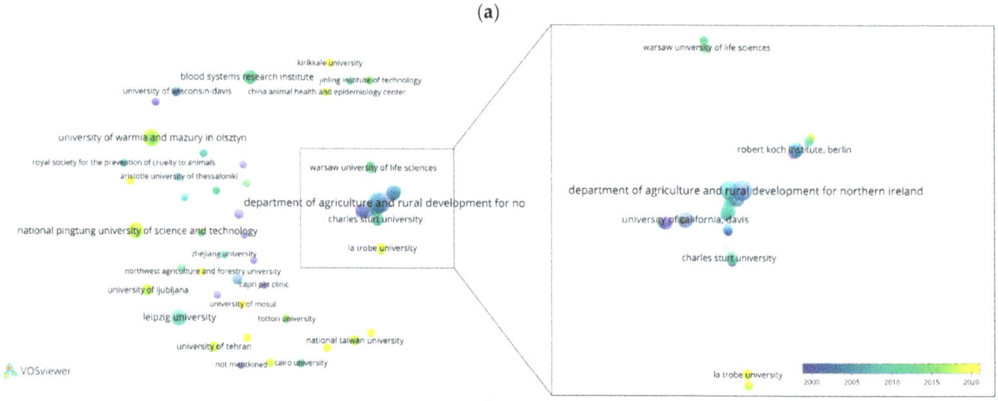

(**b**)

Figure 3. PiCV research collaboration maps among (**a**) countries and (**b**) institutions scaled by number of documents produced and colored by average publication year.

2.3. Author Collaborations and Citations

Visualized in Figure 4 is the collaboration map among authors who worked on PiCV-related research. In total, 455 individuals were identified to have authored or co-authored the retrieved records included in the current review. In general, as observed in institutional collaboration mapping above, clusters of collaborations were identified among authors. Most notable among which is the extensive collaboration cluster of Todd, Daniel of the Agri-Food and Bioscience Institute of the Department of Agriculture, Environment and Rural Affairs (DAERA), which succeeded the Department of Agriculture and Rural Development for Northern Ireland, shown above to have the most number of institutional links (Figure 3b). Previously working on other viruses with circular genomes, particularly chicken anemia virus, psittacine beak and feather disease virus, and porcine circovirus, Todd's first publication released in 2000 is also the most cited record retrieved in this review [42] based on the data gathered from the Dimensions database. Detailed in Table 2 are the top authors ranked by the number of retrieved records, with their respective citations, total link strength, and year of last publication.

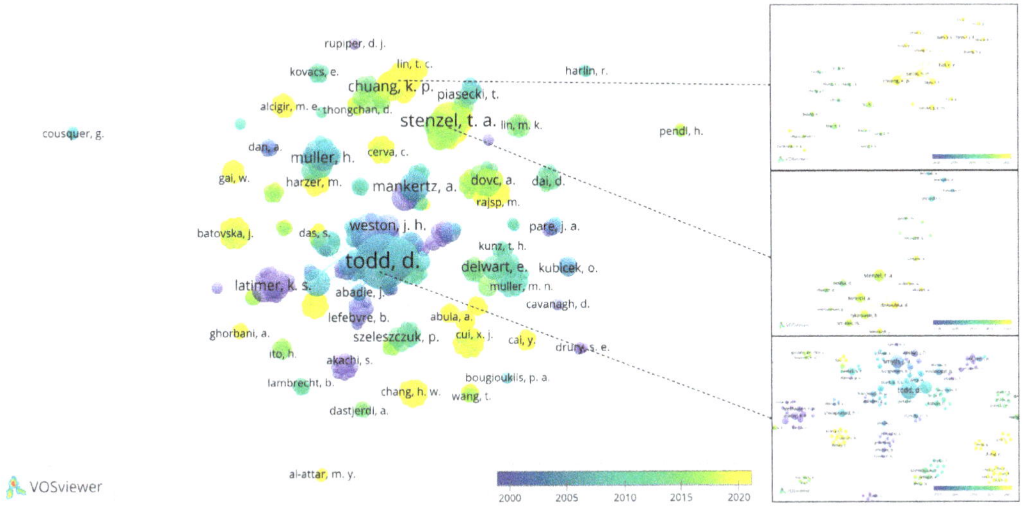

Figure 4. PiCV research collaboration map among authors scaled by number of documents produced and colored by average publication year.

Table 2. Authors with the greatest number of records.

Author	Documents	Citations [a]	Total Link Strength	Most Recent Record (Year)
Todd, D.	21	865	85	2012
Smyth, J.A.	12	460	57	2011
Duchatel, J.P.	11	169	54	2011
Stenzel, T.A. *	10	140	38	2021
Chuang, K.-P. *	6	26	39	2021
Koncicki, A. *	6	75	24	2017
Mankertz, A.	6	192	19	2011
Shivaprasad, H.L. *	5	134	33	2019
Tykalowski, B. *	5	59	23	2020
Muller, H.	5	200	22	2008
Raidal, S.R. *	5	127	18	2019
Soike, D.	5	304	15	2002

[a] as indexed in the Dimensions database; * authors with new records within the last five years.

Inspecting the details of the author collaboration networks, it is also notable that among the authors listed in Table 2, Todd is linked with Smyth, J.A., Duchatel, J.P., Raidal, S.R., and Shivaprasad, H.L., which demonstrates the tightly knitted network of the most prolific PiCV researchers (Figure 4). Important however is that this network published most of their work in the early years of PiCV research as indicated by the year of their most recent record (Table 2), except for Raidal, S.R. and Shivaprasad, H.L, who both published in 2019. While the previously mentioned cluster had links in other European countries, in two separate other networks, Mankertz, A. and Soike, D., and Muller, H., were also leading central roles in PiCV knowledge production in Germany in the first two decades of the PiCV research field. In Poland, Stenzel, T.A., Koncicki, A., and Tykalowski, B. were also working on PiCV-related research since 2012 and currently active in producing new publications. On the other hand, among the workers in Asia, only Chuang, K.-P. of Taiwan ranked within the top 10 most prolific authors (Table 2).

Cursory assessment of the most cited records showed a higher number of citations on reviews with a wider topic, i.e., not exclusively about pigeon circovirus, and those that were published earlier. Among the most important were two reviews on avian circoviruses by Todd and co-workers in 2000 and in 2004 [42,43], and the review on important viruses of pigeons by Marlier and Vindevogel [35]. Interestingly, two original articles exclusively on PiCV were also among the most cited documents. These were the papers that reported the first cloning and sequencing of the PiCV genome [15], and the paper that effectively formalized the association between PiCV and the young pigeon disease syndrome (YPDS) [44].

3. PiCV Knowledge—Current Understanding and Updates

With a wider understanding of the landscape of PiCV research across the world, this review also aims to systematically synthesize the current knowledge of the field from its historical development to locate its status and the gaps that should be addressed in the future.

3.1. The Virus

Pigeon circovirus belongs to the species *Pigeon circovirus*, genus *Circovirus* of the family *Circoviridae*. The species was formally recognized as a new member of genus *Circovirus* in 2005. According to the taxonomic report of the International Committee on Taxonomy of Viruses (ICTV) July 2019 release (EC 51, Berlin, Germany), the family *Circoviridae* is included in the order *Cirlivirales*, class *Asfiviricetes*, phylum *Cressdnaviricota*, kingdom *Shotokuvirae*, and realm *Monodnaviria*. No changes were made on these taxonomic assignments during the most recent report released in 2021 (ratified in 2022) [2].

Assignment to a given genus within the family *Circoviridae*, either as a member of genus *Circovirus* or *Cyclovirus*, is based on the location of the putative origin of replication *ori*. Circoviruses have the *ori* on the same strand encoding the replication-associated protein (Rep), while the *ori* is located on the capsid protein (CP)-encoding strand for the cycloviruses. An 80% genome-wide nucleotide sequence identity is set as the demarcation criteria for circovirus species based on the distribution of pairwise identities [45,46].

Primarily understood based on studies on porcine circoviruses, members of the family *Circoviridae* are characterized to have non-enveloped virions with icosahedral T = 1 symmetry. Pigeon circovirus virions were observed to be ranging from 14–25 nm in diameter [16,47]. Circovirus genomes are monopartite, circular, single-stranded DNA of approximately 1.7–2.1 kb in length which are replicated via rolling circle replication [46]. PiCV has approximately 2030 bp genome, relatively larger than other circoviruses [22]. A nonanucleotide motif '(T/n)A(G/t)TATTAC' marking the *ori* is found at the apex of a potential stem-loop structure located at the intergenic regions between the 5′-ends of Rep- and Cp-encoding open-reading frames, which are also known as ORFs V1 and C1, respectively. It is held that the Rep initiates the rolling circle replication by creating a nick on the virion-sense strand between the 7th and 8th positions in the conserved nonanucleotide motif. Except for the chicken anemia virus genome, all circovirus genomes have an ambisense organization with the Rep encoded on the virion-sense strand, while the Cp is encoded on the complimentary strand of the replicative form dsDNA [22,46,48]. The PiCV genome also encodes three other ORFs; the functions of the corresponding proteins are still yet to be known [22,49].

3.2. Geographic Epidemiology, Host Range, and Genetic Diversity

Prior to having the sequence of the pigeon circovirus, detection and diagnosis of infection were heavily reliant on histopathological examinations and electron microscopy findings of intracytoplasmic and/or intranuclear inclusion bodies in lymphoreticular and hepato-intestinal tissues. There were no means of verification for viral identification, and earlier studies can only rule out the possibility that the infecting agent is the Beak and Feather Disease Virus (BFDV) by employing immunologic assays, in situ hybridization, and

PCR techniques specifically targeting this previously characterized pathogen [47,50–55]. In the early 2000s, Mankertz et al. provided the first complete sequence of the PiCV genome, opening an avenue for a more accurate detection of the virus in infected pigeons [15]. Following its publication, development and evaluation of nucleic acid-based detection methods, such as PiCV-specific PCR techniques, in situ hybridization, and dot blot analysis, soon followed [18,26,27,31,34].

Later on, detection methods coupled with quantification strategies were developed into real-time quantitative and digital droplet PCR assays with the aim of correlating viral load to the clinical status of the host [56–58]. Rapid and accurate detection was also developed employing loop-mediated isothermal amplification that was reported to have no cross-reactivity against porcine circovirus 2 (PCV2), which is unlike the in situ hybridization technique that not only detects PCV2 but also the more closely related BFDV [19,59]. High-throughput sequencing is by far the most advanced technology ever used that detects PiCV, but the complexity, cost, and duration of result turn-out make this method unsuitable for diagnostic applications [60].

Columba livia domestica, either kept and bred for different purposes (sporting/racing, fancy, or utility) or feral, is known for its cosmopolitan distribution [61], and so are PiCV infections [23]. Considering that racing, the most common reason for pigeon breeding and raising [61–64], allows for the close interaction of pigeons from different local and even international lofts during transport and the actual competition, it is not surprising therefore that pigeon circovirus has also been detected in many countries across the world. Sporting events also provide opportunities for intermingling among feral and racing pigeons. Additionally, international trade for pigeons with good pedigree to be used as breeding stocks has also been suggested to facilitate the international distribution of this virus [22,65–67].

Retrieved records for this review showed that the first descriptions of case examinations demonstrating the presence of circovirus infections in pigeons were first made available in 1993 from studies in the United States and South Africa [20,50,68]. Although, based on other accounts, the virus has also been observed as early as the late 1980s in Canada and in Australia [3,54].

Within the first decade after the short report of Schmidt [1], retrieved records showed detection of the virus in several other countries including Northern Ireland [16,34,69,70], Germany [15,44], Italy [31,71], France [72], Belgium [30,33,34], and Japan [21,38]. Meanwhile, between 2003 and 2012, researchers in Czech Republic [29,32], Poland [36,58,67,73–80], Denmark [30], Slovenia [81,82], Hungary [66,83], United Arab Emirates [84], Libya [85], Nigeria [83,86], and Taiwan [59,87–93] reported first local detections of PiCV. Lastly, during the most recent decade of PiCV research, new publications and reports were also produced from Brazil [40], Iran [94,95], Iraq [39], and Turkey [96]. Other than the first local detection report for each country indicated above, succeeding reports demonstrating detections of PiCV were also cited to add to, if not complete, previous reviews.

Additionally, while no literature records were retrievable from Croatia and Algeria, sequences of PiCV detected from these two countries in 2010 (KP773230.1) and 2017 (MH932546.1), respectively, were available in the National Center for Biotechnology Information (NCBI) database. Shown in Figure 5 is a world map indicating countries with reported detection of PiCV colored according to the earliest retrieved record of detection report.

Attempts to identify infecting circoviruses in other avian species have also uncovered evidence of PiCV cross-species infection. Relying only on serological assays using hemagglutination targeting BFDV in wild Senegal doves (*Streptopelia senegalensis syn. Spilopelia senegalensis*) presenting similar symptoms to psittacine beak feather disease has proven that the infecting agent is a different circovirus [52]. In a retrospective study later on, PiCV infection was confirmed in *S. senegalensis* after performing sequence comparison [49]. Relying on a combination of clinico-histopathological observations and probe hybridization techniques, other earlier studies on circovirus-like infections in other bird species also drew suspicion of PiCV infection. However, unlike the case of PiCV infection in

Senegal doves, these other cases were later proven to be caused by entirely new circovirus species [42,53,97,98].

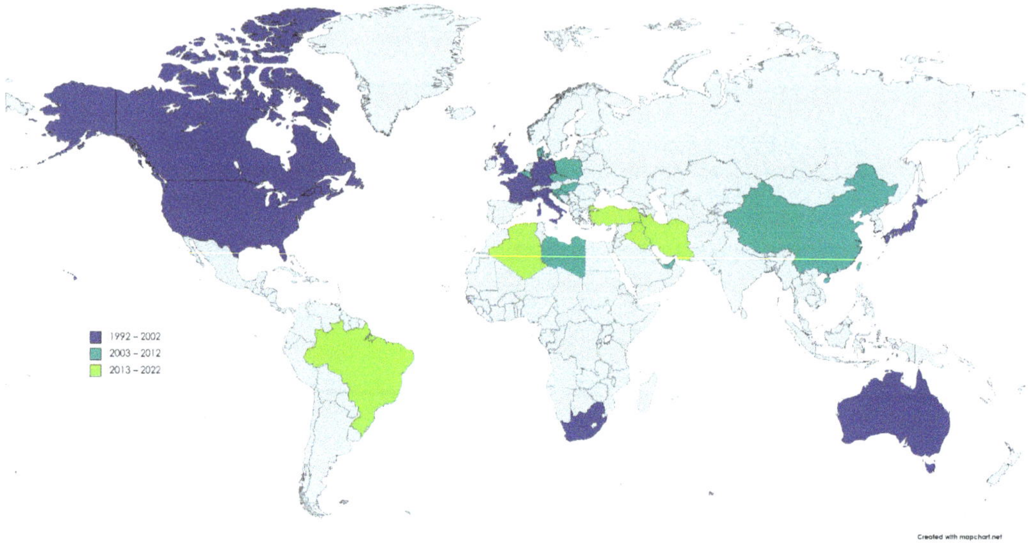

Figure 5. Countries with reported detection of PiCV colored according to the earliest retrieved record of detection report.

A shift in diagnostics occurred when the PiCV whole genome sequence was made available, allowing sequence comparison as a form of verification [15]. This was proven to be useful when PiCV was detected in Eurasian collared dove (*Streptopelia decaocto*) using the nested PCR approach targeting the partial capsid gene of the virus [32]. Meanwhile, degenerate primers targeting the conserved region of the *rep* gene in circoviruses was amplified in tissue samples from chicken obtained from Nigeria [86].

Moreover, a recent study involving outbreak of Chlamydia psittaci in aviaries in Taiwan surprisingly detected PiCV in 12 species of birds (*Aix galericulata, Cygnus melancoryphus, Caloenas nicobarica, Columba pulchricollis, Goura cristata, Ocyphaps lophotes, Spilopelia chinensis, Streptopelia orientalis, Pavo cristatus, Threskiornis aethiopicus, Phoenicopterus chilensis, Phoenicopterus ruber*) coming from five different orders by amplifying the partial cap gene of the virus. The authors have correlated the lethal outbreak of chlamydiosis to a concurrent circulation of PiCV which may have "play[ed] a key role in augmenting [the] disease progression" although further proof certainly is needed [92]. Moreover, PiCV detection is not only limited to different species of birds as it was also detected in ticks parasitizing camels and sheep from Inner Mongolia. The observation was published as a preprint. It is notable though that the whole genome sequence of PiCV (Accession Number: MN920392) detected from these ticks was recovered using high throughput and Sanger sequencing verified by PCR [60].

From these studies, partial or complete PiCV sequences were obtained from 15 avian species and one from a parasitic arachnid (Figure 6).

Although reportedly first observed in Canada and Australia, the origin and duration of the circulation of PiCV are unknown. The first whole genome sequence provided a reference point for its taxonomic classification. In relation to BFDV, PCV1, and PCV2, genomic sequence homologies between these viruses and PiCV were reported to be 57%, 36%, and 34%, respectively [15,16].

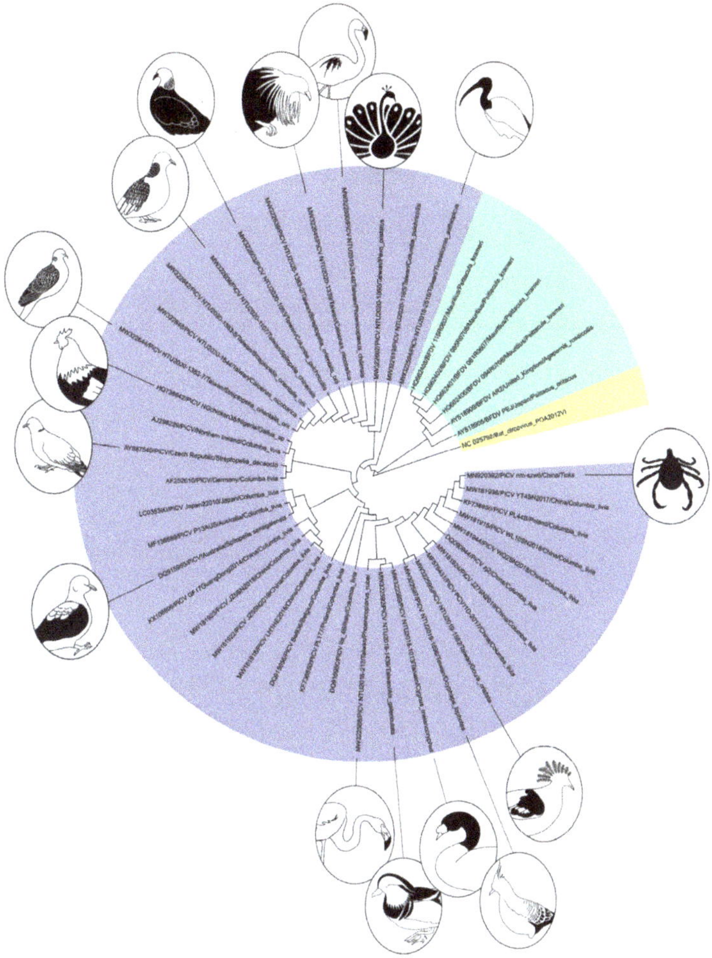

Figure 6. A neighbor-joining tree depicting the relationship among PiCV sequences from various hosts.

Using the 371 available PiCV whole genome sequences retrieved from NCBI, a pairwise identity calculation showed that the two sequences with the lowest identity similarity (83.201%) are sequences from Japan (Accession Number LC035390) and China (Accession Number KJ704801), which is consistent with the study of Khalifeh et al. (2021) that reported >83% genome-wide identity [73] (Supplementary Figure S1 and Supplementary Table S1). The same study reported as many as 95 PiCV genotypes based on a 98% pairwise identity genotype demarcation criterion [73]. This is also consistent with our observation based on the previously mentioned 371 whole genome sequences. Shown in Supplementary Figure S2a,b are the resulting phylogenetic trees from the same dataset. Topology and grouping of sequences widely reflects that of a previous report [73].

Most of the variability, according to earlier studies, is attributable to differences in the *cp* gene sequence [22,49,67,74,99–101]. Between the *rep* gene and the *cp* gene, sequence analyses have shown that the *cp* sequence undergoes more frequent mutations as an adaptive response to the hosts' immune system driving the evolution of circoviruses primarily by positive selection, although negative selection has also been observed. As a whole, these mutations in the *cp* explains the high genetic variability among PiCV sequences [22,49,66,67,75,88,99,100]. One the other hand, potential genetic admixture with other circoviruses has also been observed in Australia [102].

More recently, wider interest and more studies focused on the role of recombination as an evolutionary driver of circovirus evolution have been published [40,66,73,76,102]. Besides the observed frequent mutations in the *cp* gene, this observed high genetic diversity in PiCV sequences is also attributed as a product of constant recombination driving the evolution of this virus [75,88]. To understand the genetic recombination in PiCV, in an experimental study, infected and non-infected pigeons were housed in a single loft for a period of 23 days to observe recombination events [73]. Recombination detection analysis of the 178 whole genome sequences from this study revealed 13 recombination events, with event 2 as most common recombination event present in 109 genomes. Recombinant event 2 was shown to be located at the 5′ portion of the *rep* spanning a ~100 nt region. Additionally, localizations of the recombination hotspots were also detected at the 3′ region of *rep* and the intergenic region, while a cold spot was identified near the 5′ region of the *cp* [73]. The authors note that these observations were different from previously detected hotspots, which were located near the 3′ end of the *rep* and *cp*. It was also reasoned that the detected recombination cold spot located in the *cp* might be due to importance of the capsid protein to the transmission and infection ability of the virus.

Similarly, using the 371 whole genome sequences that we were able to retrieve, we also attempted to perform a recombination detection analysis of this global dataset. The analysis was performed as in [73] using RDP5 [103]. Our preliminary analysis revealed 130 detected hypothesized recombination events, of which 105 were accepted based on the detection of these events by at least three methods employed by the program. In our analysis, localizations of the recombination hot spots were observed at the 5′ regions of the *rep* and *cp*, as well as the 3′ region of *rep* up to the intergenic region, while recombination cold spots were observed at the central region of *cp* (Supplementary Figure S3). These observations seemingly merges together previously differing detection of recombinant hot spots, while at the same time consistent with previously reported observations of cold spots [73,75]. After removing the accepted recombinant regions, shown in Supplementary Figure S2c is the resulting phylogenetic tree. We note that the grouping of sequences into clades remain largely the same relative to Supplementary Figure S2a,b, but the positions of the clades within the tree differ, particularly the relationships of genotypes 5 and 6 to genotype 4, and genotype 7 to genotype 4. To explain this, we suggest that an assessment of each of the recombination events among these sequences be evaluated in more detail.

Up to this point, efforts to group PiCV sequences yielded varying results. Initially, two main clades were reported, however subsequent reports described phylogenetic reconstructions with as many as nine clades [38,40,49,67,101,104]. Despite multiple attempts in classifying PiCV into groups, there was no observed association between clades and geographic origin or the pathogenicity of the virus [22,49,73,100]. We however note that from the trees generated in this review, there tends to be clustering of sequences from individual studies, which might suggest association of PiCV sequence grouping to temporality, or geography or location of isolation, or an interaction of both. However, it must also be emphasized that while the current analysis utilized all available whole genome sequences, populational representation (or the lack thereof) of the "true" circulating strains might have also contributed to current observations. Of the 371 genomes used in this review, around 200 were from Western countries, particularly from Poland, which has contributed ~160 whole genome sequences that are mostly from one single study [73]. More samples of PiCV whole genomes are encouraged to be sequenced, especially from underrepresented regions and/or countries to come up with a better resolution of this observed and/or hypothesized tendencies/associations.

3.3. Transmission, Infection, and Control

Pigeon circovirus can be transmitted through both horizontal and vertical pathways. Detection of the virus in the intestine, cloaca, and faeces supports faecal–oral route transmission [18,33,50]. Although, inhalation of other faecal contaminated materials, such as feather dust, has also been suggested as a potential respiratory route of infection. The virus has previously been reported to be detected in the pharynx, trachea, and lungs, suggesting that respiratory tissues may be a major potential site for virus replication and persistence, especially in older pigeons [33]. On the other hand, vertical transmission was proven by the detection of the virus in testis and semen samples of breeding cocks, the ovary but not in the oviduct of the hens, in embryonated eggs, and chicks recovered from eggs shortly before hatching [18,30,33,56].

Detection of the virus in pigeon embryos was observed to be more common than previously thought. Reports have shown that about 11–36% of embryos may have already been infected with the virus before hatching [30,33]. No specific embryonic tissue seemed to be targeted by the virus infection as DNA of the virus has been detected in pharynx, trachea, lung, liver, spleen, intestine, kidney, heart, and bursa of Fabricius. Interestingly, viral DNA is not detected in embryonic bone marrow [33]. Observation of the status of viral infection during the first few weeks from hatching showed that while detection of the virus DNA in cloacal swabs of pre-weaning (up to 28 day old) chicks were only ranging from 1–20%, weaning and the subsequent transfer to the rearing loft coincides with the detection of the virus in 100% of the birds, suggesting high probability of virus transmission and new infection in rearing lofts [33]. In addition, noteworthy is the observation that although viral DNA was observed in crop tissues of young birds by in situ hybridization, detection of the viral DNA by PCR of crop swab samples yielded no positive result, suggesting that excretion and transmission of the virus through the crop milk is rare [30].

Pigeon circovirus infection is commonly investigated and reported in pigeons less than 1 years of age, most of which were not exhibiting any clinical sign albeit the high viral loads observed in multiple tissue samples, particularly in the bursa of Fabricius [19,27,30]. Specifically, the virus has also been detected in varying percentage in different reports using different methods in tissue samples from the spleen, liver, thymus, kidney, crop, intestine, brain, trachea, lung, heart, blood, bone marrow, esophagus, Peyer's patch, nose, and the third eyelid [19,27,30,89]. Meanwhile, among older pigeons (more than 1 year old), detection of the virus is more common in spleen, and respiratory tissue samples than in any other organs. Compared to younger pigeons, detection of the virus in the blood, pharyngeal and cloacal swabs of older pigeons is lower, while detection in other organs remains high. Taken together, these observations suggest recovery of these older pigeons from the infection. However, this also suggests that total elimination of infected pigeons from the breeding stock would not be possible [30,33].

Demonstration of the capacity of PiCV to cause clinical disease manifestation had been challenging due to both the absence of a reliable method to propagate this virus and specific pathogen free-certified pigeons. However, purified virus particles from infected tissues have been used to experimentally challenge pre-screened pigeons. The challenged birds did not display clinical disease manifestation, however, the characteristic lesions in the bursa of Fabricius and the spleen were evident in the challenged but not in the control pigeons. PiCV DNA was also detected in cloacal swab and tissue samples of the inoculated birds. These observations support the hypothesized immunosuppressive role of PiCV in pigeons [105]. More recently, examination of the lymphocyte populations among symptomatic and asymptomatic PiCV-positive, and PiCV-negative pigeons confirmed this immunosuppressive role of PiCV. It was observed that PiCV infection induces B lymphocyte apoptosis thereby suppressing humoral immunity [58]. These two studies are perhaps the best evidence to settle contentions on the immunosuppressive role of PiCV.

To date, no commercial vaccine is available to control PiCV infections. Experimental studies on the production of the recombinant capsid protein and the evaluation of its potential use as an antigen for vaccine formulations have been reported in the literature [79,93]. Moreover, probiotics and interferons have also been reportedly explored as alternative and/or supplemental modalities to control the effects of PiCV infections [90,91]. At the moment, protection against potential detrimental effects of PiCV infection relies on good biosecurity measures in the loft. However, while biosecurity principles can be easily adapted in meat pigeon farms, the same is not true for racing pigeon lofts. Importing breeding stocks, flight training, and, importantly, race events all violate biosecurity principles [22].

3.4. *Revisiting Young 'Pigeon Disease Syndrome'*

Like in many historical cases of new virus discovery, the emergence of a new condition negatively affecting the health of, in this case, pigeons, led to the discovery of PiCV. Retrieved records for this review show that Woods et al. (1993) provided the first case report of a circovirus-like infection in a pigeon from a flock that experienced 100% squab mortality three to four days following observation of anorexia and lethargy. The subject of the case report, a female racing pigeon, is described as "emaciated, with pectoral muscle atrophy [50]". The histopathologic findings of this case remain as one of the most detailed of the succeeding cases reported in literature. The report also noted distinctness of this circovirus from PBFD based on the results of immunohistochemistry, DNA in situ hybridization, and polymerase chain reaction with DNA dot-blot hybridization. The findings contained in this report were echoed by subsequent reports, often with striking similarities but with unique and appreciable differences.

For the several years that followed, reports of this circovirus infection centered on the similarities of the cases; thus, distilling unifying features observed. These features are the observation of: (1) intracytoplasmic/intranuclear inclusion bodies common in the bursa of Fabricius and the spleen, but are also sometimes observed in other tissues, that are (2) composed of paracrystalline arrays of tightly packed, nonenveloped icosahedral virions, and accompanied by (3) lymphoid depletion, evidenced, primarily, by bursal atrophy. Consistently, PiCV reports also indicated non-specific clinical signs of ill thrift, lethargy, anorexia, poor racing performance, and diarrhea. These features were often observed in conjunction with coinfections of other bacterial, fungal, and/or viral agents. These observations led to the hypothesized immunosuppressive function of PiCV that predisposes the pigeons to other infections. See [3,37,43,50,54,65,72]. Of exception to this is the paper by Tavernier et al. (2000) that investigated pigeons that were presented for necropsy after dying spontaneously or suffering from clinical disease. The report noted that the proportion of examined cases with concurrent infections and the nature of these coinfections, lesions, and clinical signs were similar among pigeons whether or not circovirus infection was detected, but the mortality is more common among circovirus infected pigeons. The paper added that clinical indications of the immunosuppressive effects of circovirus were not found in their examinations, but experimental infection and additional immunological study must be conducted to verify the contending reports [55]. Challenge and immunological studies conducted years later confirmed immunosuppression [58,105].

Curiously, in 2005, the term 'young pigeon disease syndrome' or 'YPDS' first appeared in the literature we managed to retrieve in full-text [44]. The paper also considered what was then called "swollen gut syndrome" as an equivalent designation, also adding that other non-infectious and infectious causes, "including *Spironucleus columbae*, *Escherichia coli* and avian adenoviruses, have been considered to contribute to the pathogenesis of YPDS (Dorrestein et al., 1992; De Herdt et al., 1994; Warzecha, 2002)" [44]. This epidemiological study obtained, among other information, occurrence of YPDS and the observed clinical signs [44]. Notably, the survey showed that "pigeons affected by YPDS were mainly between 4- and 12-weeks post-weaning. The clinical signs were not specific and included anorexia, depression, ruffled feathers, vomiting, diarrhoea, polyuria, and fluid-filled crop" [44].

Additionally, the abovementioned paper, also noted that *Escherichia coli* and *Klebsiella pneumoniae* were isolated more frequently from YPDS-affected pigeons, and that depletion of the splenic and bursal lymphocytes was only observed in pigeons with YPDS. All investigated pigeons with YPDS in this report were also positive for PiCV infection in at least one organ, most importantly in the bursa of Fabricius with 100% (45/45) positive detection of PiCV DNA by PCR. The report concluded that "YPDS is a multifactorial disease in which a pigeon might be a crucial factor, possibly by inducing immunosuppression in infected birds" [44]. This paper is pivotal since it provided a de facto definition of YPDS that would be cited in almost all subsequent reports.

4. Conceptualizing PiCV Diseases (PiCVDs)

4.1. Perspective—Problematizing 'Young Pigeon Disease Syndrome'

Prior to the 2005 epidemiologic report [44] that popularized the use of the term 'young pigeon disease syndrome' or YPDS, we note that the manner of reporting and the description of observations flow unidirectionally; that is, by our reading, the provided clinical picture of having non-specific clinical signs plus the three main unifying features noted above were describing PiCV infection. We also note that we did not encounter any specific terminology that referred to a conceptualization of the abovementioned set of conditions as either a disease (multifactorial or otherwise), a syndrome, a disease syndrome, or a disease complex. If any, the disease *is* PiCV infection, or, more specifically, an acquired immunosuppression caused by PiCV infection resulting to predisposition to secondary infections. Afterall, these reports were exclusively describing cases of PiCV infection. See [3,37,43,50,54,65,72].

In our attempt to trace the term 'young pigeon disease syndrome (YPDS)' prior to [44], we tried to retrieve the records cited (i.e., "Dorrestein et al., 1992; De Herdt et al., 1994; Warzecha, 2002") that, according to the paper, considered the contribution of other infectious agents in the pathogenesis of YPDS. It is unfortunate that we were not successful. It was indicated in the reference list that the first two were conference proceedings, while the latter was an article in a magazine specializing in racing/carrier pigeons [44]. However, in another published work by De Herdt and co-workers published in 1995 reporting on an epidemic of fatal hepatic necrosis, PiCV and its association with immunodeficiency were also mentioned and the paper also cited the same paper by Dorrestein et al. (1992), but never positively identifying a condition called 'young pigeon disease syndrome' or any specific terminology that bared similarity in construction [51]. The same is true with the report co-authored by De Herdt published in 2000 [55]. Considering these observations, our sense is that 'young pigeon disease syndrome (YPDS)' in this exact form or, probably, in other translational equivalents, as a coined term (as opposed to a describing phrase) were not mentioned in the cited texts, but the clinical picture of the conditions described therein and that of YPDS were similar.

With the appearance of the term YPDS in the PiCV literature [44], an apparent reversal of the flow of attribution was also observed. While reports prior to 2005 were describing observed clinical presentations and the accompanying histopathological observations of PiCV infected birds, the epidemiologic report [44] associated what was first observed as a set of clinical presentations (termed YPDS) to PiCV infection. Nevertheless, the term 'young pigeon disease syndrome (YPDS)' seems to be widely used by 2005, such that the same report was able to conduct an epidemiological survey on YPDS using a questionnaire administered to German breeders of racing pigeons. These, to us, suggesting that the term "young pigeon disease syndrome (YPDS)" might have originated in a lay characterization among fanciers of an observed illness or set of clinical manifestations that gained popular currency before it was eventually adapted by veterinary health practitioners and researchers, and reported in the scientific literature. A paper published later on supports this hypothesis stating, "In racing pigeons this combination of clinical signs (i.e., lethargy, weight loss, respiratory distress, diarrhoea and poor racing performance) is commonly described by pigeon fanciers as 'young pigeon sickness,'" [33].

As previously mentioned, this paper [44] became pivotal for its influence in the operational and functional use of the term YPDS, and the understanding and subsequent research on PiCV in the succeeding years.

We would like to note two key points about the paper [44] at this point:

i. The phrase "might be a crucial factor" in the concluding statement, nuanced to the limitations of their methodology, did not set PiCV infection as a precondition/requirement to YPDS. As such, it logically follows that YPDS as, in the words of the authors, "a multifactorial disease" might also be possible without PiCV infection.
ii. The concluding statement did not categorically identify its use of the term YPDS as either referring to the set of clinical, histopathological, and molecular observations reported, or the colloquial use of the term which only involved, in their words, "clinical signs [that] were not specific." Hence, to our reading of the text, there exist virtually two conceptualizations of YPDS: one that is colloquial, and the other that is more technical in its requirements.

We contend that these two points are crucial to the developments in PiCV research that followed the abovementioned report, and to the studies that are still to be conducted in the future. As such, we also investigated how this paper and, in particular, the two key points noted above, were referred to, and propagated in the literature. Our underpinning questions are: (1) How is PiCV described to be related to YPDS?; and (2) What is 'young pigeon disease syndrome (YPDS)' as operationalized in the literature?

Examination of the retrieved texts showed that the relationship of PiCV infection to YPDS has been described in variable terms of associatedness, importance, and/or cruciality. Specifically, the role of PiCV infection to YPDS has been described in some papers as remaining "uncertain" and that it "has not yet been conclusively clarified", while also described as "speculated as crucial", "might be crucial", "crucial", "probably associated", "associated", "strongly associated", "important", "central", "one of the putative factors", "a key factor", "one of the causative agents", "among the major causative agents", "the likely aetiological agent", and "likely causal agent" in other records. Likewise, other texts also described PiCV infection to be "caus[al of] YPDS", or "responsible for YPDS". While these descriptors are not mutually exclusive of each other, these are also not equivalents. Additionally, the mechanism by which PiCV infection contributes to the development of YPDS has also been described in various descriptions of certainties, ranging from "assumed to induce immunosuppression", "possibly by immunosuppression," to "causes profound immunosuppression". See [22,23,33,40,44,48,58,59,66,67,75,77,79,84,88–90,93–95,100–102,104,106–117]. To us, these degrees of variability among these descriptors do not seem to correlate with time. In other words, these variable descriptors, and the timing of their usages in the literature do not suggest increasing or decreasing acceptance of the importance of PiCV infection to YPDS, or the perception of certainty of immunosuppression as the mechanism for the PiCV association with YPDS.

Unsurprisingly, we also found that the descriptions, definitions, and/or operationalized usage of the term "young pigeon disease syndrome (YPDS)" also vary, corresponding to the above noted colloquial and technical conceptualizations. Indeed, in most texts, YPDS is referred to as a "multifactorial" or "complex disease" characterized by a combination of symptoms (often described as non-specific), with no to variable level of attributed importance to histopathological markers. Consistent with this is our observation that YPDS is used to describe cases (or loft histories) prior to the detection of PiCV infection in numerous reports. In others, YPDS is a category of viral infections. These are all also while other records explicitly state that YPDS is a consequence of PiCV infection and subsequent immunosuppression, or categorically require specific conditions for YPDS diagnosis, such as the observation of both clinical symptoms and PiCV-specific histological lesions, presence of viral particles, and a demonstration of the presence of a large amount of viral DNA. See [22,23,33,40,56–59,67,77,79,80,90,93,96,100,105,109–111,113,117].

Problematizing PiCV infection and its pathology, YPDS, and their correlation with each other is not new to this review. These have been, in fact, subjects of several discussions and opinions before. For instance, Dr. Henk de Weerd pointed out, "As you know, about 10 years ago 'everyone' (except me) was extremely alarmed about the circovirus. According to many, that was the cause of 'just about everything' in terms of pigeon outbreaks, including 'adeno-coli'," which, still according to Dr. de Weerd, is internationally known as the "Young Bird disease" [118]. Similarly, in another perspective piece, Dr. Colin Walker expressed that, 'young bird disease' clusters numerous diseases with similar symptoms together such that "[f]anciers run the risk of labelling any young pigeon with these symptoms simply as having 'young bird disease' when, in fact, all they are acknowledging is that the young pigeon is sick with wasting and diarrhoea" [119]. Indeed, many other agents were considered to be associated with YPDS, including fowl adenovirus, pigeon herpes virus, pigeon adenovirus, paramyxovirus, *Chlamydia*, *Escherichia coli*, *Salmonella*, and other parasites in varying combinations with varying degrees of support [74,96,108,115,120]. Dr. Walker also opined, "the term 'young bird disease' is a poor one and one that I think should be abandoned" [119].

Most recently, experimental infection with the newly discovered and isolated pigeon Rotavirus A belonging to a novel clade also produced clinical manifestations that were also described as YPDS-like [117,121]. Some workers in the field expressed that the discovery of pigeon Rotavirus A and the demonstration of its capacity to cause a disease that is like YPDS seem to depreciate the role of PiCV in the etiology of YPDS [58].

Therefore, the questions of what YPDS is, what its relation to PiCV infection is, and how this term has been used and propagated in the literature, which reflects how veterinary health practitioners and researchers perceive YPDS, need critical assessment. The current review revealed that usages of the term YPDS in the literature are neither standard nor equivalent. YPDS in the literature is a diverse range of characterizations.

In consideration of the complex history of YPDS and PiCV infection, our current opinion is that YPDS, as colloquially conceptualized pointed above—that is, a combination of non-specific clinical symptoms, may still be useful as a putative diagnosis pending identification of the causative and/or confounding agents. We are also under the impression that PiCV must be categorically identified as a causative agent of the diseases that we propose below to be identified as PiCV subclinical infection (PiCV-SI) and PiCV systemic disease (PiCV-SD).

4.2. Proposal—PiCV Diseases Case Definition

From the early days of PiCV research, this virus has been consistently compared and likened to other related pathogens, most remarkably to porcine circovirus (PCV) and the associated disease formerly referred to as post-weaning multisystemic wasting syndrome (PMWS), in terms of basic biology, to genetic diversity, pathobiology, and approaches to treatment and control [23,42,57,79,88,90,93,99]. While the first case of PMWS was first recognized in 1991, years later than the observation of clinical cases in pigeons which may be categorized as YPDS, swine health practitioners were able to come up with diagnostic criteria fairly early. Notably, this was after much controversy and debate. Regardless, as early as 1999, three major criteria were set to establish diagnosis of PWMS, which was later renamed as PCV2-systemic disease (PCV2-SD), one of the porcine circovirus diseases (PCVDs) [122–124]. These requirements are "(1) presence of compatible clinical signs, mainly wasting, (2) observation of moderate-to-severe histological lesions in lymphoid tissues (lymphocytic infiltration and histiocytic infiltration) and (3) detection of moderate to high amount of PCV-2 within such lesions" [124]. Such strict definition of a case was considered acceptable by the veterinary and scientific communities worldwide considering a general reluctance to accept that PCV2 was truly pathogenic. This is in sharp contrast to the discovery of PCV3 and the acceptance of its causality in associated diseases, which was accepted much faster probably due to the previous experience with PCV2 [124]. The case of YPDS and its relationship with PiCV infection has not been similar, which as pointed above still necessitates a cohesive conceptualization and associated terminologies.

Additional parallels between PiCV infection and PCV2 infection can be long, but we would like to highlight several important points of similarities related to pathology, and disease. Like PiCV, early experimental studies with PCV2 sought to demonstrate the capacity of the virus to produce clinical signs compatible to PMWS. In PCV2, successes were achieved mostly after inoculation with another infectious or non-infectious agent.

Together with other field observations, these studies lead to the conclusion that PCV2 is immunosuppressive and a "necessary but not sufficient factor to develop the clinical disease" [122,123]. While no artificial/controlled co-infection studies on PiCV and other suspected pathogens had been reported to date, there is preponderance of evidence from case studies of PiCV infection and YPDS to demonstrate that PiCV infected birds with clinical symptoms were coinfected with other infectious agents. Additionally, analogously applied, PiCV infection cases displaying clinical symptoms fully satisfy even the strict definition of PCV2-SD. Important to point out here is the difference in the characterization of YPDS and PCV2-SD. While PCV2-SD explicitly necessitates PCV2 and its detection in associated lesions, YPDS, in most cases of usage, does not require or expresses no same certainty to the importance of PiCV. This is except in one paper by Duchatel and Szeleszczuk (2011), which, as noted in the previous section, requires three major requirements for YPDS diagnosis [23]. Remarkably, this paper has previously been cited by other workers, but rarely about its specification of YPDS diagnostic criteria. In one recent review, this diagnostic criteria was cited, but what Duchatel and Szeleszczuk categorically defines as YPDS, the review identifies as circovirosis [22,23].

From this perspective of the current understanding of PiCV and YPDS research, and guided by the lessons from PCV2 and PCVD, we draw our proposed PiCV diseases, as shown in Table 3.

Table 3. Proposed terminology for pigeon circovirus diseases together with their case definition based on clinical and laboratorial findings.

PiCVD Proposed Name (Acronym)	Main Clinical Sign	Individual Diagnostic Criteria
PiCV-subclinical infection (PiCV-SI)	No evident clinical sign	1. Lack of overt clinical signs 2. No or minimal histologic lesions in the lymphoid organs, mainly bursa of Fabricius and/or spleen 3. Detection of PiCV at least from fecal or cloacal swab samples by standard or quantitative PCR
PiCV-systemic disease (PiCV-SD)	Lethargy, depression, weight loss, diarrhea, vomiting	1. Clinical signs 2. Histologic lesions with characteristic intranuclear and/or intracytoplasmic viral inclusions mainly in the bursa and/or in the spleen 3. Moderate to high amount of PiCV in damaged tissue, demonstrable by electron microscopy, ISH, and/or quantitative PCR

Based on epidemiologic and case studies conducted in multiple countries across the world, PiCV infection is quite common among both domestic and feral pigeons, with some reports identifying that most young pigeons in an affected loft would have been infected before reaching 1 year old. Previously performed experimental infections of pigeons with PiCV infected tissue homogenates, as well as documented naturally infected pigeons without clinical symptoms, confirm that while no clinical manifestations were observed, characteristic histopathologic lesions were present in challenged pigeons [79,93,105]. Therefore, PiCV-subclinical infection (PiCV-SI) is a common disease.

On other hand, PiCV-SD, like PCV2-SD, is a multifactorial disease in which PiCV is the strictly required factor. Like PCVDs, as PiCV has also been noted for its wide distribution similar to PCV2, cases of PiCV-SD must be considered to be of complex causality. This proposal, therefore, functionally reclassifies Duchatel's and Szeleszczuk's conceptualization of YPDS as PiCV-SD. As such, the previously reviewed clinico-pathological picture of PiCV infection with clinical manifestations and characteristic histological lesions would also be descriptive of PiCV-SD [22,23]. In this sense, many cases previously identified as YPDS can be classified as cases of PiCV-SD. In contrast, the proportion of the diagnosed YPDS cases that are not PiCV-SD cases cannot be assessed, due to, as pointed out earlier, variable characterizations of YPDS.

Furthermore, the current proposal is not to deny the involvement of other infectious agents in PiCV-SD. For instance, consistent with Schmidt et al. (2020), the discovery of pigeon Rotavirus A does not contradict to its potential importance in the development of what we are currently proposing as PiCV-SD. What this proposal is trying to resolve and formalize is the importance of and necessity for PiCV infection to pigeon health and the development of a particular disease. In extension, the proposal recognizes the pathogenicity of PiCV, perhaps, best evidenced by the demonstration of the virus' apoptotic ability to B lymphocytes confirming its immunosuppressive function [52,92].

Guided by our examination of the literature, we are of the view that this proposal is the best step moving forward as an attempt to lodge a more cohesive approach to dealing with matters important to pigeon health management, and in extension, to scientific research.

5. Research Gaps

Perhaps, among the most important research gaps in PiCV research is still the absence of a reliable and replicated method for isolating this viral agent. We, however, note two records reporting propagation of PiCV [96,112] achieved by inoculating PiCV-positive tissue homogenates into specific pathogen free (SPF) embryonated chicken eggs.

In Sahindokuyucu et al. (2022), inoculation of tissue homogenates positive for PiAdV-A and PiCV into the chorioallantoic cavity and yolk sack of the SPF embryonated eggs yielded PiCV-positive allantoic fluid at the fifth passage. None of the other tested viruses were positive by PCR. Additionally, cultivation of the virus from the same tissue homogenate using primary chicken embryo fibroblast also yielded PiCV- and PiADV-A-positive supernatant. It was, however, unclear how many cell culture passages were performed [96]. Meanwhile, in Van Borm (2013), SPF embryonated eggs were inoculated in the allantoic cavity. Notably, the goal of their investigation is to determine the genetic diversity of pigeon paramyxovirus type 1 (PPMV-1) in their collection using next generation sequencing approach. However, their sequence data did not only provide information on their PPMV-1 isolates, but also revealed contaminating PiCV. RT-PCR analysis of their virus stocks, even those that were passaged in embryonated chicken eggs, were positive for PiCV [112]. Both these reports provide hope for possible isolation, purification, and cultivation of PiCV isolates. Replicating these successes would be essential as this would open various opportunities to investigate this virus in a more controlled and more detailed manner.

Particularly, the establishment of a reliable isolation protocol for PiCV would enable future research on the more basic understanding of the biology of this virus, including its specific mechanisms for entry, replication, and release, as well as the functions of the other proteins encoded its genome. In particular, the role of ORF C3 protein in PiCV is still unclear. In PCV2 and duck circovirus, ORF C3 has been confirmed to induce lymphocyte depletion by apoptosis [15,58,125,126]. Additional understanding of these basic facets may be critical to the understanding of the viral pathogenicity, and to the development of new approaches for treatment and control. Furthermore, propagation of pure isolates of this virus would also enable studies related to pathogenesis, virus–host interactions, especially during coinfections, host immunity, and recovery, among others.

On the other hand, with the reported detection of PiCV in other animals, particularly in ticks, it is tempting to speculate whether all or some of these species are hosts (definitive or intermediate) or a vector of PiCV. With most of these reports relying on partial sequences, it remains to be shown if these detected infections are of a different subtype or strain, or perhaps a completely new circovirus species that is a very close relative sharing a common origin with PiCV. Nevertheless, these reports seem to be a prelude to studies needed to uncover what might be a cross-species jumping in play in PiCV evolution, together with its implication to PiCV epidemiology, pathobiology, and genetic diversity.

More than thirty years after the discovery of PiCV, there is still a lot to learn about this viral agent.

Supplementary Materials: The following supporting information can be downloaded at: https://www.mdpi.com/article/10.3390/v14071498/s1, Figure S1: Pairwise identity matrix of 371 PiCV whole genome sequences available from NCBI.; Figure S2: Phylogenetic trees of the 371 PiCV whole genome sequences available from NCBI.; Figure S3: Recombination breakpoint analysis of the 371 PiCV whole genome sequences available from NCBI. Table S1: Nucleotide pairwise identity analysis of the 371 PiCV whole genome sequences available from NCBI.

Author Contributions: Conceptualization, K.-P.C., J.R.E., B.B.I.S. and M.L.R.U.; methodology, B.B.I.S. and M.L.R.U.; formal analysis, K.-P.C., J.R.E., B.B.I.S., M.L.R.U., S.A.S., Y.-C.T., K.-C.H., C.-C.C. and K.-C.H.; investigation, B.B.I.S., M.L.R.U., J.R.E., A.M.S. and A.J.D.L.; data curation, B.B.I.S., M.L.R.U., J.R.E., A.M.S. and A.J.D.L.; writing—original draft preparation, B.B.I.S., M.L.R.U., J.R.E., A.M.S. and A.J.D.L.; writing—review and editing, K.-P.C., Y.-C.T., J.R.E., B.B.I.S., M.L.R.U., S.A.S., C.-C.C. and K.-C.H.; visualization, B.B.I.S., M.L.R.U., A.M.S. and A.J.D.L.; supervision, K.-P.C. and J.R.E.; project administration, K.-P.C.; funding acquisition, K.-P.C. and Y.-C.T. All authors have read and agreed to the published version of the manuscript.

Funding: This research was funded by the National Pingtung University of Science and Technology and the Kaohsiung Chang Gung Memorial Hospital and National Pingtung University of Science and Technology Joint Research Program (Grant Number NPUST-KMU-111-P002), and the Higher Education Sprout Project by the Ministry of Education and the Ministry of Science and Technology (MOST 107-3017-F-020-003).

Institutional Review Board Statement: Not applicable.

Informed Consent Statement: Not applicable.

Data Availability Statement: Not applicable.

Acknowledgments: Grateful acknowledgments to Paulyne Gonzales for the expert assistance in preparing important visuals for this paper, and to Angelo Miguel Elijah S. De Guzman for assisting in the initial text mining. Additionally, deepest gratitude to Cristian Tablazon for the critical discussions and comments that immensely improved this paper.

Conflicts of Interest: The funders had no role in the design of the study; in the collection, analyses, or interpretation of data; in the writing of the manuscript, or in the decision to publish the results.

References

1. Schmidt, R.E. Circovirus in Pigeons. *J. Assoc. Avian Vet.* **1992**, *6*, 204. [CrossRef]
2. International Committee on the Taxonomy of Viruses (ICTV). Available online: https://talk.ictvonline.org/taxonomy/p/taxonomy_releases (accessed on 13 June 2022).
3. Woods, L.W.; Latimer, K.S.; Niagro, F.D.; Riddell, C.; Crowley, A.M.; Anderson, M.L.; Daft, B.M.; Moore, J.D.; Campagnoli, R.P.; Nordhausen, R.W. A retrospective study of circovirus infection in pigeons: Nine cases (1986–1993). *J. Vet. Diagn. Investig.* **1994**, *6*, 156–164. [CrossRef]
4. Tian, D. Bibliometric analysis of pathogenic organisms. *Biosaf. Health* **2020**, *2*, 95–103. [CrossRef]
5. Gray Neils, M.E.; Pfaeffle, H.O.I.; Kulatti, A.T.; Titova, A.; Lyles, G.S.; Plotnikova, Y.; Zorkaltseva, E.; Ogarkov, O.B.; Vitko, S.M.; Dillingham, R.A.; et al. A geospatial Bibliometric review of the HIV/AIDS epidemic in the Russian federation. *Front. Public Health* **2020**, *8*, 75. [CrossRef] [PubMed]
6. Deng, Z.; Chen, J.; Wang, T. Bibliometric and Visualization Analysis of Human Coronaviruses: Prospects and Implications for COVID-19 Research. *Front. Cell. Infect. Microbiol.* **2020**, *10*, 581404. [CrossRef]
7. Furstenau, L.; Rabaioli, B.; Sott, M.; Cossul, D.; Bender, M.; Farina, E.; Filho, F.; Severo, P.; Dohan, M.; Bragazzi, N. A Bibliometric Network Analysis of Coronavirus during the First Eight Months of COVID-19 in 2020. *Int. J. Environ. Res. Public Health* **2021**, *18*, 952. [CrossRef]
8. Raman, R.; Vinuesa, R.; Nedungadi, P. Bibliometric Analysis of SARS, MERS, and COVID-19 Studies from India and Connection to Sustainable Development Goals. *Sustainability* **2021**, *13*, 7555. [CrossRef]
9. Nejad, A.S.M.; Noor, T.; Munim, Z.H.; Alikhani, M.Y.; Ghaemi, A. A bibliometric review of oncolytic virus research as a novel approach for cancer therapy. *Virol. J.* **2021**, *18*, 98. [CrossRef]
10. Turatto, F.; Mazzalai, E.; Pagano, F.; Migliara, G.; Villari, P.; De Vito, C. A Systematic Review and Bibliometric Analysis of the Scientific Literature on the Early Phase of COVID-19 in Italy. *Front. Public Health* **2021**, *9*, 776. [CrossRef]
11. Ogunsakin, R.E.; Ebenezer, O.; Ginindza, T.G. A Bibliometric Analysis of the Literature on Norovirus Disease from 1991–2021. *Int. J. Environ. Res. Public Health* **2022**, *19*, 2508. [CrossRef]
12. Xu, C.; Zhang, W.; Pan, Y.; Wang, G.; Yin, Q.; Fu, S.; Li, F.; He, Y.; Xu, S.; Wang, Z.; et al. A Bibliometric Analysis of Global Research on Japanese Encephalitis From 1934 to 2020. *Front. Cell. Infect. Microbiol.* **2022**, *12*, 8. [CrossRef] [PubMed]

13. Brookes, V.J.; Barrett, T.E.; Ward, M.P.; Roby, J.A.; Hernandez-Jover, M.; Cross, E.M.; Donnelly, C.M.; Barnes, T.S.; Wilson, C.S.; Khalfan, S. A scoping review of African swine fever virus spread between domestic and free-living pigs. *Transbound. Emerg. Dis.* **2021**, *68*, 2643–2656. [CrossRef] [PubMed]
14. Van Eck, N.J.; Waltman, L. Software survey: VOSviewer, a computer program for bibliometric mapping. *Scientometrics* **2009**, *84*, 523–538. [CrossRef] [PubMed]
15. Mankertz, A.; Hattermann, K.; Ehlers, B.; Soike, D. Cloning and sequencing of columbid circovirus (CoCV), a new circovirus from pigeons. *Arch. Virol.* **2000**, *145*, 2469–2479. [CrossRef]
16. Todd, D.; Weston, J.; Soike, D.; Smyth, J. Genome Sequence Determinations and Analyses of Novel Circoviruses from Goose and Pigeon. *Virology* **2001**, *286*, 354–362. [CrossRef]
17. Todd, D.; Weston, J.; Ball, N.W.; Borghmans, B.J.; Smyth, J.A.; Gelmini, L.; Lavazza, A. Nucleotide sequence-based identification of a novel circovirus of canaries. *Avian Pathol.* **2001**, *30*, 321–325. [CrossRef]
18. Soike, D.; Hattermann, K.; Albrecht, K.; Segalés, J.; Domingo, M.; Schmitt, C.; Mankertz, A. A diagnostic study on columbid circovirus infection. *Avian Pathol.* **2001**, *30*, 605–611. [CrossRef]
19. Smyth, J.A.; Weston, J.; Moffett, D.A.; Todd, D. Detection of circovirus infection in pigeons by in situ hybridization using cloned DNA probes. *J. Vet. Diagn. Investig.* **2001**, *13*, 475–482. [CrossRef]
20. Gerdes, G.H. Two very small viruses—A presumptive identification. *J. S. Afr. Vet. Assoc.* **1993**, *64*, 2.
21. Sato, K.; Shoyama, T.; Ikemachi, Y.; Nakanishi, K.; Asahi, S.; Akachi, S.; Yamauchi, K.; Higashiyama, Y.; Obata, H.; Yamanaka, S.; et al. A Case of Racing Pigeons Concomitantly Infected with a Circovirus-like Virus and Chlamydia psittaci in Mie Prefecture. *J. Japan Vet. Med. Assoc.* **2000**, *53*, 446–450. [CrossRef]
22. Stenzel, T.; Koncicki, A. The epidemiology, molecular characterization and clinical pathology of circovirus infections in pigeons—Current knowledge. *Vet. Q.* **2017**, *37*, 166–174. [CrossRef] [PubMed]
23. Duchatel, J.P.; Szeleszczuk, P. Young pigeon disease syndrome. *Med. Weter.* **2011**, *67*, 291–294.
24. Duchatel, J.P. L'infection du pigeon par le circovirus. *Ann. Med. Vet.* **2009**, *153*, 211–218.
25. Cavanagh, D. Innovation and discovery: The application of nucleic acid-based technology to avian virus detection and characterization. *Avian Pathol.* **2001**, *30*, 581–598. [CrossRef] [PubMed]
26. Hattermann, K.; Soike, D.; Grund, C.; Mankertz, A. A method to diagnose Pigeon circovirus infection in vivo. *J. Virol. Methods* **2002**, *104*, 55–58. [CrossRef]
27. Todd, D.; Duchatel, J.P.; Weston, J.H.; Ball, N.W.; Borghmans, B.J.; Moffett, D.A.; Smyth, J.A. Evaluation of polymerase chain reaction and dot blot hybridisation tests in the diagnosis of pigeon circovirus infections. *Vet. Microbiol.* **2002**, *89*, 1–16. [CrossRef]
28. Roy, P.; Dhillon, A.S.; Lauerman, L.; Shivaprasad, H.L. Detection of Pigeon Circovirus by Polymerase Chain Reaction. *Avian Dis.* **2003**, *47*, 218–222. [CrossRef]
29. Taras, L.; Kubíček, O.; Juranová, R.; Jurajda, V. The First Demonstration of Pigeon Circovirus Infection in the Czech Republic Based on Histology and Nested PCR. *Acta Vet. Brno* **2003**, *72*, 577–582. [CrossRef]
30. Duchatel, J.P.; Todd, D.; Curry, A.; Smyth, J.A.; Bustin, J.C.; Vindevogel, H. New data on the transmission of pigeon circovirus. *Vet. Rec.* **2005**, *157*, 413–415. [CrossRef]
31. Franciosini, M.P.; Fringuelli, E.; Tarhuni, O.; Guelfi, G.; Todd, D.; Proietti, P.C.; Falocci, N.; Asdrubali, G. Development of a polymerase chain reaction-based in vivo method in the diagnosis of subclinical pigeon circovirus infection. *Avian Dis.* **2005**, *49*, 340–343. [CrossRef]
32. Kubíček, O.; Taras, L. Incidence of pigeon circovirus in Eurasian collared-dove (Streptopelia decaocto) detected by nested PCR. *Acta Vet. Brno* **2005**, *74*, 361–368. [CrossRef]
33. Duchatel, J.P.; Todd, D.; Smyth, J.A.; Bustin, J.C.; Vindevogel, H. Observations on detection, excretion and transmission of pigeon circovirus in adult, young and embryonic pigeons. *Avian Pathol.* **2006**, *35*, 30–34. [CrossRef] [PubMed]
34. Todd, D.; Duchatel, J.P.; Bustin, J.C.; Scullion, F.T.; Scullion, M.G.; Scott, A.N.J.; Curry, A.; Ball, N.W.; Smyth, J.A. Detection of pigeon circovirus in cloacal swabs: Implications for diagnosis, epidemiology and control. *Vet. Rec.* **2006**, *159*, 314–317. [CrossRef] [PubMed]
35. Marlier, D.; Vindevogel, H. Viral infections in pigeons. *Vet. J.* **2006**, *172*, 40–51. [CrossRef] [PubMed]
36. Wieliczko, A.; Piasecki, T.; Houszka, M. Infection of circovirus in pigeons. *Med. Weter.* **2005**, *61*, 94–97.
37. Lumeij, J.T. Diseases of racing pigeon: An update. *Vet. Q.* **1996**, *18*, 66–67. [CrossRef]
38. Yamamoto, E.; Ito, H.; Kitamoto, E.; Morinishi, K.; Yano, A.; Miyoshi, S.; Ito, T. Complete genome sequence of pigeon circovirus detected in racing pigeons in western Japan. *Virus Genes* **2015**, *51*, 140–143. [CrossRef]
39. Al-Baroodi, S.Y.; Al-Attar, M.Y. Isolation and identification of circovirus in pigeon. *Iraqi J. Vet. Sci.* **2021**, *35*, 207–210. [CrossRef]
40. Loiko, M.R.; Junqueira, D.M.; Varela, A.P.M.; Tochetto, C.; Scheffer, C.M.; Lima, D.A.; Morel, A.P.; Cerva, C.; Paim, W.P.; Mayer, F.Q.; et al. Columbid circoviruses detected in free ranging pigeons from Southern Brazil: Insights on PiCV evolution. *Arch. Virol.* **2018**, *163*, 3083–3090. [CrossRef]
41. van Eck, N.J.; Waltman, L. VOSviewer Manual. Available online: https://www.vosviewer.com/documentation/Manual_VOSviewer_1.6.7.pdf (accessed on 3 February 2022).
42. Todd, D. Circoviruses: Immunosuppressive threats to avian species: A review. *Avian Pathol.* **2000**, *29*, 373–394. [CrossRef]
43. Todd, D. Avian circovirus diseases: Lessons for the study of PMWS. *Vet. Microbiol.* **2004**, *98*, 169–174. [CrossRef] [PubMed]

44. Raue, R.; Schmidt, V.; Freick, M.; Reinhardt, B.; Johne, R.; Kamphausen, L.; Kaleta, E.F.; Müller, H.; Krautwald-Junghanns, M.E. A disease complex associated with pigeon circovirus infection, young pigeon disease syndrome. *Avian Pathol.* **2005**, *34*, 418–425. [CrossRef]
45. Rosario, K.; Breitbart, M.; Harrach, B.; Segalés, J.; Delwart, E.; Biagini, P.; Varsani, A. Revisiting the taxonomy of the family Circoviridae: Establishment of the genus Cyclovirus and removal of the genus Gyrovirus. *Arch. Virol.* **2017**, *162*, 1447–1463. [CrossRef] [PubMed]
46. Breitbart, M.; Delwart, E.; Rosario, K.; Segalés, J.; Varsani, A. ICTV virus taxonomy profile: Circoviridae. *J. Gen. Virol.* **2017**, *98*, 1997–1998. [CrossRef] [PubMed]
47. Shivaprasad, H.L.; Chin, R.P.; Jeffrey, J.S.; Latimer, K.S.; Nordhausen, R.W.; Niagro, F.D.; Campagnoli, R.P. Case Report—Particles Resembling Circovirus in the Bursa of Fabricius of Pigeons. *Avian Dis.* **1994**, *38*, 635–641. [CrossRef] [PubMed]
48. Mankertz, A. Circoviruses. In *Encyclopedia of Virology*; Elsevier: Amsterdam, The Netherlands, 2008; pp. 513–519.
49. Todd, D.; Fringuelli, E.; Scott, A.; Borghmans, B.; Duchatel, J.; Shivaprasad, H.; Raidal, S.; Abadie, J.; Franciosini, M.; Smyth, J. Sequence comparison of pigeon circoviruses. *Res. Vet. Sci.* **2008**, *84*, 311–319. [CrossRef]
50. Woods, L.W.; Latimer, K.S.; Barr, B.C.; Niagro, F.D.; Campagnoli, R.; Nordhausen, R.W.; Castro, A.E. Circovirus-Like Infection in a Pigeon. *J. Vet. Diagn. Investig.* **1993**, *5*, 609–612. [CrossRef]
51. De Herdt, P.; Ducatelle, R.; Lepoudre, C.; Charlier, G.; Nauwynck, H. An epidemic of fatal hepatic necrosis of viral origin in racing pigeons(*Columba livia*). *Avian Pathol.* **1995**, *24*, 475–483. [CrossRef]
52. Raidal, S.R.; Riddoch, P.A. A feather disease in Senegal doves (*Streptopelia senegalensis*) morphologically similar to psittacine beak and feather disease. *Avian Pathol.* **1997**, *26*, 829–836. [CrossRef] [PubMed]
53. Soike, D.; Köhler, B.; Albrecht, K. A circovirus-like infection in geese related to a runting syndrome. *Avian Pathol.* **1999**, *28*, 199–202. [CrossRef]
54. Paré, J.A.; Brash, M.L.; Hunter, D.B.; Hampson, R.J. Observations on pigeon circovirus infection in Ontario. *Can. Vet. J.* **1999**, *40*, 659–662. [PubMed]
55. Tavernier, P.; De Herdt, P.; Thoonen, H.; Ducatelle, R. Prevalence and pathogenic significance of circovirus-like infections in racing pigeons (*Columba livia*). *Vlaams Diergeneeskd. Tijdschr.* **2000**, *69*, 338–341.
56. Duchatel, J.P.; Todd, D.; Willeman, C.; Losson, B. Quantification of pigeon circovirus in serum, blood, semen and different tissues of naturally infected pigeons using a real-time polymerase chain reaction. *Avian Pathol.* **2009**, *38*, 143–148. [CrossRef]
57. Duchatel, J.P.; Todd, D.; Smyth, J.; Costes, B.; Jauniaux, T.; Farnir, F.; Losson, B.; Vanderplasschen, A. Pigeon circovirus: Baculovirus expression of the capsid protein gene, specific antibody and viral load measured by real time polymerase chain reaction. *Isr. J. Vet. Med.* **2011**, *66*, 26–31.
58. Stenzel, T.; Dziewulska, D.; Tykałowski, B.; Koncicki, A. The Clinical Infection with Pigeon Circovirus (PiCV) Leads to Lymphocyte B Apoptosis But Has No Effect on Lymphocyte T Subpopulation. *Pathogens* **2020**, *9*, 632. [CrossRef]
59. Tsai, S.S.; Chang, Y.L.; Huang, Y.L.; Liu, H.J.; Ke, G.M.; Chiou, C.J.; Hsieh, Y.C.; Chang, T.C.; Cheng, L.T.; Chuang, K.P. Development of a loop-mediated isothermal amplification method for rapid detection of pigeon circovirus. *Arch. Virol.* **2014**, *159*, 921–926. [CrossRef]
60. Kong, Y.; Yan, C.; Zhang, G.; Cai, Y.; He, B.; Li, Y. Detection of pigeon circoviruses in ticks of sheep and camels in Inner Mongolia, China. *bioRxiv* **2021**. [CrossRef]
61. Jerolmack, C. Animal archeology: Domestic pigeons and the nature-culture dialectic. *Qual. Sociol. Rev.* **2007**, *3*, 74–95. [CrossRef]
62. Traxler, B.; Brem, G.; Müller, M.; Achmann, R. Polymorphic DNA microsatellites in the domestic pigeon, Columba livia var. domestica. *Mol. Ecol.* **2000**, *9*, 366–368. [CrossRef]
63. Jedrzejczak-Silicka, M.; Yu, Y.; Cheng, Y.; Dybus, A. the Influence of LDHA Gene Polymorphism on Relative Level of Its Expression in Racing Pigeons. *Acta Sci. Pol. Zootech.* **2018**, *17*, 9–16. [CrossRef]
64. Chang, C.-C.; Silva, B.B.I.; Huang, H.-Y.; Tsai, C.-Y.; Flores, R.J.D.; Tayo, L.L.; Tyan, Y.-C.; Tsai, M.-A.; Catulin, G.E.M.; Chuang, K.-P.; et al. Development and Validation of KASP Assays for the Genotyping of Racing Performance-Associated Single Nucleotide Polymorphisms in Pigeons. *Genes* **2021**, *12*, 1383. [CrossRef] [PubMed]
65. Pare, J.A.; Robert, N. Circovirus. In *Infectious Diseases of Wild Birds*; Blackwell Publishing Professional: Ames, IA, USA, 2004; pp. 194–205.
66. Cságola, A.; Lőrincz, M.; Tombacz, K.; Wladár, Z.; Kovács, E.; Tuboly, T. Genetic diversity of pigeon circovirus in Hungary. *Virus Genes* **2011**, *44*, 75–79. [CrossRef] [PubMed]
67. Stenzel, T.; Pestka, D. Occurrence and genetic diversity of pigeon circovirus strains in Poland. *Acta Vet. Hung.* **2014**, *62*, 274–283. [CrossRef] [PubMed]
68. Shivaprasad, H.L.; Chin, R.P.; Jeffrey, J.S.; Nordhausen, R.W.; Latimer, K.S. A new viral disease of pigeons? Particles resembling circovirus in the bursa of fabricius. In Proceedings of the 42nd Western Poultry Disease Conference, Sacramento, CA, USA, 28 February–2 March 1993; pp. 99–100.
69. Smyth, J.; Carroll, B. Circovirus infection in European racing pigeons. *Vet. Rec.* **1995**, *136*, 173–174. [CrossRef] [PubMed]
70. Begg, N.; Kennedy, S. VSD-AFBI Northern Ireland disease surveillance, July to September 2010. *Vet. Rec.* **2010**, *167*, 924–927. [CrossRef]
71. Coletti, M.; Franciosini, M.P.; Asdrubali, G.; Passamonti, F. Atrophy of the primary lymphoid organs of meat pigeons in Italy associated with circoviruslike particles in the bursa of Fabricius. *Avian Dis.* **2000**, *44*, 454–459. [CrossRef] [PubMed]

72. Abadie, J.; Nguyen, F.; Groizeleau, C.; Amenna, N.; Fernandez, B.; Guereaud, C.; Guigand, L.; Robart, P.; Lefebvre, B.; Wyers, M. Pigeon circovirus infection: Pathological observations and suggested pathogenesis. *Avian Pathol.* **2001**, *30*, 149–158. [CrossRef] [PubMed]
73. Khalifeh, A.; Kraberger, S.; Dziewulska, D.; Varsani, A.; Stenzel, T. A Pilot Study Investigating the Dynamics of Pigeon Circovirus Recombination in Domesticated Pigeons Housed in a Single Loft. *Viruses* **2021**, *13*, 964. [CrossRef]
74. Stenzel, T.A.; Pestka, D.; Tykałowski, B.; Śmiałek, M.; Koncicki, A. Epidemiological investigation of selected pigeon viral infections in Poland. *Vet. Rec.* **2012**, *171*, 562. [CrossRef]
75. Stenzel, T.; Piasecki, T.; Chrząstek, K.; Julian, L.; Muhire, B.M.; Golden, M.; Martin, D.P.; Varsani, A. Pigeon circoviruses display patterns of recombination, genomic secondary structure and selection similar to those of beak and feather disease viruses. *J. Gen. Virol.* **2014**, *95*, 1338–1351. [CrossRef]
76. Stenzel, T.; Pestka, D.; Choszcz, D. The prevalence and genetic characterization of Chlamydia psittaci from domestic and feral pigeons in Poland and the correlation between infection rate and incidence of pigeon circovirus. *Poult. Sci.* **2014**, *93*, 3009–3016. [CrossRef] [PubMed]
77. Stenzel, T.; Woźniakowski, G.; Pestka, D.; Choszcz, D.; Tykałowski, B.; Śmiałek, M.; Koncicki, A. Application of pigeon circovirus recombinant capsid protein for detecting anti-PiCV antibodies in the sera of asymptomatic domestic pigeons and the potential use of a combination of serological and molecular tests for controlling circovirus infections in pigeon breeding flocks. *Poult. Sci.* **2016**, *96*, 303–308. [CrossRef] [PubMed]
78. Dolka, B.; Dolka, I.; Ledwoń, A.; Sapierzyński, R.; Koralewski, A.; Szeleszczuk, P. Investigations into feather abnormalities in racing pigeons. *Med. Weter.* **2016**, *72*, 693–698. [CrossRef]
79. Stenzel, T.; Dziewulska, D.; Tykałowski, B.; Śmiałek, M.; Kowalczyk, J.; Koncicki, A. Immunogenicity of Pigeon Circovirus Recombinant Capsid Protein in Pigeons. *Viruses* **2018**, *10*, 596. [CrossRef] [PubMed]
80. Stenzel, T.; Dziewulska, D.; Śmiałek, M.; Tykałowski, B.; Kowalczyk, J.; Koncicki, A. Comparison of the immune response to vaccination with pigeon circovirus recombinant capsid protein (PiCV rCP) in pigeons uninfected and subclinically infected with PiCV. *PLoS ONE* **2019**, *14*, e0219175. [CrossRef] [PubMed]
81. Krapež, U.; Slavec, B.; Steyer, A.F.; Pintarič, S.; Dobeic, M.; Rojs, O.Z.; Dovc, A. Prevalence of Pigeon Circovirus Infections in Feral Pigeons in Ljubljana, Slovenia. *Avian Dis.* **2012**, *56*, 432–435. [CrossRef]
82. Dovč, A.; Jereb, G.; Krapež, U.; Gregurić-Gračner, G.; Pintarič, Š.; Slavec, B.; Knific, R.L.; Kastelic, M.; Kvapil, P.; Mićunović, J.; et al. Occurrence of Bacterial and Viral Pathogens in Common and Noninvasive Diagnostic Sampling from Parrots and Racing Pigeons in Slovenia. *Avian Dis.* **2016**, *60*, 487–492. [CrossRef]
83. Phan, T.; Vo, N.P.; Boros, A.; Pankovics, P.; Reuter, G.; Li, O.T.W.; Wang, C.; Deng, X.; Poon, L.; Delwart, E. The Viruses of Wild Pigeon Droppings. *PLoS ONE* **2013**, *8*, e72787. [CrossRef]
84. Ledwoń, A.; Bailey, T.; O'Donovan, D.; McKeown, S.; Lloyd, C.; Wieckowski, T.; Kinne, J.; Silvanose, C.; Szeleszczuk, P.; Wernery, U. Prevalence of circovirus and adenovirus in pigeons in Dubai. *Med. Weter.* **2011**, *67*, 752–756.
85. El-Rahman, S.S.A.; El-Dakhly, A.T. Demonstration of pigeon circovirus infection in Western-Libya, Based on histopathology, electron microscopy and apoptosis. *Glob. Vet.* **2010**, *4*, 242–248.
86. Li, L.; Kapoor, A.; Slikas, B.; Bamidele, O.S.; Wang, C.; Shaukat, S.; Alam Masroor, M.; Wilson, M.L.; Ndjango, J.-B.N.; Peeters, M.; et al. Multiple Diverse Circoviruses Infect Farm Animals and Are Commonly Found in Human and Chimpanzee Feces. *J. Virol.* **2010**, *84*, 1674–1682. [CrossRef] [PubMed]
87. Lin, Y.J. *Epidemiological Analysis of Circovirus Isolated from Pigeon and the Order Passeriformes in Taiwan*; National Taiwan University: Taipei, Taiwan, 2012.
88. Liao, P.-C.; Wang, K.-K.; Tsai, S.-S.; Liu, H.-J.; Huang, B.-H.; Chuang, K.-P. Recurrent positive selection and heterogeneous codon usage bias events leading to coexistence of divergent pigeon circoviruses. *J. Gen. Virol.* **2015**, *96*, 2262–2273. [CrossRef] [PubMed]
89. Huang, Y.-L.; Castaneda, O.A.; Thongchan, D.; Khatri-Chhetri, R.; Tsai, S.-S.; Wu, H.-Y. Pigeon circovirus infection in disqualified racing pigeons from Taiwan. *Avian Pathol.* **2017**, *46*, 359–366. [CrossRef]
90. Santos, H.M.; Chen, C.C.; Tsai, C.-Y.; Hsish, Y.C.; Chung, F.C.; Tyan, Y.-C.; Tayo, L.L.; Chuang, K.P. Influence of pigeon interferon alpha (PiIFN-α) on pigeon circovirus (PiCV) replication and cytokine expression in Columba livia. *Vet. Microbiol.* **2020**, *242*, 108591. [CrossRef] [PubMed]
91. Tsai, C.; Hu, S.; Santos, H.; Catulin, G.; Tayo, L.; Chuang, K. Probiotic supplementation containing *Bacillus velezensis* enhances expression of immune regulatory genes against pigeon circovirus in pigeons (*Columba livia*). *J. Appl. Microbiol.* **2020**, *130*, 1695–1704. [CrossRef]
92. Chen, W.-T.; Teng, C.-A.; Shih, C.-H.; Huang, W.-H.; Jiang, Y.-F.; Chang, H.-W.; Jeng, C.-R.; Lai, Y.-H.; Guo, J.-C.; Wang, P.-J.; et al. Investigation of Lethal Concurrent Outbreak of Chlamydiosis and Pigeon Circovirus in a Zoo. *Animals* **2021**, *11*, 1654. [CrossRef] [PubMed]
93. Huang, H.-Y.; Silva, B.; Tsai, S.-P.; Tsai, C.-Y.; Tyan, Y.-C.; Lin, T.-C.; Flores, R.; Chuang, K.-P. Immunogenicity and Protective Activity of Pigeon Circovirus Recombinant Capsid Protein Virus-Like Particles (PiCV rCap-VLPs) in Pigeons (*Columba livia*) Experimentally Infected with PiCV. *Vaccines* **2021**, *9*, 98. [CrossRef]
94. Mahzounieh, M.; Khoei, H.H.; Shamsabadi, M.G.; Dastjerdi, A. Detection and phylogenetic characterization of Columbid circoviruses in Chaharmahal va Bakhtiari province, Iran. *Avian Pathol.* **2014**, *43*, 524–528. [CrossRef]

95. Haddadmarandi, M.R.; Madani, S.A.; Nili, H.; Ghorbani, A. Molecular survey of avian circoviruses in some non-psittacine birds and detection of a novel canary circovirus in a pigeon. *Iran. J. Vet. Res.* **2020**, *21*, 57–60.
96. Sahindokuyucu, I.; Turkmen, M.B.; Sumer, T.; Elhag, A.E.; Alcigir, M.E.; Yazici, Z.; Barry, G.; Gulbahar, M.Y.; Kul, O. First detection and molecular characterisation of a pigeon aviadenovirus A and pigeon circovirus co-infection associated with Young Pigeon Disease Syndrome (YPDS) in Turkish pigeons (*Columba livia domestica*). *Vet. Med. Sci.* **2022**, *8*, 139–149. [CrossRef]
97. Twentyman, C.; Alley, M.; Meers, J.; Cooke, M.; Duignan, P. Circovirus-like infection in a southern black-backed gull (*Larus dominicanus*). *Avian Pathol.* **1999**, *28*, 513–516. [CrossRef] [PubMed]
98. Smyth, J.A.; Todd, D.; Scott, A.; Beckett, A.; Twentyman, C.M.; Bröjer, C.; Uhlhorn, H.; Gavier-Widen, D. Identification of circovirus infection in three species of gull. *Vet. Rec.* **2006**, *159*, 212–214. [CrossRef] [PubMed]
99. Zhang, Z.; Lu, C.; Wang, Y.; Wang, S.; Dai, D.; Chen, Z.; Fan, H. Molecular characterization and epidemiological investigation of pigeon circovirus isolated in eastern china. *J. Vet. Diagn. Investig.* **2011**, *23*, 665–672. [CrossRef] [PubMed]
100. Zhang, Z.; Dai, W.; Wang, S.; Dai, D. Epidemiology and genetic characteristics of pigeon circovirus (PiCV) in eastern China. *Arch. Virol.* **2014**, *160*, 199–206. [CrossRef]
101. Wang, K.C.; Zhuang, Q.Y.; Qiu, Y.; Wang, T.; Chen, J.M. Genome sequence characterization of pigeon circoviruses in China. *Virus Res.* **2017**, *233*, 1244. [CrossRef]
102. Sarker, S.; Das, S.; Ghorashi, S.; Forwood, J.; Raidal, S.R. Pigeon circoviruses from feral pigeons in Australia demonstrate extensive recombination and genetic admixture with other circoviruses. *Avian Pathol.* **2019**, *48*, 512–520. [CrossRef]
103. Martin, D.P.; Varsani, A.; Roumagnac, P.; Botha, G.; Maslamoney, S.; Schwab, T.; Kelz, Z.; Kumar, V.; Murrell, B. RDP5: A computer program for analyzing recombination in, and removing signals of recombination from, nucleotide sequence datasets. *Virus Evol.* **2021**, *7*, veaa087. [CrossRef]
104. Wang, H.; Gao, H.; Jiang, Z.; Shi, L.; Zhao, P.; Zhang, Y.; Wang, C. Molecular detection and phylogenetic analysis of pigeon circovirus from racing pigeons in China during 2016–2019. *Authorea* **2021**, 299–308. [CrossRef]
105. Schmidt, V.; Schlömer, J.; Lüken, C.; Johne, R.; Biere, B.; Muller, H.; Krautwald-Junghanns, M.-E. Experimental Infection of Domestic Pigeons with Pigeon Circovirus. *Avian Dis.* **2008**, *52*, 380–386. [CrossRef]
106. Johne, R.; Fernández-De-Luco, D.; Höfle, U.; Müller, H. Genome of a novel circovirus of starlings, amplified by multiply primed rolling-circle amplification. *J. Gen. Virol.* **2006**, *87*, 1189–1195. [CrossRef]
107. Cousquer, G.; Parsons, D. Veterinary care of the racing pigeon. *Practice* **2007**, *29*, 344–355. [CrossRef]
108. Freick, M.; Müller, H.; Raue, R. Rapid detection of pigeon herpesvirus, fowl adenovirus and pigeon circovirus in young racing pigeons by multiplex PCR. *J. Virol. Methods* **2008**, *148*, 226–231. [CrossRef] [PubMed]
109. de Herdt, P.; Pasmans, F. Pigeons. In *Handbook of Avian Medicine*; Saunders: Philadelphia, PA, USA, 2009; pp. 350–376, ISBN 9780702028748.
110. Daum, I.; Finsterbusch, T.; Härtle, S.; Göbel, T.W.; Mankertz, A.; Korbel, R.; Grund, C. Cloning and expression of a truncated pigeon circovirus capsid protein suitable for antibody detection in infected pigeons. *Avian Pathol.* **2009**, *38*, 135–141. [CrossRef]
111. Duchatel, J.P.; Jauniaux, T.; Smyth, J.; Habsch, I.; De Bournonville, M.; Losson, B.; Todd, D. Effect of a Commercial Paratyphus Vaccine on the Development of Pigeon Circovirus Infection in Young Pigeons (*Columba livia domestica*). *J. Avian Med. Surg.* **2010**, *24*, 107–114. [CrossRef] [PubMed]
112. Van Borm, S.; Rosseel, T.; Steensels, M.; Berg, T.V.D.; Lambrecht, B. What's in a strain? Viral metagenomics identifies genetic variation and contaminating circoviruses in laboratory isolates of pigeon paramyxovirus type 1. *Virus Res.* **2013**, *171*, 186–193. [CrossRef] [PubMed]
113. Lai, G.-H.; Lin, Y.-C.; Tsai, Y.-L.; Lien, Y.-Y.; Lin, M.-K.; Chen, H.-J.; Chang, W.-T.; Tzen, J.T.C.; Lee, M.-S. High yield production of pigeon circovirus capsid protein in the E. coliby evaluating the key parameters needed for protein expression. *BMC Vet. Res.* **2014**, *10*, 115. [CrossRef] [PubMed]
114. Gai, W.; Zheng, W.; Zhao, Z.; Wong, G.; Sun, P.; Yan, L.; He, H.; Zheng, X. Assembly of pigeon circovirus-like particles using baculovirus expression system. *Microb. Pathog.* **2020**, *139*, 103905. [CrossRef] [PubMed]
115. Santos, H.M.; Tsai, C.-Y.; Catulin, G.E.M.; Trangia, K.C.G.; Tayo, L.L.; Liu, H.-J.; Chuang, K.P. Common bacterial, viral, and parasitic diseases in pigeons (*Columba livia*): A review of diagnostic and treatment strategies. *Vet. Microbiol.* **2020**, *247*, 108779. [CrossRef]
116. Harzer, M.; Heenemann, K.; Sieg, M.; Vahlenkamp, T.; Freick, M.; Rückner, A. Prevalence of pigeon rotavirus infections: Animal exhibitions as a risk factor for pigeon flocks. *Arch. Virol.* **2020**, *166*, 65–72. [CrossRef]
117. Schmidt, V.; Kümpel, M.; Cramer, K.; Sieg, M.; Harzer, M.; Rückner, A.; Heenemann, K. Pigeon rotavirus A genotype G18P [17]-associated disease outbreaks after fancy pigeon shows in Germany—A case series. *Tierärztliche Prax. Ausg. K Kleintiere/Heimtiere* **2021**, *49*, 22–27. [CrossRef]
118. de Weerd, H. Dead Pigeons, A Nightmare for Every Pigeon Fancier. Available online: https://www.belgicadeweerd.com/en/dead-pigeons-a-nightmare-for-every-pigeon-fancier/ (accessed on 10 June 2022).
119. Walker, C. Circo Virus. Available online: https://www.auspigeonco.com.au/circo-virus.html (accessed on 10 June 2022).
120. Teske, L.; Rubbenstroth, D.; Meixner, M.; Liere, K.; Bartels, H.; Rautenschlein, S. Identification of a novel aviadenovirus, designated pigeon adenovirus 2 in domestic pigeons (*Columba livia*). *Virus Res.* **2017**, *227*, 15–22. [CrossRef] [PubMed]

121. Rubbenstroth, D.; Ulrich, R.; Wylezich, C.; Rautenschlein, S.; Beer, M.; Mohr, L. First experimental proof of Rotavirus A (RVA) genotype G18P[17] inducing the clinical presentation of 'young pigeon disease syndrome' (YPDS) in domestic pigeons (*Columba livia*). *Transbound. Emerg. Dis.* **2020**, *67*, 1507–1516. [CrossRef] [PubMed]
122. Segalés, J.; Domingo, M. Postweaning mulstisystemic wasting syndrome (PMWS) in pigs. A review. *Vet. Q.* **2002**, *24*, 109–124. [CrossRef] [PubMed]
123. Segalés, J. Porcine circovirus type 2 (PCV2) infections: Clinical signs, pathology and laboratory diagnosis. *Virus Res.* **2012**, *164*, 10–19. [CrossRef]
124. Saporiti, V.; Franzo, G.; Sibila, M.; Segalés, J. Porcine circovirus 3 (PCV-3) as a causal agent of disease in swine and a proposal of PCV-3 associated disease case definition. *Transbound. Emerg. Dis.* **2021**, *68*, 2936–2948. [CrossRef]
125. Liu, J.; Chen, I.; Kwang, J. Characterization of a Previously Unidentified Viral Protein in Porcine Circovirus Type 2-Infected Cells and Its Role in Virus-Induced Apoptosis. *J. Virol.* **2005**, *79*, 8262–8274. [CrossRef]
126. Xiang, Q.-W.; Wang, X.; Xie, Z.-J.; Sun, Y.-N.; Zhu, Y.-L.; Wang, S.-J.; Liu, H.-J.; Jiang, S.-J. ORF3 of duck circovirus: A novel protein with apoptotic activity. *Vet. Microbiol.* **2012**, *159*, 251–256. [CrossRef]

viruses

MDPI

Article

Genome-Wide Reassortment Analysis of Influenza A H7N9 Viruses Circulating in China during 2013–2019

Dongchang He [1,†], Xiyue Wang [1,†], Huiguang Wu [1,†], Xiaoquan Wang [1,2,3], Yayao Yan [1], Yang Li [1], Tiansong Zhan [1], Xiaoli Hao [1], Jiao Hu [1,2,3], Shunlin Hu [1,2,3], Xiaowen Liu [1,2,3], Chan Ding [2,4], Shuo Su [5], Min Gu [1,2,3,*] and Xiufan Liu [1,2,3,*]

1 Animal Infectious Disease Laboratory, College of Veterinary Medicine, Yangzhou University, Yangzhou 225009, China; virolog@outlook.com (D.H.); bioyusy@outlook.com (X.W.); hgwu80@163.com (H.W.); wxq@yzu.edu.cn (X.W.); bioyyy@outlook.com (Y.Y.); liyang1631@163.com (Y.L.); biocreater@outlook.com (T.Z.); xlhao@yzu.edu.cn (X.H.); hujiao@yzu.edu.cn (J.H.); slhu@yzu.edu.cn (S.H.); xwliu@yzu.edu.cn (X.L.)
2 Jiangsu Co-innovation Center for Prevention and Control of Important Animal Infectious Diseases and Zoonosis, Yangzhou University, Yangzhou 225009, China; shoveldeen@shvri.ac.cn
3 Jiangsu Key Laboratory of Zoonosis, Yangzhou University, Yangzhou 225009, China
4 Department of Avian Diseases, Shanghai Veterinary Research Institute, Chinese Academy of Agricultural Sciences, Shanghai 200241, China
5 College of Veterinary Medicine, Nanjing Agricultural University, Nanjing 210095, China; shuosu@njau.edu.cn
* Correspondence: gumin@yzu.edu.cn (M.G.); xfliu@yzu.edu.cn (X.L.)
† These authors have contributed equally to this work.

Abstract: Reassortment with the H9N2 virus gave rise to the zoonotic H7N9 avian influenza virus (AIV), which caused more than five outbreak waves in humans, with high mortality. The frequent exchange of genomic segments between H7N9 and H9N2 has been well-documented. However, the reassortment patterns have not been described and are not yet fully understood. Here, we used phylogenetic analyses to investigate the patterns of intersubtype and intrasubtype/intralineage reassortment across the eight viral segments. The H7N9 virus and its progeny frequently exchanged internal genes with the H9N2 virus but rarely with the other AIV subtypes. Before beginning the intrasubtype/intralineage reassortment analyses, five Yangtze River Delta (YRD A-E) and two Pearl River Delta (PRD A-B) clusters were divided according to the HA gene phylogeny. The seven reset segment genes were also nomenclatured consistently. As revealed by the tanglegram results, high intralineage reassortment rates were determined in waves 2–3 and 5. Additionally, the clusters of PB2 c05 and M c02 were the most dominant in wave 5, which could have contributed to the onset of the largest H7N9 outbreak in 2016–2017. Meanwhile, a portion of the YRD-C cluster (HP H7N9) inherited their PB2, PA, and M segments from the co-circulating YRD-E (LP H7N9) cluster during wave 5. Untanglegram results revealed that the reassortment rate between HA and NA was lower than HA with any of the other six segments. A multidimensional scaling plot revealed a robust genetic linkage between the PB2 and PA genes, indicating that they may share a co-evolutionary history. Furthermore, we observed relatively more robust positive selection pressure on HA, NA, M2, and NS1 proteins. Our findings demonstrate that frequent reassortment, particular reassorted patterns, and adaptive mutations shaped the H7N9 viral genetic diversity and evolution. Increased surveillance is required immediately to better understand the current state of the HP H7N9 AIV.

Keywords: avian influenza virus; H7N9; highly pathogenic; diversity; reassortment; tanglegram

Citation: He, D.; Wang, X.; Wu, H.; Wang, X.; Yan, Y.; Li, Y.; Zhan, T.; Hao, X.; Hu, J.; Hu, S.; et al. Genome-Wide Reassortment Analysis of Influenza A H7N9 Viruses Circulating in China during 2013–2019. *Viruses* **2022**, *14*, 1256. https://doi.org/10.3390/v14061256

Academic Editor: Chi-Young Wang

Received: 4 May 2022
Accepted: 6 June 2022
Published: 9 June 2022

Publisher's Note: MDPI stays neutral with regard to jurisdictional claims in published maps and institutional affiliations.

Copyright: © 2022 by the authors. Licensee MDPI, Basel, Switzerland. This article is an open access article distributed under the terms and conditions of the Creative Commons Attribution (CC BY) license (https://creativecommons.org/licenses/by/4.0/).

1. Introduction

Influenza A virus carries enveloped genomes consisting of eight gene segments of single-stranded ribonucleic acid molecules. When two viruses co-infect the same cell, the progeny virions may contain heterogeneous segments from different genomic sources. This process is known as reassortment, which is crucial for viral evolution [1] and adaptation [2]. Reassortment creates epidemiologically significant variants that can influence virus propagation, illness severity, antiviral resistance, and vaccine efficacy [3–5]. The combination of reassortment and genomic mutations increases the viral diversity while also shaping their short-term evolution [6–8]. Particularly, when a reassortant contains novel antigens against the naive population and is able to achieve cross-host transmission from other species to humans (host jump), it meets the critical requirements for a pandemic with high potential. However, most reassortants are less fit than either parent and then cleaned by the purifying selection [9]. Occasionally, the reassortment leads to the formation of a gene combination with mutations under a particular set of selection pressures, resulting in improved fitness [9] and the potential of a pandemic. Almost all influenza viruses responsible for human pandemics are thought to have evolved through reassortment [2,3,5,10,11].

The reassorted avian influenza H7N9 virus emerged in 2013 with antigenic novelty, resulting in substantial human mortality (30–50%) and posing a significant threat to public health [12]. The six internal genes of the H7N9 virus were derived from at least two different H9N2 virus lineages, and the H7 and N9 genes came from wild birds [13–17]. Nonetheless, the H7N9 virus has continually undergone reassortment with other subtypes of AIVs since its emergence, such as seasonal influenza A virus H1N1, H3N2, and even influenza B virus [18–20], H5N6 [21,22], H6Ny [22,23], H9N2 [23–28], and even H7N9 itself [29]. As a result, multiple novel viruses were generated, including H7N2 [30], H7N3 [31], H7N4 [32], and H7N6 [33]. However, most of these reassortants were transient except for the progeny viruses with the H9N2 (especially the G57 or S genotype H9N2 [34]). The H9N2 virus facilitated the genesis, adaptation, and prevalence of the H7N9 virus [15,35]. Significantly, reassortment with H9N2 has contributed to the onset of the fifth epidemic wave, which is the largest H7N9 outbreak in humans [36]. Due to the ease of notifying the intersubtype reassortment with NA, the actual reassortment events and patterns involving the internal segments may largely remain unknown.

In China, two lineages have been identified and named based on the outbreak sources: the Yangtze River Delta (YRD) and the Pearl River Delta (PRD). The YRD lineage was found to be responsible for the majority of H7N9 outbreaks [17]. Based on this existing nomenclature system, we further divided the H7N9 into eight genotypes. Then, we performed systemic genomic analyses on the eight segments of the H7N9 virus from wave 1 to wave 7. Our study enabled us to estimate the frequency of reassortment and particular patterns among the H7N9 virus. Therefore, we conclude that reassortment and mutation co-shaped the evolution and epidemiology of the H7N9 virus. Monitoring reassortment and mutations in H7N9 virus predominance in chicken flocks is critical for preventing the virus from spreading to humans.

2. Materials and Methods

2.1. Sequences Collection

We downloaded the H5, H6, and H9 subtypes of avian influenza viruses circulating in Asia from 2010 to 2019 from the online databases of the *Influenza Virus Resource, NCBI Influenza Research Database*, and *Global Initiative on Sharing All Influenza Data* (accessed on 1 September 2019). We downsampled the collected sequences with a 0.99 threshold by CD-HIT (http://weizhong-lab.ucsd.edu/cdhit-web-server/cgi-bin/index.cgi, accessed on 1 October 2019) to minimize the computation resource consumption. Then, sequences were aligned using MAFFT (v7.453) [37] and trimmed to keep the coding region. As a result, polymerase basic protein 2 (PB2) ($n = 1125$), polymerase basic protein 1 (PB1) ($n = 1118$), polymerase acidic protein (PA) ($n = 1069$), matrix protein (M) ($n = 883$), nucleoprotein (NP) ($n = 1000$), and non-structural protein (NS) ($n = 884$) gene sequences were obtained. For the genomic sequences of H7N9, the cd99 and cd999 datasets were adopted from our previous study [38].

2.2. Phylogenetic Analyses and Diversity Estimates

We used Bayesian evolutionary analysis sampling trees (BEAST, v1.10.4) [39] program to interpret the (cd99 dataset) phylogenetic tree and evolutionary rate (ucld.mean) of eight segments. The substitution model was determined by ModelFinder in PhyloSuite [40] according to the optimality Bayesian information criterion (BIC). The uncorrelated relaxed molecular clock model was used, and the Bayesian SkyGrid coalescent model was set as the tree prior [41]. Bayesian Markov chain Monte Carlo (MCMC) chains were set to 200 million generations with samples for every 20,000 steps to create the XML files. Each XML file was executed at least three times. Subsequently, the log files were assessed by Tracer (v1.7.1), and the effective sample size (ESS) values greater than 200 were accepted. Lastly, the maximum clade credibility (MCC) trees with median node heights were generated after the first 10% burn-in by TreeAnnotator (v1.10.4). The HA and NA MCC phylogenetic trees were adopted from our previous study [38]. Meanwhile, we performed the diversity test with the H7N9 cd999 dataset by the MEGA X (v10.2) to estimate the nucleotide diversity.

2.3. *Phylogenetic Clusters Classification*

Phylogenetic clusters were firstly classified based on the MCC trees of eight segments. Phylopart (v2.1) is a methodology for large-scale phylogeny partition [42]. We used it to define monophyletic clades with posterior probabilities of $\geq$90% and a median genetic distance threshold for clusters of 20% [15]. The assignment to these clusters was subsequently manually merged based on cluster tips containing $\leq$ 3 and failed to cluster. Then, each cluster was assigned a unique number based on the increasing tree topology. The remaining 7 MCC trees were assigned to clusters using the same processes. The H7N9 genotypes were classified using the HA clusters result and a previous investigation [17]. Phylogenetic tree and heatmap were plotted and visualized using ggtree [43,44].

2.4. *Reassortment Detections*

2.4.1. Intersubtype Reassortment

To investigate intersubtype reassortments across the six internal segments (PB2, PB1, PA, NP, M, and NS), we integrated the main subtypes (H5, H6, H7, and H9) of avian influenza viruses that circulated from 2010 to 2019. Multiple sequence alignments were performed with MAFFT [37]. Then, phylogenetic tree construction was performed by Mrbayes (v3.2.7a) [45]. The GTR+G+I model parameters were incorporated into the nexus file. Then, the nexus files were implemented in MrBayes by running 20 million MCMC chains with a sampling frequency of 5000. The phylogenetic trees were visualized and plotted using ggtree and ggtreeExtra [43,44].

2.4.2. Entanglegram and Untanglement

Tanglegram is a visual method to draw an association between two phylogenetic trees with identical tip labels. Tanglegram is also known as the incongruence tree. Theoretically, when the reassortment event is absent, the twines connect the tanglegram in a noncrossing or horizontally way [46]. To uncover the H7N9 AIV intrasubtype reassortment, we used the backronymed adaptable lightweight tree import code (Baltic) to generate incongruence visualization of eight H7N9 AIV MCC trees. The phylogenetic position of each strain was traced and colored according to the HA clusters across all segments. Entanglement was generated from an adapted script (https://github.com/dven42/phylogenetic-incongruence, accessed on 1 June 2021) and modified to visualize the positions of specific isolates for our study.

Untanglegram is a visual method to minimize crossings of hybridization networks between the tanglegram by rotating their branches around ancestral nodes [47]. There is an arrangement of their branches such that the association edges do not intersect if the topologies of the two phylogenies are entirely concordant [47]. The untanglegram phylogeny reveals the extent of reassortment between the HA and seven other segments (PB2, PB1, PA, NA, NP, and M). The identical tips were connected across the trees. Tips and connecting lines were colored according to the HA clusters. Untanglements were plotted with Dendextend using the step1side untangle method [48] by fixing the left tree (HA) and rotating the right tree (the other segments) to generate the minimization hybridization tanglegram in R (v4.0.5). The bigger the untanglement value is, the worse the reassortment between the paired tree.

2.4.3. Quantified Investigation Reassortment Events

We investigated the reassortment events over the entire genome by combined approaches of Graph Incompatibility based Reassortment Finder (GiRaF, v1.02) [49] and Recombination Detection Program (RDP, v5.05) [50]. In brief, Mrbayes assessed multiple trees per segment for incompatible splits using the GiRaF. After twenty million generations and sampling every 5000 steps, a burn-in of 25% samples was established to obtain trees and subsequently used for input files. Reassortment events with a confidence level greater than 0.9 were deemed accurate. Meanwhile, we also investigated the reassortment events using the RDP from the concatenated segments [51]. The algorithms incorporate the RDP, GENECONV, BOOTSCAN, MAXCHI, CHIMAERA, SISCAN, and 3SEQ techniques. At least four of the seven detection methods with a p-value of 10^{-6} were acknowledged as recombination events. Only strains confirmed concurrently by GiRaF and RDP were considered putative reassortment events.

BEAST2/CoalRe (v0.05) [51] estimated intralineage reassortment networks between H7 and N9. Three runs of 200 million MCMC sampling steps were performed following the tutorial (https://taming-the-beast.org/tutorials/Reassortment-Tutorial/, accessed on 1 June 2021). The embed segment tree was drawn using icytree (https://icytree.org/, accessed on 1 October 2021) to depict the reassortment network between HA and NA.

2.5. *Multidimensional Scaling*

We used multidimensional scaling (MDS) to analyze each viral segment's tree-to-tree branch variation and investigate the overall level of cross-correlation between all segments in a two-dimensional space. Generally, the last 500 phylogenetic trees for each segment were obtained from independently sampled trees in BEAST. The statistical tree-to-tree variation in branch lengths was calculated. Then, the MDS statistics analysis was performed in the R (v4.0.5) to determine the cross-correlation of all segments in two-dimensional space. Only the first two scaling dimensions were plotted using the ggplot2 [52] package for visualization. Theoretically, the viral segments that share similar evolutionary histories occupy similar locations in the two-dimensional Euclidean space where the cloud of points should overlap. Clouds of points in the MDS plot indicate phylogenetic uncertainty based on 500 sampled trees. In contrast, segments are expected to exhibit uncorrelated features due to their unlinked evolutionary histories in response to reassortment. Scripts for the MDS calculation were obtained from Doctor Maude Jacquot et al. [4,53] and modified for this study.

2.6. *Selection Pressure Analysis*

We quantified the selection pressures acting on the 10 major protein-coding regions (M and NS, respectively, encode at least two proteins) under different models. Before selection analysis, the best-fit model of nucleotide substitution was obtained from Modelfinder [54] based on Bayesian information criterion (BIC) for each segment. The maximum likelihood (ML) phylogenetic tree was interpreted using IQtree in PhyloSuite (v1.2.2). Subsequently, single-likelihood ancestor counting (SLAC) [55], fast unconstrained Bayesian approximation (FUBAR) [56], fixed-effects likelihood (FEL), and mixed-effects model of evolution (MEME) [57] were used to infer sites under episodic or pervasive natural selection on each coding protein. Finally, we recommend the significance levels for FEL ($p < 0.1$), SLAC ($p < 0.05$), MEME ($p < 0.05$), and FUBAR (posterior probability > 0.9). All methods were implemented in the HyPhy (v2.5.2) [58] on a high-performance computing cluster. The selection results of HA and NA were adopted from our previous study [38].

3. Results

3.1. The Surface Glycoproteins Have Faster Evolutionary Rates but Less Genetic Diversity

Nucleotide analyses revealed that the internal segments have higher diversity than the surface glycoproteins (Figure 1). PB2 (3.86×10^{-3}) has the highest genetic diversity among all segments, exhibiting a highly heterogeneous genome. The HA gene (1.83×10^{-3}) has slightly more diversity than the NA gene (1.80×10^{-3}), which is smaller than internal genes.

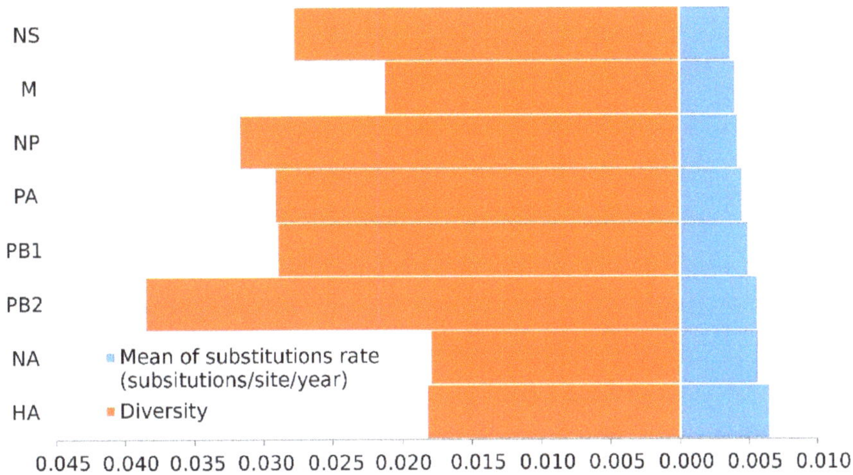

Figure 1. Diversity and evolutionary rates of H7N9 eight gene segments.

3.2. PB2 c05 and M c08 Were the Predominant Clusters in Wave 5

To study H7N9 reassortment, we investigated the clusters of each segment for H7N9 viruses recovered from all hosts in China from 2013 to 2019 (Figure 2). Segment cluster numberings were designated, including HA (c1-c8), NA (c1-c8), PB2 (c1-c5), PB1 (c1-c9), PA (c1-c7), NP (c1-c6), M (c1-c8), and NS (c1-c5) (Figures S1–S7 and Table S1). Cluster 0 (c0) represented the phylogenetic tips that could not be assigned to a cluster. The same number in different segments is not correlative. Eight genotypes were allocated based on HA clusters: the early genotype cluster was designated as W1; the Pearl River Delta (PRD) lineage was divided into two genotypes: PRD-A and PRD-B (Figure 2); and the Yangtze River Delta (YRD) lineage designated as five genotypes: YRD-A, YRD-B, YRD-C, YRD-D, and YRD-E. LP H7N9 viruses of the YRD lineage discovered in wave 5 were mainly assigned within the YRD-E cluster, which was the dominant cluster. In the YRD-E cluster, the dominating segment clusters were c05 PB2 (79.03%) and c08 M (78.23%) (Figure 3). Viruses simultaneously containing PB2 c05 and M c08 were determined in 62.10% (77/124 in the cd99 dataset) of the YRD-E cluster. The circulating HP H7N9 AIVs were assigned to the YRD-C cluster. In the HP H7N9 virus found in waves 6 and 7, the cluster of M c03 was replaced by the c08. Reassortment has a more significant impact on the diversity of viral genotypes.

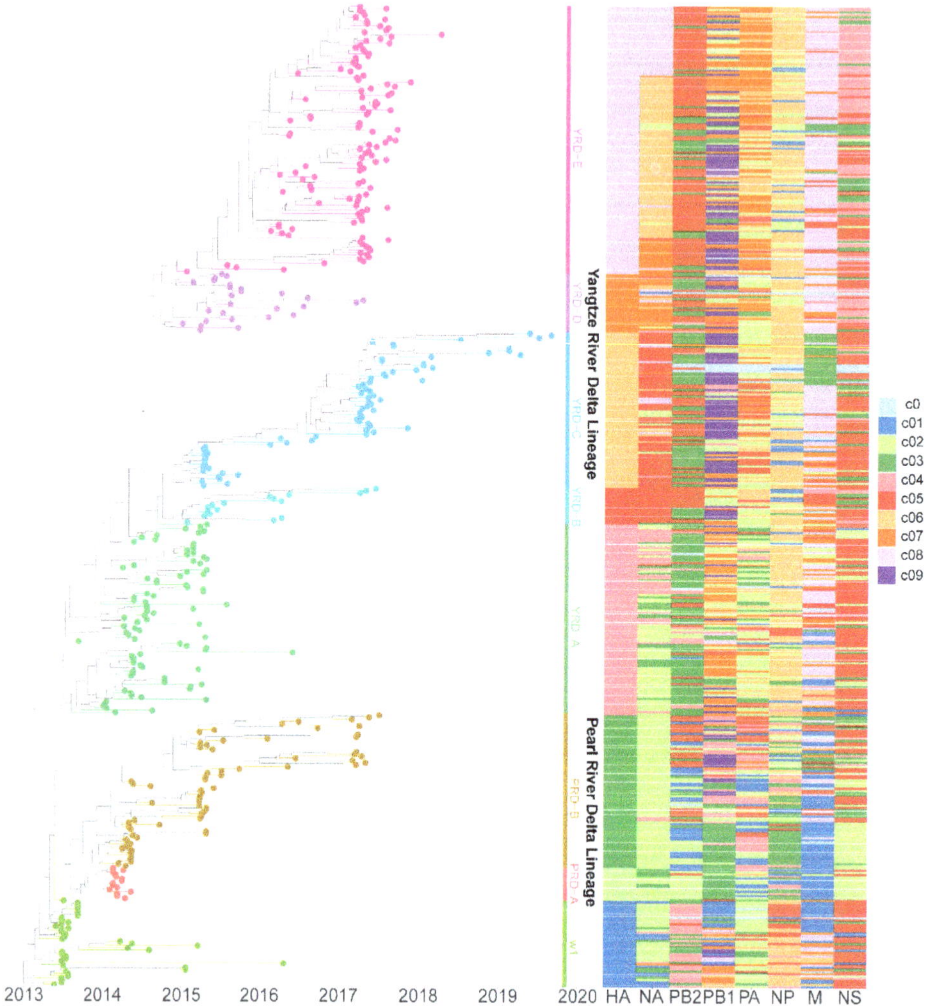

Figure 2. The genotypes heatmap of H7N9 viruses. Each segment's cluster assignment is based on MCC trees with a median branch length distance threshold of 0.20 identified using PhyloPart. The left panel is the time-scaled HA tree with two lineages (Yangtze River Delta and Pearl River Delta), and tips are colored based on clusters. The right panel displays the heatmap of each segment cluster. Shared colors and numbers indicate sequences of the same segment assigned to the same cluster. The same color and number in different segments are not correlative. Detailed clusters are available in the Supplementary Table S1.

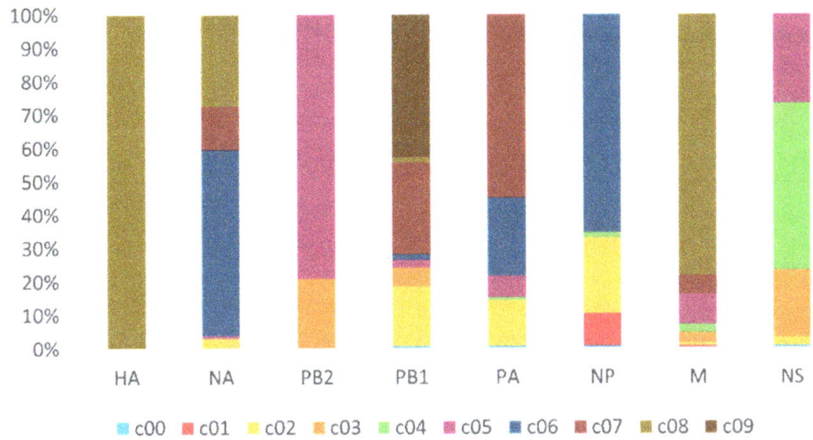

Figure 3. The gene clusters proportion of YRD-E. The same color and number in different segments are not correlative. PB2 c05 (79.03%) and M c08 (78.23%) are dominant clusters in the epidemic wave 5. Viruses have PB2 c05, and M c08 account for 62.10%.

3.3. Intensive Intersubtype Reassortments between H7N9 and H9N2

Phylogenetic tree analyses evaluated intersubtype reassortments between cocirculating subtypes of H5, H6, and H9. Surprisingly, the topology of H7N9 and H9N2 was clustered together in many branches on each tree of internal segments (Figure 4). H7N9 viruses were reassorted in rare cases with other subtypes of AIVs instead of H9N2. Duck-origin viruses in particular have a more complicated genetic background, with internal segments derived primarily from wild bird viruses, such as H5N6, H6N1, H6N2, H6N6, H7N2, H7N3, and H7N7. For example, the M gene of A/chicken/Guangdong/Q1/2016, the early-raised HP H7N9 obtained on 20 June 2016, is closely clustered with the M gene of the H5N6 subtype (A/duck/Yunnan/07.15 DQNPH129/2015, EPI668788). However, all these intersubtype reassortment events were predominantly within the H9N2 subtype (G57 or genotype S).

3.4. Dynamic and Intricate Intrasubtype Reassortment

Baltic was used to visualize inconsistencies between the MCC phylogenetic trees (tanglegram). An abundance of intralineage reassortment events was observed from the tanglegram among the 7 waves (Figure 5). Internal segments in the YRD-E (LP in wave 5) cluster were highly diverse during the wave 5 outbreak. They usually came from 2–3 main clusters on each MCC tree topology (Figure S8). Notably, part of HP H7N9 viruses in the YRD-C inherited their PB2, PA, and M segments from the YRD-E cluster (Figure S9), suggesting the LP H7N9 contributed to the genesis of HP H7N9 and frequent intralineage reassortments. There were also intralineage reassortments between the YRD and PRD lineages. For instance, some YRD-C HP H7N9 viruses obtained the NA gene from the cocirculating PRD-B cluster. Intralineage reassortment events between the YRD and PRD lineages were also commonly and consistently identified throughout all H7N9 epidemic waves by BEAST2/CoalRe (Figure 6). Especially, many intralineage reassortment instances were also determined in waves 2–3 and 5. Although the H7N9 population was severely decreased at the time, the HP H7N9 virus isolated in waves 6 and 7 continued to reassort. For instance, the M gene of A/chicken/Shanxi/SX0256/2019 came from c07 rather than the c03. The M c03 has been the dominant cluster in the previous prevalent viruses since 2018 (Figure 2). Of note, none of the genotypes was predominant in any of the seven epidemic waves. Our results demonstrated that China's H7N9 virus has a dynamic and intricate intrasubtype reassortment pattern.

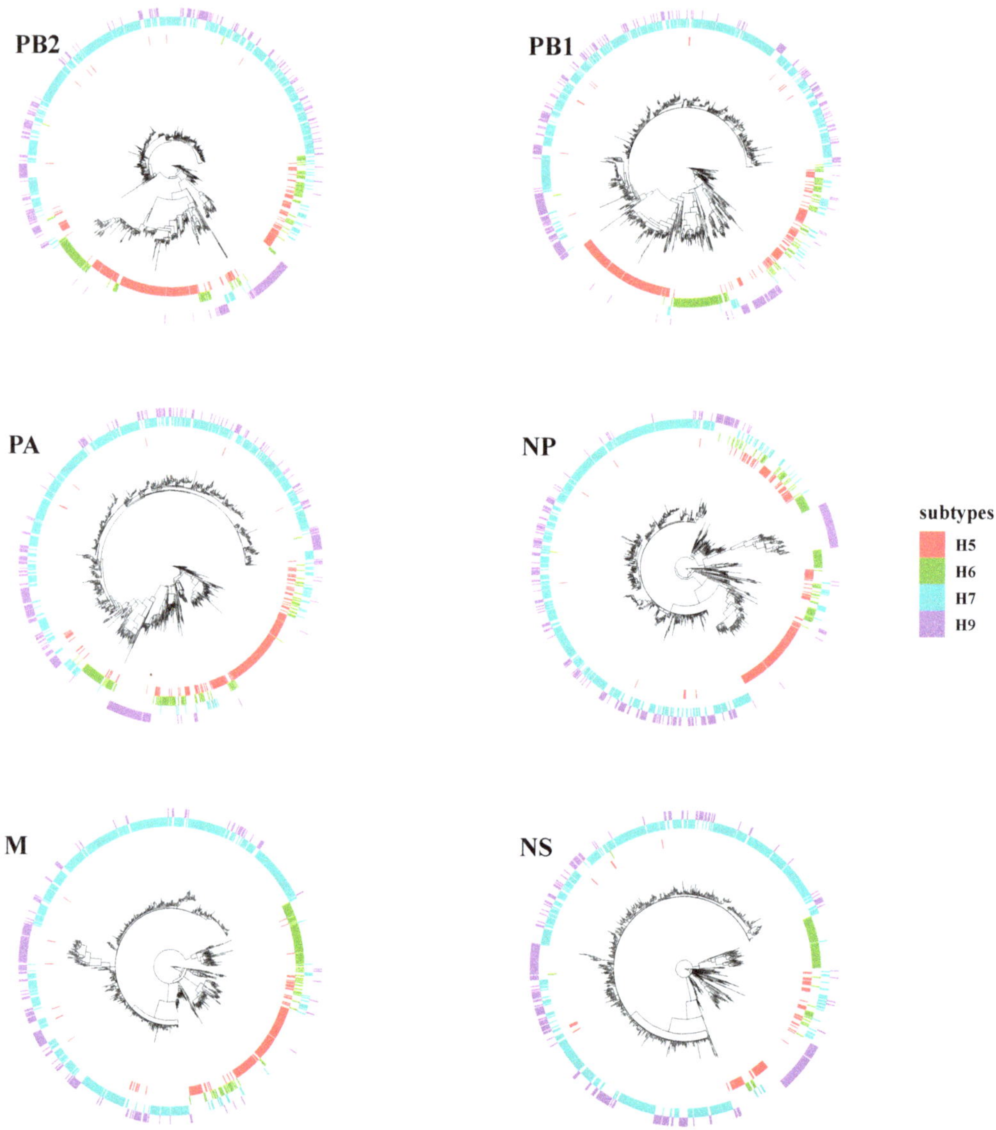

Figure 4. Intersubtype reassortment between H7N9 and H5, H6, and H9 subtypes of avian influenza viruses.

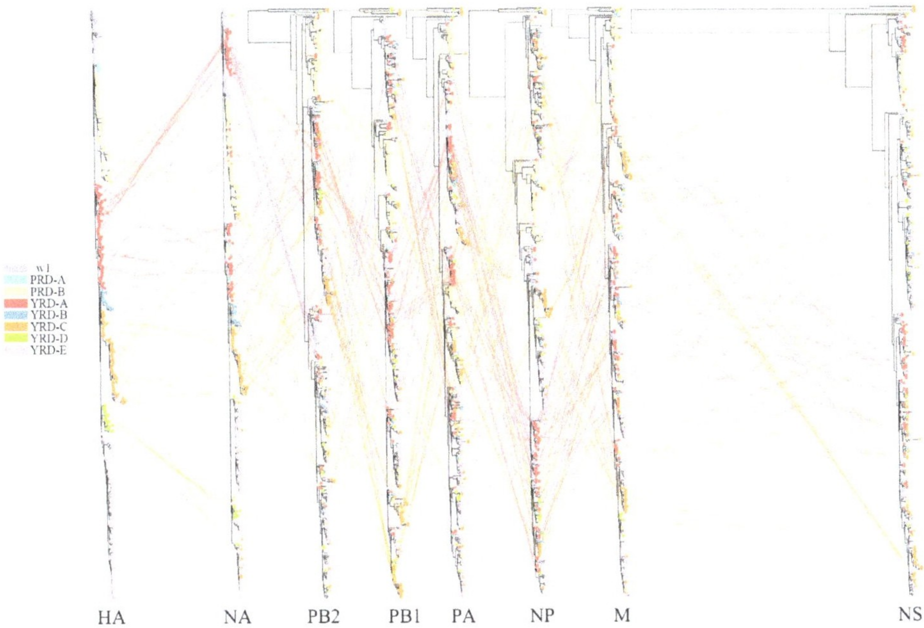

Figure 5. Phylogenetic incongruence analysis. MCC trees for the HA segment and all internal genes NA, PB2, PB1, PA, NP, and M from equivalent strains connect across the trees. Tips and connecting lines are colored according to the HA clusters.

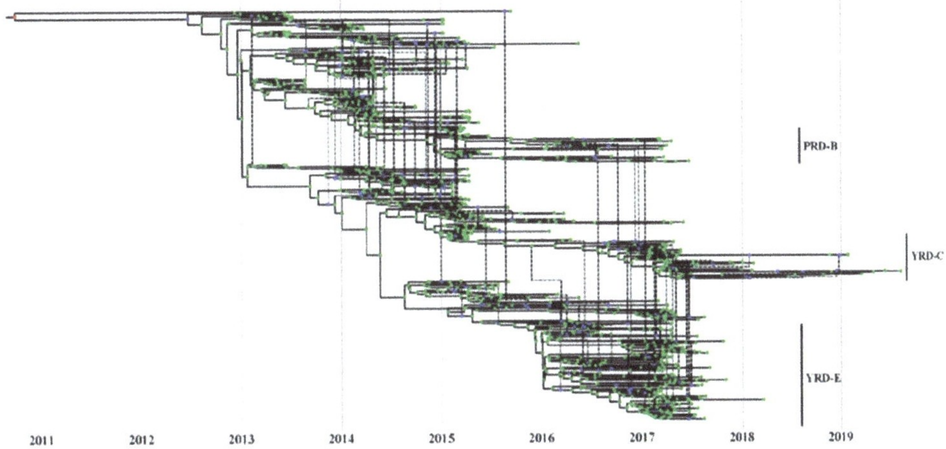

Figure 6. Estimate of MCC reassortment network between HA and NA genes of H7N9 viruses. Icytree visualizes the network as a base tree connected by dotted branches, indicating a reassortment event. The green dots indicate the tips of the tree.

3.5. HA and NA Have the Lowest Reassortment Rate

To measure the severity of reassortment over the entire genome, we untangled the tanglegram between HA and the paired trees shown as untanglement values (the bigger, the worse. Figure 7). The unentanglement values between HA and paired segments (NA,

PB2, PB1, PA, NP, M, and NS) were 0.0381, 0.1997, 0.2340, 0.0591, 0.3234, 0.0672, and 0.4894, respectively. The lowest untanglement value (0.0381) was found between HA and NA, whereas the highest value was found between HA and NS (0.4894).

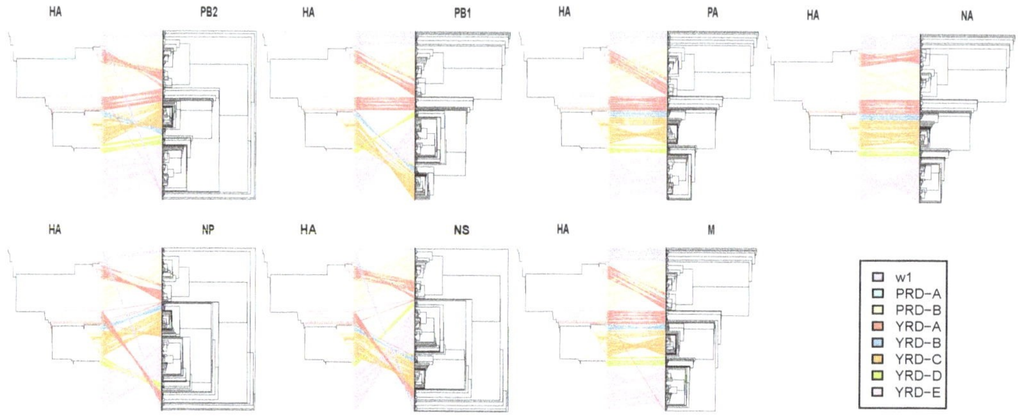

Figure 7. Evolutionary relationships of each gene segment with HA. Incongruence phylogenetic analysis shows interclade reassortment between the HA segment and seven additional genes (NA, PB2, PB1, PA, NP, and M). Equivalent strains connect across the trees. Tips and connecting lines are colored according to the HA clusters. Unentanglement is used to minimize crossings of the hybridization network between the paired trees. The values are determined by the degree of inter-segment reassortment.

3.6. *High Reassortment Rate in Waves 3 and 5*

GiRaF and RDP analyses found that 217 of 454 strains H7N9 (cd99 dataset) were involved with reassortment events (Table S2). Following that, we computed the reassortment rates in different waves. High reassortment rates were discovered in waves 3 (0.76) and 5 (0.79). Wave 1 is notable for having the lowest reassortment rate (0.10).

3.7. *PB2 and PA Shared Coevolutionary History*

MDS allows for the two-dimensional depiction of the total degree of cross-correlation between all segments, with overlap between observations indicating shared evolutionary history (i.e., linkage) between segments. In comparison, segments that split up due to reassortment are expected to occupy separate plot areas. We found that, except for the PB2 and PA segments, the rest of the H7N9 segments were very distinct (Figure 8). The segments PB1, M, NP, and NS did not show any association, indicating that they did not have any coevolutionary relationship. The PB2 and PA genes almost completely overlapped, indicating they shared a strong evolutionary history (i.e., linkage). In comparison, the capsid proteins of HA and NA only had a weak association.

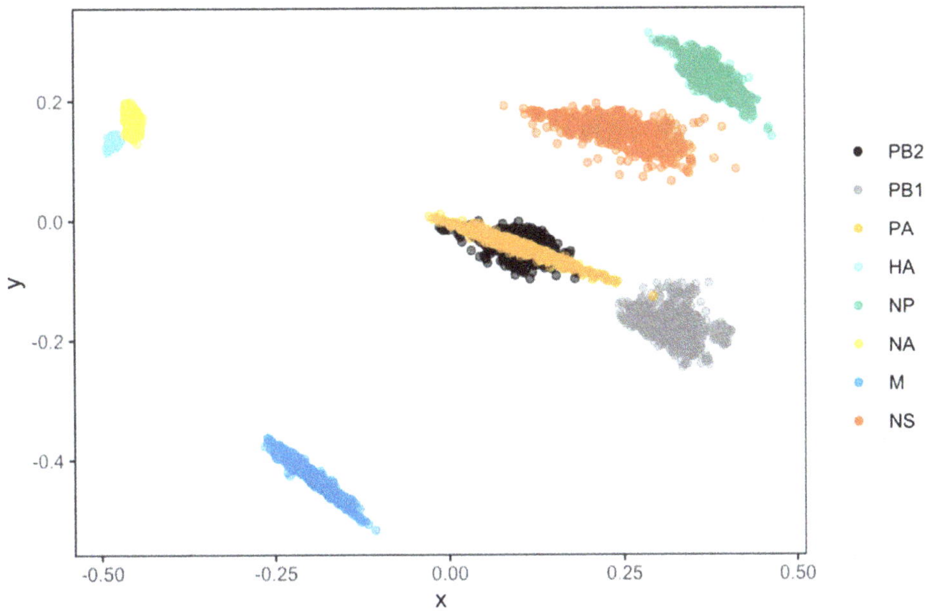

Figure 8. Correlations in time to the most recent common ancestor (tMRCA) among H7N9 viral segments as depicted by a multidimensional scaling (MDS) graphic. MDS enables a two-dimensional representation of the total level of cross-correlation between all segments. Cloud points represent phylogenetic uncertainty based on 500 trees for each segment sampled in the program of BEAST, with pairwise comparisons to other segments limited to viruses sampled in the same year. In the absence of reassortment, segments are likely to have highly correlated TMRCAs due to shared evolutionary history showing as overlapping dots. Reversely, segments split up by reassortment are predicted to inhabit various plot regions. Only the first two scaling dimensions are visible.

3.8. High Selection Pressure Acting on HA, NA, NS1, and M2

The natural selection acting on all coding regions was estimated using the dN/dS ratio. Four site-level detection methods (FEL, SLAC, FUBAR, and MEME) were used to assess positive and negative selection codons. M2 (0.5398), NS1 (0.3087), HA (0.2656), and NA (0.2946) underwent more substantial selection pressure than other segments, whereas M1 protein (0.0752) was the lowest, according to the overall dN/dS values estimated using SLAC (Table 1). Except for M2, NS1, and NEP, the other internal proteins were unaffected by higher purifying selectivity. MEME was used to find more positive selection sites by identifying fixed and sporadic positively chosen codons. As a result, HA (14 codons) and NA (13 codons) had more sites under positive selection pressure than other coding regions. PB2 (10 codons), NS1(10 codons), PB1 (7 codons), and PA (6 codons) also detected many positive selection sites.

Table 1. Selection pressures and positively and negatively selected codons of coding regions of H7N9 viruses circulating between 2013 and 2019.

Coding Region	No. of Codons	dN/dS	No. of Selected Sites (% of Codons)				
			Positively Selected				Negatively Selected (SLAC)
			SLAC [a]	FUBAR [b]	FEL [c]	MEME [d]	
PB2	759	0.1200	4	4	4	10	670 (88.16%)
PB1	757	0.0977	5	6	6	7	657 (75.60%)
PA	716	0.1137	7	4	8	6	573 (79.92%)
HA [38]	564	0.2656	16	13	16	14	374 (66.19%)
NP	498	0.1026	2	1	4	3	413 (82.77%)
NA [38]	465	0.2946	13	8	15	13	280 (60.09%)
M1	252	0.0752	0	1	1	1	190 (75.10%)
M2	97	0.5398	6	7	8	3	26 (26.53%)
NS1	217	0.3087	13	6	14	10	115 (52.75%)
NEP	121	0.2766	0	1	1	2	55 (45.08%)

[a] p-value < 0.05; [b] posterior probability of $\geq$0.9; [c] p-value < 0.1; [d] p-value < 0.05

4. Discussion

We estimated the inter- and intra- reassortment, reassortment patterns, and adaptative evolution of H7N9 viruses from waves 1–7. Our study found that H7N9 viruses have undergone considerable changes by reassortment, which had a significant impact on the H7N9 genomic composition since its emergence. Notably, the internal genes presented more diversified features than surface genes, and even the surface genes had higher substitution rates. Numerous reassortment events were determined in our study, resulting in a high level of genetic diversity, especially among the internal genes. The dynamic and intricate reassortments may shape the epidemiology and genomic evolution of the H7N9 virus and contribute to its genetic diversity.

The H7N9 haemagglutinin gene evolved faster (6.51×10^{-3} substitutions/site/year) than the seasonal influenza virus, with an average rate of 3.41×10^{-3} [59]. This substitution rate is similar to the H5N1 under vaccination pressure (6.13×10^{-3}–8.87×10^{-3}) but faster than H5N1 without vaccination pressure [60]. According to recent studies, the nucleotide substitution rate of the H7N9 HA gene under vaccination pressure increased to 1.963×10^{-2} during 2018–2019 [61]. Considering that reassortment usually leads to a transient increase in the substitution rate [62], the frequency of reassortments may also contribute to the rapid evolution of H7N9. According to other research, external genes of the influenza virus usually have evolved at faster evolutionary rates than its internal genes [63,64], resulting in more genetic diversity in surface genes. However, the external genes of the H7N9 virus showed less genetic diversity than its internal segments. This contradiction implies that H7N9 internal genes were reassorted frequently rather than in a single event since its emergence.

In the study of intersubtype reassortments, we found that the H7N9 virus was primarily associated with the cocirculating H9N2 virus (genotype S or G57). In contrast, they were rarely reassorted with other AIV subtypes. The H9N2 and H7N9 viruses have been cocirculating in China since 2013. Therefore, numerous opportunities existed for their coinfections. Several investigations in the first five epidemic waves also confirmed the high prevalence of H7N9/H9N2 coinfection in chickens [15,25,27,65]. Influenza virus reassortment occurs with high frequency without segment mismatch [66]. However, most reassortment viruses produced from divergent parental strains are also frequently outcompeted by one or both parental viruses [2]. Specific reassortment genotypes may be inefficient in forming due to incompatibility between the heterologous RNA packaging signals [2,67]. Considering the high genetic similarity between the internal segments of H7N9 and H9N2 AIVs, it may be the biological basis for the high frequency of intersubtype reassortment in the absence of mismatches between their packaging signals. Heterogeneous genome packaging signals combinations other than H7N9/H9N2 viruses may be deleterious or decrease fitness under natural conditions [68], forcing the reassorted progeny virus particles to be eliminated through purifying selection. Therefore, the high frequency of intersubtype reassortants between H7N9 and H9N2 AIVs could be expected.

The LP H7N9 virus, unlike the other LP AIV, may cause serious illness in humans and other mammalian species, demonstrating its exceptional fitness in mammalian hosts [69]. The fifth wave of 2016–2017 was the biggest epidemic to date, with nearly the same number of human cases (n = 758) as the sum of the previous four outbreaks [36]. Nonetheless, most human infections in wave 5 were caused by the LP H7N9 virus, which is primarily associated with YRD-E. Our investigation found that gene clusters PB2 c05 and M c08 have a large proportion of YRD-E cluster strains, which mainly circulated in wave 5 despite a highly diverse genetic background. Our previous studies showed that the reassortments with PB2 and M genes from genotype S H9N2 caused attenuated progeny of H5Nx and H7N9, resulting in optimizing viral fitness viruses in mice [70] and chicken [71]. Therefore, we assume that PB2 c05 and M c08 might optimize the viral fitness and contribute to the wave 5 outbreak. Aside from the PB2 and M, NP c06 and PA c07 have a certain ratio in YRD-E. It is also impossible to overlook their contribution to viral fitness.

In terms of intralineage reassortment, we found that the intralineage reassortment occurs considerably frequently rather than as a single event. Usually, intralineage reassortment is subjected to severe negative selection, which becomes more pronounced as the genetic distance between donor strains increases [72]. Due to the remarkable similarity in their internal genes and genomic packaging signals between the H9N2 and H7N9 viruses, the high compatibility between their internal genes makes the reassorts rise rather naturally. According to our findings, the YRD-E lineage contributed its PB2, PA, and M gene to the YRD-C lineage, dominated by the HP H7N9 viruses. The frequent reassortment leads to such a diversity of genotypes of H7N9. Similarly, Cui et al. documented 27 H7N9 genotypes within 3 months of H7N9 emergence [35]. Nonetheless, it should be highlighted that intralineage reassortment between pretty similar internal gene cassettes is likely to be underestimated. Using tanglegram to estimate the reassortment in phylogenies from high similarity sequences might underestimate or exaggerate reassortment events [46]. Combined with additional pieces of evidence [27–29], we confirmed that the intricate intralineage genetic reassortment of the H7N9 virus had occurred frequently since its introduction.

Usually, wild birds are the primary source of AIV reassortant [73]. However, we found that H7N9 reassortment occurred mainly in the chicken host rather than in wild birds. The most likely reason was that H7N9 and H9N2 were primarily cocirculated in poultry, while H7N9 only had sporadic spillover to wild birds [74]. The other reason is that national surveillance relied predominantly on passive reporting systems and less active and systemic surveillance in wild birds [75]. Therefore, the "sampling strategy" in the H7N9 surveillance may affect the reassortment interpretation.

Many intersubtype reassortments took place between HA and NA in waves 2–3 and 5, coinciding with the periods of extensive intrasubtype and intralineage reassortment. Theoretically, H7N2 and H9N9 should have a high probability of detection since the extensive interlineage reassortment between H7N9 and H9N2. However, the detection of the H7N2 and H9N9 virus subtypes was limited. Even the reassorted H9N9 virus in the laboratory has shown increased fitness features in poultry [76]. Although intersubtype reassortments between H7 and N9 were frequently detected in waves 2–3 and 5, the frequency of reassortment between HA and NA remained low compared to HA with other segments, demonstrating a robust functional balance between H7 and N9. Similar to other avian influenza viruses in poultry and wild bird populations, the NS gene encoding the non-structural protein has a more divergent phylogeny and high reassortment rates [73,77].

Reassortment in the influenza virus is not a random process, which is consistent with observations from other segmented viruses [4,67]. Except for the PB2 and PA segments, the other H7N9 segments did not share their evolutionary history segment. However, the HA-NA capsid proteins and the PB1-PB2/PA RNA polymerase complex proteins only had a weak association. The times to most recent common ancestor (tMRCA) were consistent in PB2 and PA, indicating they were less amenable to reassortment than the other segments in history. This inherited linkage also implies physical or biochemical interactions between their encoded proteins since epistatic interactions result in tighter evolutionary connections. According to a recent study, reintroducing PB2 and PA with adaptive mutations from cocirculating H9N2 in 2015 resulted in a novel H7N9 genotype, which increased polymerase activity and became dominant in the fifth H7N9 virus epidemic [36]. Because the internal segments primarily arose from genotype S H9N2 [34] (similarly with G57 genotype H9N2 [36]), this linkage PB2-PA pattern most likely originated from H9N2. A previous study found that a coadapted PB1-PB2-HA gene pattern was established during influenza B interlineage reassortment, which has been demonstrated to be critical for whole-genome fitness [78]. As a result, we assume that the particular PB2-PA pattern might confer a specific advantage to the H7N9/H9N2-like gene cassette. Further studies are warranted.

In addition to M2, NS1, and NEP, high dN/dS values were found in the HA and NA proteins (0.5398, 0.3087, 0.2766, 0.2656, and 0.2946, respectively), which were generally in agreement with the previous study in H3N2 [46]. Surface proteins (HA, NA, and M2) are typically more sensitive to positive selection and evolve more rapidly than internal genes, especially when the influenza virus circulates in a naive population [63,64]. The reassortment events lead to a transient increase in the rate of amino acid replacements on the descendant phylogenetic branches [79]. Therefore, the extensive reassortment might increase the H7N9 evolutionary rate and result in rapid adaptive evolution at the molecular level.

In summary, our phylogenetic analyses revealed the comprehensive genetic evolution of the H7N9 viruses. The particular reassortment patterns indicate that the H9N2-original internal gene constellation has superior compatibility to the genesis and evolution of H7N9 over the other AIV subtypes. The H7N9 viruses were further diversified by frequent inter- and intra-lineage reassortment events with adaptative mutation, leading to successful H7N9 genotypes. The evolution and epidemiology of the H7N9 virus in China may be shaped by reassortment and adaptive mutations. Further research is needed to understand the dynamics of reassortment of the circulating H7N9 virus.

Supplementary Materials: The following supporting information can be downloaded at: https://www.mdpi.com/article/10.3390/v14061256/s1. Table S1, PhyloPart clusters; Figures S1 to S7, H7N9 segments phylopart cluster tree; Table S2, Girad and RDP; Figure S8, Phylogenetic incongruence analysis of YRD-E; Figure S9, Phylogenetic incongruence analysis of YRD-C.

Author Contributions: Conceptualization, M.G., S.S. and C.D.; methodology, D.H., X.W. (Xiyue Wang) and H.W.; formal analysis, D.H. and X.W. (Xiyue Wang); resources, X.W. (Xiaoquan Wang), S.H. and J.H.; writing—original draft preparation, D.H.; writing—review and editing, T.Z., Y.Y., Y.L. and X.H.; visualization, D.H. and X.W. (Xiyue Wang); supervision, M.G.; project administration, X.L. (Xiufan Liu) and M.G.; funding acquisition, X.L. (Xiufan Liu), M.G., X.L. (Xiaowen Liu) and X.W. (Xiaoquan Wang) All authors have read and agreed to the published version of the manuscript.

Funding: This work was supported by the National Natural Science Foundation of China (32072892), the National Key Research and Development Project of China (2021YFD1800202), the Earmarked Fund for China Agriculture Research System (No. CARS-40), and the Priority Academic Program Development of Jiangsu Higher Education Institutions (PAPD).

Institutional Review Board Statement: Not applicable.

Informed Consent Statement: Not applicable.

Data Availability Statement: Data are derived from the public database and 10.1016/j.meegid.2021.104993.

Acknowledgments: We are thankful to Maude Jacquot (Free University of Brussels), who shared the scripts for the MDS calculation. This research was supported by the High-Performance Computing Cluster of the College of Veterinary Medicine, Yangzhou University.

Conflicts of Interest: The authors declare that the research was conducted without any commercial or financial relationships that could be construed as a potential conflict of interest.

References

1. Steel, J.; Lowen, A.C. Influenza A virus reassortment. *Curr. Top. Microbiol. Immunol.* **2014**, *385*, 377–401. [CrossRef] [PubMed]
2. Lowen, A.C. Constraints, Drivers, and Implications of Influenza A Virus Reassortment. *Annu. Rev. Virol.* **2017**, *4*, 105–121. [CrossRef] [PubMed]
3. Maljkovic Berry, I.; Melendrez, M.C.; Li, T.; Hawksworth, A.W.; Brice, G.T.; Blair, P.J.; Halsey, E.S.; Williams, M.; Fernandez, S.; Yoon, I.-K.; et al. Frequency of influenza H3N2 intra-subtype reassortment: Attributes and implications of reassortant spread: Attributes and implications of reassortant spread. *BMC Biol.* **2016**, *14*, 117. [CrossRef] [PubMed]
4. Nomikou, K.; Hughes, J.; Wash, R.; Kellam, P.; Breard, E.; Zientara, S.; Palmarini, M.; Biek, R.; Mertens, P. Widespread Reassortment Shapes the Evolution and Epidemiology of Bluetongue Virus following European Invasion. *PLoS Pathog.* **2015**, *11*, e1005056. [CrossRef] [PubMed]
5. White, M.C.; Lowen, A.C. Implications of segment mismatch for influenza A virus evolution. *J. Gen. Virol.* **2018**, *99*, 3–16. [CrossRef] [PubMed]
6. Nelson, M.I.; Simonsen, L.; Viboud, C.; Miller, M.A.; Taylor, J.; George, K.S.; Griesemer, S.B.; Ghedin, E.; Ghedi, E.; Sengamalay, N.A.; et al. Stochastic processes are key determinants of short-term evolution in influenza a virus. *PLoS Pathog.* **2006**, *2*, e125. [CrossRef]
7. Wolf, Y.I.; Viboud, C.; Holmes, E.C.; Koonin, E.V.; Lipman, D.J. Long intervals of stasis punctuated by bursts of positive selection in the seasonal evolution of influenza A virus. *Biol. Direct.* **2006**, *1*, 34. [CrossRef]
8. Rambaut, A.; Pybus, O.G.; Nelson, M.I.; Viboud, C.; Taubenberger, J.K.; Holmes, E.C. The genomic and epidemiological dynamics of human influenza A virus. *Nature* **2008**, *453*, 615–619. [CrossRef]
9. Lowen, A.C. It's in the mix: Reassortment of segmented viral genomes. *PLoS Pathog.* **2018**, *14*, e1007200. [CrossRef]
10. de Silva, U.C.; Tanaka, H.; Nakamura, S.; Goto, N.; Yasunaga, T. A comprehensive analysis of reassortment in influenza A virus. *Biol. Open* **2012**, *1*, 385–390. [CrossRef]
11. Wille, M.; Holmes, E.C. The Ecology and Evolution of Influenza Viruses. *Cold Spring Harb. Perspect. Med.* **2020**, *10*, a038489. [CrossRef] [PubMed]
12. Li, C.; Chen, H. H7N9 Influenza Virus in China. *Cold Spring Harb. Perspect. Med.* **2021**, *11*, a038349. [CrossRef] [PubMed]
13. Gao, R.; Cao, B.; Hu, Y.; Feng, Z.; Wang, D.; Hu, W.; Chen, J.; Jie, Z.; Qiu, H.; Xu, K.; et al. Human infection with a novel avian-origin influenza A (H7N9) virus. *N. Engl. J. Med.* **2013**, *368*, 1888–1897. [CrossRef] [PubMed]
14. Lam, T.T.-Y.; Wang, J.; Shen, Y.; Zhou, B.; Duan, L.; Cheung, C.-L.; Ma, C.; Lycett, S.J.; Leung, C.Y.-H.; Chen, X.; et al. The genesis and source of the H7N9 influenza viruses causing human infections in China. *Nature* **2013**, *502*, 241–244. [CrossRef] [PubMed]
15. Pu, J.; Wang, S.; Yin, Y.; Zhang, G.; Carter, R.A.; Wang, J.; Xu, G.; Sun, H.; Wang, M.; Wen, C.; et al. Evolution of the H9N2 influenza genotype that facilitated the genesis of the novel H7N9 virus. *Proc. Natl. Acad. Sci. USA* **2015**, *112*, 548–553. [CrossRef] [PubMed]
16. Wu, A.; Su, C.; Wang, D.; Peng, Y.; Liu, M.; Hua, S.; Li, T.; Gao, G.F.; Tang, H.; Chen, J.; et al. Sequential reassortments underlie diverse influenza H7N9 genotypes in China. *Cell Host Microbe* **2013**, *14*, 446–452. [CrossRef]
17. Wang, D.; Yang, L.; Zhu, W.; Zhang, Y.; Zou, S.; Bo, H.; Gao, R.; Dong, J.; Huang, W.; Guo, J.; et al. Two Outbreak Sources of Influenza A (H7N9) Viruses Have Been Established in China. *J. Virol.* **2016**, *90*, 5561–5573. [CrossRef]
18. Zhu, Y.; Qi, X.; Cui, L.; Zhou, M.; Wang, H. Human co-infection with novel avian influenza A H7N9 and influenza A H3N2 viruses in Jiangsu province, China. *Lancet* **2013**, *381*, 2134. [CrossRef]
19. Li, J.; Kou, Y.; Yu, X.; Sun, Y.; Zhou, Y.; Pu, X.; Jin, T.; Pan, J.; Gao, G.F. Human co-infection with avian and seasonal influenza viruses, China. *Emerg. Infect. Dis.* **2014**, *20*, 1953–1955. [CrossRef]
20. Pu, Z.; Yang, J.; Shen, X.; Irwin, D.M.; Shen, Y. Human-isolated H7N9 obtained internal genes from duck and human influenza viruses. *J. Infect.* **2019**, *78*, 491–503. [CrossRef]
21. Yang, L.; Xie, J.; Zhang, Y.; Zhu, W.; Li, X.; Wei, H.; Li, Z.; Zhao, L.; Bo, H.; Liu, J.; et al. Emergence of waterfowl-originated gene cassettes in HPAI H7N9 viruses caused severe human infection in Fujian, China. Influenza Other Respir. *Viruses* **2019**, *13*, 496–503. [CrossRef]
22. Jin, Y.; Ren, H.; Teng, Y.; Hu, M.; Peng, X.; Yue, J.; Liang, L. Novel reassortment of avian influenza A(H7N9) virus with subtype H6N6 and H5N6 viruses circulating in Guangdong Province, China. *J. Infect.* **2017**, *75*, 179–182. [CrossRef] [PubMed]
23. Quan, C.; Shi, W.; Yang, Y.; Yang, Y.; Liu, X.; Xu, W.; Li, H.; Li, J.; Wang, Q.; Tong, Z.; et al. New Threats from H7N9 Influenza Virus: Spread and Evolution of High- and Low-Pathogenicity Variants with High Genomic Diversity in Wave Five: Spread and Evolution of High- and Low-Pathogenicity Variants with High Genomic Diversity in Wave Five. *J. Virol.* **2018**, *92*, e00301-18. [CrossRef] [PubMed]
24. Li, C.; Wang, S.; Bing, G.; Carter, R.A.; Wang, Z.; Wang, J.; Wang, C.; Wang, L.; Wu, G.; Webster, R.G.; et al. Genetic evolution of influenza H9N2 viruses isolated from various hosts in China from 1994 to 2013. *Emerg. Microbes Infect.* **2017**, *6*, e106. [CrossRef] [PubMed]

25. Liu, W.; Fan, H.; Raghwani, J.; Lam, T.T.-Y.; Li, J.; Pybus, O.G.; Yao, H.-W.; Wo, Y.; Liu, K.; An, X.-P.; et al. Occurrence and reassortment of avian influenza A (H7N9) viruses derived from coinfected birds in China. *J. Virol.* **2014**, *88*, 13344–13351. [CrossRef]
26. Ke, C.; Lu, J.; Wu, J.; Guan, D.; Zou, L.; Song, T.; Yi, L.; Zeng, X.; Liang, L.; Ni, H.; et al. Circulation of reassortant influenza A(H7N9) viruses in poultry and humans, Guangdong Province, China, 2013. *Emerg. Infect. Dis.* **2014**, *20*, 2034–2040. [CrossRef]
27. Qi, W.; Shi, W.; Li, W.; Huang, L.; Li, H.; Wu, Y.; Yan, J.; Jiao, P.; Zhu, B.; Ma, J.; et al. Continuous reassortments with local chicken H9N2 virus underlie the human-infecting influenza A (H7N9) virus in the new influenza season, Guangdong, China. *Protein Cell* **2014**, *5*, 878–882. [CrossRef]
28. Bai, R.; Sikkema, R.S.; Munnink, B.B.O.; Li, C.R.; Wu, J.; Zou, L.; Jing, Y.; Lu, J.; Yuan, R.Y.; Liao, M.; et al. Exploring utility of genomic epidemiology to trace origins of highly pathogenic influenza A/H7N9 in Guangdong. *Virus Evol.* **2020**, *6*, veaa097. [CrossRef]
29. Ma, M.-J.; Yang, Y.; Fang, L.-Q. Highly Pathogenic Avian H7N9 Influenza Viruses: Recent Challenges: Recent Challenges. *Trends Microbiol.* **2019**, *27*, 93–95. [CrossRef]
30. Qiu, Y.; Sun, R.; Hou, G.; Yu, X.; Li, Y.; Li, J.; Zhang, Q.; Zou, F.; Liu, H.; Jiang, W. Novel reassortant H7N2 originating from the H7N9 highly pathogenic avian influenza viruses in China, 2019. *J. Infect.* **2019**, *79*, 462–470. [CrossRef]
31. Nakayama, M.; Uchida, Y.; Shibata, A.; Kobayashi, Y.; Mine, J.; Takemae, N.; Tsunekuni, R.; Tanikawa, T.; Harada, R.; Osaka, H.; et al. A novel H7N3 reassortant originating from the zoonotic H7N9 highly pathogenic avian influenza viruses that has adapted to ducks. *Transbound. Emerg. Dis.* **2019**, *66*, 2342–2352. [CrossRef] [PubMed]
32. Qu, B.; Li, X.; Cardona, C.J.; Xing, Z. Reassortment and adaptive mutations of an emerging avian influenza virus H7N4 subtype in China. *PLoS ONE* **2020**, *15*, e0227597. [CrossRef] [PubMed]
33. Zhao, Z.; Liu, L.; Guo, Z.; Zhang, C.; Wang, Z.; Wen, G.; Zhang, W.; Shang, Y.; Zhang, T.; Jiao, Z.; et al. A Novel Reassortant Avian H7N6 Influenza Virus Is Transmissible in Guinea Pigs via Respiratory Droplets. *Front. Microbiol.* **2019**, *10*, 18. [CrossRef] [PubMed]
34. Gu, M.; Chen, H.; Li, Q.; Huang, J.; Zhao, M.; Gu, X.; Jiang, K.; Wang, X.; Peng, D.; Liu, X. Enzootic genotype S of H9N2 avian influenza viruses donates internal genes to emerging zoonotic influenza viruses in China. *Vet. Microbiol.* **2014**, *174*, 309–315. [CrossRef] [PubMed]
35. Cui, L.; Liu, D.; Shi, W.; Pan, J.; Qi, X.; Li, X.; Guo, X.; Zhou, M.; Li, W.; Li, J.; et al. Dynamic reassortments and genetic heterogeneity of the human-infecting influenza A (H7N9) virus. *Nat. Commun.* **2014**, *5*, 3142. [CrossRef]
36. Pu, J.; Yin, Y.; Liu, J.; Wang, X.; Zhou, Y.; Wang, Z.; Sun, Y.; Sun, H.; Li, F.; Song, J.; et al. Reassortment with dominant chicken H9N2 influenza virus contributed to the fifth H7N9 virus human epidemic. *J. Virol.* **2021**, *95*, e01578-20. [CrossRef]
37. Katoh, K.; Standley, D.M. MAFFT multiple sequence alignment software version 7: Improvements in performance and usability. *Mol. Biol. Evol.* **2013**, *30*, 772–780. [CrossRef]
38. He, D.; Gu, J.; Gu, M.; Wu, H.; Li, J.; Zhan, T.; Chen, Y.; Xu, N.; Ge, Z.; Wang, G.; et al. Genetic and antigenic diversity of H7N9 highly pathogenic avian influenza virus in China. *Infect. Genet. Evol.* **2021**, *93*, 104993. [CrossRef]
39. Drummond, A.J.; Suchard, M.A.; Xie, D.; Rambaut, A. Bayesian phylogenetics with BEAUti and the BEAST 1.7. *Mol. Biol. Evol.* **2012**, *29*, 1969–1973. [CrossRef]
40. Zhang, D.; Gao, F.; Jakovlić, I.; Zou, H.; Zhang, J.; Li, W.X.; Wang, G.T. PhyloSuite: An integrated and scalable desktop platform for streamlined molecular sequence data management and evolutionary phylogenetics studies. *Mol. Ecol. Resour.* **2020**, *20*, 348–355. [CrossRef]
41. Hill, V.; Baele, G. Bayesian estimation of past population dynamics in BEAST 1.10 using the Skygrid coalescent model. *Mol. Biol. Evol.* **2019**, *36*, 2620–2628. [CrossRef] [PubMed]
42. Prosperi, M.C.F.; Ciccozzi, M.; Fanti, I.; Saladini, F.; Pecorari, M.; Borghi, V.; Di Giambenedetto, S.; Bruzzone, B.; Capetti, A.; Vivarelli, A.; et al. A novel methodology for large-scale phylogeny partition. *Nat. Commun.* **2011**, *2*, 321. [CrossRef] [PubMed]
43. Yu, G. Using ggtree to Visualize Data on Tree-Like Structures. *Curr. Protoc. Bioinform.* **2020**, *69*, e96. [CrossRef]
44. Xu, S.; Dai, Z.; Guo, P.; Fu, X.; Liu, S.; Zhou, L.; Tang, W.; Feng, T.; Chen, M.; Zhan, L.; et al. ggtreeExtra: Compact Visualization of Richly Annotated Phylogenetic Data. *Mol. Biol. Evol.* **2021**, *38*, 4039–4042. [CrossRef]
45. Altekar, G.; Dwarkadas, S.; Huelsenbeck, J.P.; Ronquist, F. Parallel Metropolis coupled Markov chain Monte Carlo for Bayesian phylogenetic inference. *Bioinformatics* **2004**, *20*, 407–415. [CrossRef]
46. Westgeest, K.B.; Russell, C.A.; Lin, X.; Spronken, M.I.J.; Bestebroer, T.M.; Bahl, J.; van Beek, R.; Skepner, E.; Halpin, R.A.; de Jong, J.C.; et al. Genomewide analysis of reassortment and evolution of human influenza A(H3N2) viruses circulating between 1968 and 2011. *J. Virol.* **2014**, *88*, 2844–2857. [CrossRef]
47. Avino, M.; Ng, G.T.; He, Y.; Renaud, M.S.; Jones, B.R.; Poon, A.F.Y. Tree shape-based approaches for the comparative study of cophylogeny. *Ecol. Evol.* **2019**, *9*, 6756–6771. [CrossRef]
48. Galili, T. dendextend: An R package for visualizing, adjusting and comparing trees of hierarchical clustering. *Bioinformatics* **2015**, *31*, 3718–3720. [CrossRef]
49. Nagarajan, N.; Kingsford, C. GiRaF: Robust, computational identification of influenza reassortments via graph mining. *Nucleic Acids Res.* **2011**, *39*, e34. [CrossRef]
50. Martin, D.P.; Murrell, B.; Golden, M.; Khoosal, A.; Muhire, B. RDP4: Detection and analysis of recombination patterns in virus genomes. *Virus Evol.* **2015**, *1*, vev003. [CrossRef]

51. Müller, N.F.; Stolz, U.; Dudas, G.; Stadler, T.; Vaughan, T.G. Bayesian inference of reassortment networks reveals fitness benefits of reassortment in human influenza viruses. *Proc. Natl. Acad. Sci. USA* **2019**, *117*, 17104–17111. [CrossRef] [PubMed]
52. Wickham, H. *Ggplot2: Elegrant graphics for data analysis/Hadley Wickham; with contributions by Carson Sievert*, 2nd ed.; Springer: Cham, Switzerland, 2016; pp. 9–199.
53. Jacquot, M.; Rao, P.P.; Yadav, S.; Nomikou, K.; Maan, S.; Jyothi, Y.K.; Reddy, N.; Putty, K.; Hemadri, D.; Singh, K.P.; et al. Contrasting selective patterns across the segmented genome of bluetongue virus in a global reassortment hotspot. *Virus Evol.* **2019**, *5*, vez027. [CrossRef] [PubMed]
54. Kalyaanamoorthy, S.; Minh, B.Q.; Wong, T.K.F.; von Haeseler, A.; Jermiin, L.S. ModelFinder: Fast model selection for accurate phylogenetic estimates. *Nat. Methods* **2017**, *14*, 587–589. [CrossRef] [PubMed]
55. Kosakovsky Pond, S.L.; Frost, S.D.W. Not so different after all: A comparison of methods for detecting amino acid sites under selection. *Mol. Biol. Evol.* **2005**, *22*, 1208–1222. [CrossRef] [PubMed]
56. Murrell, B.; Moola, S.; Mabona, A.; Weighill, T.; Sheward, D.; Kosakovsky Pond, S.L.; Scheffler, K. FUBAR: A fast, unconstrained bayesian approximation for inferring selection. *Mol. Biol. Evol.* **2013**, *30*, 1196–1205. [CrossRef] [PubMed]
57. Murrell, B.; Wertheim, J.O.; Moola, S.; Weighill, T.; Scheffler, K.; Kosakovsky Pond, S.L. Detecting individual sites subject to episodic diversifying selection. *PLoS Genet.* **2012**, *8*, e1002764. [CrossRef] [PubMed]
58. Kosakovsky Pond, S.L.; Poon, A.F.Y.; Velazquez, R.; Weaver, S.; Hepler, N.L.; Murrell, B.; Shank, S.D.; Magalis, B.R.; Bouvier, D.; Nekrutenko, A.; et al. HyPhy 2.5-A Customizable Platform for Evolutionary Hypothesis Testing Using Phylogenies. *Mol. Biol. Evol.* **2020**, *37*, 295–299. [CrossRef]
59. Chen, R.; Holmes, E.C. Avian influenza virus exhibits rapid evolutionary dynamics. *Mol. Biol. Evol.* **2006**, *23*, 2336–2341. [CrossRef]
60. Cattoli, G.; Fusaro, A.; Monne, I.; Coven, F.; Joannis, T.; El-Hamid, H.S.A.; Hussein, A.A.; Cornelius, C.; Amarin, N.M.; Mancin, M.; et al. Evidence for differing evolutionary dynamics of A/H5N1 viruses among countries applying or not applying avian influenza vaccination in poultry. *Vaccine* **2011**, *29*, 9368–9375. [CrossRef]
61. Wu, Y.; Hu, J.; Jin, X.; Li, X.; Wang, J.; Zhang, M.; Chen, J.; Xie, S.; Qi, W.; Liao, M.; et al. Accelerated Evolution of H7N9 Subtype Influenza Virus under Vaccination Pressure. *Virol. Sin.* **2021**, *36*, 1124–1132. [CrossRef]
62. Zhang, L.; Zhang, Z.; Weng, Z.; Shi, W. Substitution rates of the internal genes in the novel avian H7N9 influenza virus. *Clin. Infect. Dis.* **2013**, *57*, 1213–1215. [CrossRef] [PubMed]
63. Fitch, W.M.; Leiter, J.M.; Li, X.Q.; Palese, P. Positive Darwinian evolution in human influenza A viruses. *Proc. Natl. Acad. Sci. USA* **1991**, *88*, 4270–4274. [CrossRef] [PubMed]
64. Suzuki, Y. Natural selection on the influenza virus genome. *Mol. Biol. Evol.* **2006**, *23*, 1902–1911. [CrossRef] [PubMed]
65. Yu, X.; Jin, T.; Cui, Y.; Pu, X.; Li, J.; Xu, J.; Liu, G.; Jia, H.; Liu, D.; Song, S.; et al. Influenza H7N9 and H9N2 viruses: Coexistence in poultry linked to human H7N9 infection and genome characteristics. *J. Virol.* **2014**, *88*, 3423–3431. [CrossRef]
66. Marshall, N.; Priyamvada, L.; Ende, Z.; Steel, J.; Lowen, A.C. Influenza virus reassortment occurs with high frequency in the absence of segment mismatch. *PLoS Pathog.* **2013**, *9*, e1003421. [CrossRef]
67. Gerber, M.; Isel, C.; Moules, V.; Marquet, R. Selective packaging of the influenza A genome and consequences for genetic reassortment. *Trends Microbiol.* **2014**, *22*, 446–455. [CrossRef]
68. Essere, B.; Yver, M.; Gavazzi, C.; Terrier, O.; Isel, C.; Fournier, E.; Giroux, F.; Textoris, J.; Julien, T.; Socratous, C.; et al. Critical role of segment-specific packaging signals in genetic reassortment of influenza A viruses. *Proc. Natl. Acad. Sci. USA* **2013**, *110*, E3840-8. [CrossRef]
69. Sun, X.; Belser, J.A.; Pappas, C.; Pulit-Penaloza, J.A.; Brock, N.; Zeng, H.; Creager, H.M.; Le, S.; Wilson, M.; Lewis, A.; et al. Risk Assessment of Fifth-Wave H7N9 Influenza A Viruses in Mammalian Models. *J. Virol.* **2019**, *93*, e01740-18. [CrossRef]
70. Hao, X.; Hu, J.; Wang, X.; Gu, M.; Wang, J.; Liu, D.; Gao, Z.; Chen, Y.; Gao, R.; Li, X.; et al. The PB2 and M genes are critical for the superiority of genotype S H9N2 virus to genotype H in optimizing viral fitness of H5Nx and H7N9 avian influenza viruses in mice. *Transbound. Emerg. Dis.* **2020**, *67*, 758–768. [CrossRef]
71. Hao, X.; Wang, X.; Hu, J.; Gu, M.; Wang, J.; Deng, Y.; Jiang, D.; He, D.; Xu, H.; Yang, Y.; et al. The PB2 and M genes of genotype S H9N2 virus contribute to the enhanced fitness of H5Nx and H7N9 avian influenza viruses in chickens. *Virology* **2019**, *535*, 218–226. [CrossRef]
72. Villa, M.; Lässig, M. Fitness cost of reassortment in human influenza. *PLoS Pathog.* **2017**, *13*, e1006685. [CrossRef] [PubMed]
73. Lu, L.; Lycett, S.J.; Leigh Brown, A.J. Reassortment patterns of avian influenza virus internal segments among different subtypes. *BMC Evol. Biol.* **2014**, *14*, 16. [CrossRef] [PubMed]
74. Yao, Y.; Zhang, T.; Yang, W.; Shao, Z.; He, B.; Chen, X.; Wu, L.; Jin, E.; Liu, H.; Chen, J.; et al. Avian Influenza A (H7N9) Virus in a Wild Land Bird in Central China, Late 2015. *Virol. Sin.* **2018**, *33*, 96–99. [CrossRef] [PubMed]
75. Bui, C.; Bethmont, A.; Chughtai, A.A.; Gardner, L.; Sarkar, S.; Hassan, S.; Seale, H.; MacIntyre, C.R. A Systematic Review of the Comparative Epidemiology of Avian and Human Influenza A H5N1 and H7N9 - Lessons and Unanswered Questions. *Transbound. Emerg. Dis.* **2016**, *63*, 602–620. [CrossRef]
76. Bhat, S.; James, J.; Sadeyen, J.-R.; Mahmood, S.; Everest, H.J.; Chang, P.; Walsh, S.; Byrne, A.M.P.; Mollett, B.; Lean, F.; et al. Co-infection of chickens with H9N2 and H7N9 avian influenza viruses leads to emergence of reassortant H9N9 virus with increased fitness for poultry and enhanced zoonotic potential. *J. Virol.* **2021**, *96*, e0185621. [CrossRef]

77. Dugan, V.G.; Chen, R.; Spiro, D.J.; Sengamalay, N.; Zaborsky, J.; Ghedin, E.; Nolting, J.; Swayne, D.E.; Runstadler, J.A.; Happ, G.M.; et al. The evolutionary genetics and emergence of avian influenza viruses in wild birds. *PLoS Pathog.* **2008**, *4*, e1000076. [CrossRef]
78. Dudas, G.; Bedford, T.; Lycett, S.; Rambaut, A. Reassortment between influenza B lineages and the emergence of a coadapted PB1-PB2-HA gene complex. *Mol. Biol. Evol.* **2015**, *32*, 162–172. [CrossRef]
79. Neverov, A.D.; Lezhnina, K.V.; Kondrashov, A.S.; Bazykin, G.A. Intrasubtype reassortments cause adaptive amino acid replacements in H3N2 influenza genes. *PLoS Genet.* **2014**, *10*, e1004037. [CrossRef]

viruses

MDPI

Article

Gga-miR-30c-5p Enhances Apoptosis in Fowl Adenovirus Serotype 4-Infected Leghorn Male Hepatocellular Cells and Facilitates Viral Replication through Myeloid Cell Leukemia-1

Areayi Haiyilati [1,2], Linyi Zhou [1,2], Jiaxin Li [1,2], Wei Li [1,2], Li Gao [1,2], Hong Cao [1,2], Yongqiang Wang [1,2], Xiaoqi Li [2,*] and Shijun J. Zheng [1,2,*]

1 Key Laboratory of Animal Epidemiology of the Ministry of Agriculture, China Agricultural University, Beijing 100193, China; aray713@cau.edu.cn (A.H.); zlyi123321@126.com (L.Z.); lijiaxinswq@163.com (J.L.); liwei19940506@163.com (W.L.); gaoli194@cau.edu.cn (L.G.); caohong@edu.cau.cn (H.C.); vetwyq@cau.edu.cn (Y.W.)

2 College of Veterinary Medicine, China Agricultural University, Beijing 100193, China

* Correspondence: leexiaoqi@cau.edu.cn (X.L.); sjzheng@cau.edu.cn (S.J.Z.); Tel./Fax: +86-(10)-6273-4681 (S.J.Z.)

Abstract: Fowl adenovirus serotype 4 (FAdV-4) is the primary causative agent responsible for the hepatitis-hydropericardium syndrome (HHS) in chickens, leading to considerable economic losses to stakeholders. Although the pathogenesis of FAdV-4 infection has gained attention, the underlying molecular mechanism is still unknown. Here, we showed that the ectopic expression of gga-miR-30c-5p in leghorn male hepatocellular (LMH) cells enhanced apoptosis in FAdV-4-infected LMH cells by directly targeting the myeloid cell leukemia-1 (Mcl-1), facilitating viral replication. On the contrary, the inhibition of endogenous gga-miR-30c-5p markedly suppressed apoptosis and viral replication in LMH cells. Importantly, the overexpression of Mcl-1 inhibited gga-miR-30c-5p or FAdV-4-induced apoptosis in LMH cells, reducing FAdV-4 replication, while the knockdown of Mcl-1 by RNAi enhanced apoptosis in LMH cells. Furthermore, transfection of LMH cells with gga-miR-30c-5p mimics enhanced FAdV-4-induced apoptosis associated with increased cytochrome *c* release and caspase-3 activation. Thus, gga-miR-30c-5p enhances FAdV-4-induced apoptosis by directly targeting Mcl-1, a cellular anti-apoptotic protein, facilitating FAdV-4 replication in host cells. These findings could help to unravel the mechanism of how a host responds against FAdV-4 infection at an RNA level.

Keywords: FAdV-4; chicken microRNA; gga-miR-30c-5p; apoptosis; Mcl-1

Citation: Haiyilati, A.; Zhou, L.; Li, J.; Li, W.; Gao, L.; Cao, H.; Wang, Y.; Li, X.; Zheng, S.J. Gga-miR-30c-5p Enhances Apoptosis in Fowl Adenovirus Serotype 4-Infected Leghorn Male Hepatocellular Cells and Facilitates Viral Replication through Myeloid Cell Leukemia-1. *Viruses* **2022**, *14*, 990. https://doi.org/10.3390/v14050990

Academic Editor: Chi-Young Wang

Received: 28 February 2022
Accepted: 27 April 2022
Published: 7 May 2022

Publisher's Note: MDPI stays neutral with regard to jurisdictional claims in published maps and institutional affiliations.

Copyright: © 2022 by the authors. Licensee MDPI, Basel, Switzerland. This article is an open access article distributed under the terms and conditions of the Creative Commons Attribution (CC BY) license (https://creativecommons.org/licenses/by/4.0/).

1. Introduction

Fowl adenoviruses (FAdVs), belonging to the genus *Aviadenovirus*, the family of *Adenoviridae*, are generally divided into five species (*FAdV-A* to *FAdV-E*) based on their restriction enzyme digestion patterns [1], including 12 serotypes (FAdV1-7, FAdV8a/b, and FAdV9-11) as determined by a serum cross-neutralization test [2]. Although FAdVs could be isolated from healthy chickens, some serotypes of FAdVs were associated with several notable diseases, such as inclusion body hepatitis (IBH) [3], hepatitis-hydropericardium syndrome (HHS) [4], and adenoviral gizzard erosion (AGE) [5] in chickens and other birds. FAdV-4 is the primary causative agent responsible for HHS at mortality of 30–80% [6–8]. The predominant gross lesions of HHS in chicken are hydropericardium, accumulation of clear or yellowish jelly-like fluid in the pericardial sac, as well as a yellow brown-colored and swollen, enlarged, congested, and friable liver with foci of hemorrhages and necrosis, and in a few cases, hemorrhages in spleens, lungs, and kidneys [9,10], and enlarged bursa of Fabricius could be observed [11]. In the past few years, the occurrence of HHS in Asia, particularly in China, has gained attention, leading to considerable economic losses to stakeholders [12]. FAdVs, highly resistant to inactivation, persistently exist in the environment for a long period of time and can be transmitted both horizontally and vertically [13]. Once flocks were infected with FAdVs, the virus would circulate in flocks for a long time. Thus, FAdVs pose severe threats to the poultry industry around the world.

FAdV-4, a non-enveloped and double-stranded DNA (dsDNA) virus, potentially contains 46 ORFs encoding 11 structural proteins and approximately 32 non-structural proteins [14]. Among these viral proteins, Hexon, a structural protein with a mass of 107 kD, is the most abundant capsid protein, containing group, type, and subtype-specific antigenic determinants [15,16] and determining the FAdV-4 pathogenicity [17]. It was found that the FAdV-4 Hexon interacted with chaperonin containing TCP-1 subunit eta (CCT7), enhancing viral replication [18]. PX, another structural protein of FAdV, is involved in attaching the linear double-stranded DNA genome to the capsid during replication [19] and induces apoptosis in FAdV-4-infected LMH cells, serving as a virulence factor for FAdV-4 infection [20]. In addition, other viral proteins were also involved in the pathogenesis of FAdV-4 infection [21].

MicroRNAs (miRNAs), small non-coding RNAs, are abundant classes of post-transcriptional regulators with the length of 20–24 nucleotides that are involved in the control of a broad range of cellular activities, such as cell proliferation [22], differentiation [23], apoptosis [24], and metabolism [25], and play a role in the replication and propagation of viruses [26,27]. The nucleotides 2–7 of the miRNA 5′ end, which is called the seed-sequence, usually target the 3′ untranslated region (3′-UTR), 5′ untranslated region (5′-UTR), or the coding sequence (CDS) of mRNAs either fully or partially to degrade mRNA or inhibit translation [28–30].

However, the role of miRNAs in the host response to FAdV-4 infection remains elusive. In the present study, we found that the infection of LMH cells with FAdV-4 altered the expressions of miRNAs in host cells, among which gga-miR-30c-5p expression was downregulated and attracted our attention. We found that the transfection of host cells with gga-miR-30c-5p enhanced apoptosis in FAdV-4-infected LMH cells by directly targeting the myeloid cell leukemia-1 (Mcl-1), a cellular anti-apoptotic protein, favoring FAdV-4 growth in LMH cells, while the inhibition of gga-miR-30c-5p in FAdV-4-infected cells slows down viral replication. These data indicate that gga-miR-30c-5p serves as a pro-apoptotic factor in the host response to FAdV-4 infection, uncovering a novel mechanism of host response against FAdV-4 infection at an RNA level.

2. Materials and Methods

2.1. Cells and Virus

LMH cells, an immortalized chicken liver cell line, were kindly provided by Dr. Jinhua Liu (CAU, Beijing, China). The LMH cells were cultured in Waymouth MB 752/1 Medium (MACGENE Technology, Beijing, China) supplemented with 1× Penicillin-Streptomycin (MACGENE Technology) and 10% fetal bovine serum (Gibco, Grand Island, NE, USA) in a 5% CO_2 incubator at 37 °C. The cell culture plates were coated with 0.1% gelatin solution (Cat. ES-006-B, Millipore, Billerica, MA, USA) and incubated at 4 °C for 10 min before splitting cells. The FAdV4 HuBWH strain was originally isolated from the liver of a diseased chicken [18] and was stocked at −80 °C till use.

2.2. Reagents, Chemicals, and Antibodies

OPTI-MEM™ I and Lipofectamine 3000 transfection reagents were purchased from Invitrogen (Carlsbad, CA, USA). The jetPRIMETM transfection reagent was obtained from Polyplus-transfection Biotechnology Company (Strasbourg, France). Mouse anti-GAPDH antibodies (60004-1-Ig) were purchased from Proteintech (Wuhan, China), anti-Tubulin antibodies from Abcam (Cambridge, UK), anti-Hexon monoclonal antibodies and anti-PX monoclonal antibodies from CAEU Biological Company (Beijing, China), anti-Mcl-1 monoclonal antibodies (ab32087) and anti-cytochrome *c* antibodies (ab133504) from Abcam (Cambridge, UK), and horseradish peroxidase (HRP)-conjugated goat anti-mouse/rabbit IgG antibodies from DingGuoShengWu (Beijing, China). An annexin V-PE/7-AAD apoptosis detection kit was purchased from BD Pharmingen (Franklin Lakes, NJ, USA). A dual-specific luciferase assay kit was purchased from 122 Promega (Madison, WI, USA). Caspase-3 colorimetric assay kits were obtained from BioVision (San Francisco, CA, USA). The pRK5-Flag plasmid was obtained from Clontech (Mountain View, CA, USA). The CCK-8 solution was purchased from Beyotime Biotechnology (Beijing, China).

2.3. Plasmid Construction

Gallus gallus Mcl-1 (GenBank accession No. NM_001257283.4) was cloned from LMH cells with the primer pairs shown in Table 1. The primers were synthesized by Sangon Company (Shanghai, China). pRK5-Flag-Mcl-1 expression plasmids were constructed by standard molecular biology techniques.

Table 1. The primers of PCR for Mcl-1sequences.

	Sequences (5′–3′)
Mcl-1 sense primer	AGGACGACGATGACAAGGGATCCATGTTTGCAGTCAAGCGGA
Mcl-1 antisense primer	CGGCCAAGCTTCTGCAGGTCGACTCACCGGATCATGTAGGCCAAGCTC

2.4. Sequences of miRNA Mimics or Inhibitors

Mimics/inhibitors for miRNA were synthesized by Genepharma Company (Shanghai, China). The sense sequences for miRNAs are shown in Table 2.

Table 2. Sheet of miRNAs sequences.

The Name of miRNA	Sequences (5′–3′)
gga-miR-30c-5p mimics	UGUAAACAUCCUACACUCUCAGCU
mimics negative control	UUCUCCGAACGUGUCACGUTT
gga-miR-30c-5p inhibitors	AGCUGAGAGUGUAGGAUGUUUACA
inhibitors negative control	CAGUACUUUUGUGUAGUACAA

2.5. miRNA Target Prediction

The miRNA targets in host cells were predicted by RNA22.v2 (https://cm.jefferson.edu/rna22, accessed on 15 January 2022), miRanda (http://www.microrna.org/microrna/home.do, accessed on 15 January 2022), and Targetscan (http://www.targetscan.org/vert_70/, accessed on 15 January 2022).

2.6. RNA Isolation and Quantitative Real-Time PCR (qRT-PCR) Analysis

Total RNA and miRNA were prepared from LMH cells using the EASYspin Plus kit and RNA misi microRNA kit (Aidlab, Beijing, China), respectively, as per the manufacturer's instructions. Quantitative reverse transcription-PCR (qRT-PCR) was performed using a PrimeScript RT reagent kit (TaKaRa) on a Light Cycler 480 II (Roche, Basel, Switzerland). The primers used for qRT-PCR are shown in Table 3. All primers were designed and synthesized by Sangon Company. The thermal cycling parameters were as follows: 94 °C for 2 min; 40 cycles of 94 °C for 20 s, 56 °C for 20 s, and 72 °C for 20 s; and 1 cycle of 95 °C for 30 s, 60 °C for 30 s, and 95 °C for 30 s. qRT-PCR analysis of gga-miR-30c-5p was performed with an RT-PCR Quantitation Kit (GenePharma, Suzhou, China). The thermal cycling parameters for miRNA were as follows: 95 °C for 3 min; 40 cycles of 95 °C for 12 s, 62 °C for 40 s; and 1 cycle of 95 °C for 30 s, 60 °C for 30 s, and 95 °C for 30 s. The final step was to obtain a melt curve for the PCR products to determine the specificity of the amplification, and the U6 snRNA was utilized as the reference gene. The expression levels of gga-miR-30c-5p were calculated relative to that of U6 snRNA and presented as fold increases or decreases relative to the control samples. All samples were carried out in triplicate on the same plate.

Table 3. The primers of qRT-PCR for Mcl-1 and GAPDH sequences.

The Name of Gene	Sense Primers (5′–3′)	Antisense Primers (5′–3′)
Mcl-1	GGGATCATCACGGACGCATTGG	TCCTCAACTCGGAAGAAGTCAACAAAG
GAPDH	CAACTACATGGTTTACATGTTCC	GGACTGTGGTCATGAGTCCT

2.7. Apoptosis Assay

LMH cells were seeded on 12-well plates and cultured for 24 h, followed by transfection with miRNA controls, gga-miR-30c-5p mimics, or gga-miR-30c-5p inhibitors. At different time points (24 and 48 h) post-transfection, cells were harvested, double stained with 7-AAD and Annexin V-PE using an apoptosis detection kit (BD Pharmingen™), and examined by flow cytometry. To determine the effect of gga-miR-30c-5p on FAdV-4-induced apoptosis, LMH cells were cultured and transfected with miRNA controls, gga-miR-30c-5p mimics, or gga-miR-30c-5p inhibitors. Twenty-four hours after transfection, the cells were infected with FAdV-4 at an MOI of 1 and harvested at different time points (24 and 48 h) post-infection for further analysis. To confirm the effect of Mcl-1 on cell apoptosis, LMH cells were transfected with pRK5-Flag-Mcl-1 or empty vectors. Twenty-four hours after transfection, the cells were harvested. Similarly, LMH cells were seeded onto 12-well plates and cultured for 24 h before transfection with siRNA against Mcl-1 or RNAi controls using Lipofectamine 3000. Double transfections were performed at 24 h intervals. Twenty-four hours after the second transfection, cells were harvested for further analysis. The samples were subjected to flow cytometry analysis as described above. The data were then analyzed with CellQuest Pro software (version 5.1, BD) and presented as means ± standard deviations (SD) of three independent experiments.

2.8. Measurement of FAdV-4 Growth in LMH Cells

Normal cells or cells receiving gga-miR-30c-5p mimics, miRNA controls (80 nM per well), gga-miR-30c-5p inhibitors, or miRNA inhibitor controls (200 nM per well) were infected with FAdV-4 at an MOI of 1. Cell culture samples were harvested at different time points (12, 24, 48, and 72 h) post-FAdV-4 infection. To confirm the effect of Mcl-1 on FAdV-4 replication, LMH cells were transfected with pRK5-Flag-Mcl-1 or an empty vector. Twenty-four hours after transfection, the cells were infected with FAdV-4 at an MOI of 1 and harvested after 24 h. The samples were subjected to three rounds of freeze-thawed treatment. The viral contents in the total cell lysates were titrated using 50% tissue culture infective doses ($TCID_{50}$) in LMH cells. Briefly, the viral solution was diluted 10-fold in Waymouth MB 752/1 Medium. A 100 μL aliquot of each diluted sample was added to the wells of 96-well plates, followed by the addition of 100 μL of LMH cells at a density of 3×10^5 cells/mL. The cells were cultured at 37 °C in 5% CO_2 for 5 days. Tissue culture wells with cytopathic effect (CPE) were determined as positive. The titer was calculated based on a previously described method [31].

2.9. Western Blot Analysis

All cell lysates were prepared using a lysis buffer (50 mM Tris-HCl, pH 8.0, 150 mM NaCl, 5 mM EDTA, 1% NP-40, 10% glycerol, 1× complete cocktail protease inhibitor), boiled with SDS loading buffer for 10 min, and fractionated by electrophoresis on 10% SDS-PAGE gels. The resolved proteins were then transferred onto polyvinylidene difluoride (PVDF) membranes. After being blocked with 5% skimmed milk, the membranes were incubated with anti-Hexon, anti-PX, anti-Mcl-1, anti-cytochrome *c*, anti-GAPDH, or anti-Tubulin antibodies, followed by incubation with HRP-conjugated secondary antibodies. Western Blots were developed using an enhanced chemiluminescence (ECL) kit (Kangwei Biological Company, Suzhou, China) per the manufacturer's instructions.

2.10. Luciferase Reporter Gene Assays

LMH cells were seeded on 24-well plates and cultured overnight, followed by transfection with luciferase reporter gene plasmids (pGL3-target-Mcl-1-wt or pGL3-target-Mcl-1-mutant) and miRNA mimics, miRNA controls, miRNA inhibitors, or miRNA inhibitor controls. To normalize for transfection efficiency, another plasmid pRL-TK expressing Renilla luciferase reporter gene was added to each transfection as a control. Forty-eight hours post-transfection, luciferase reporter gene assays were performed with a dual-luciferase reporter gene assay system. Firefly luciferase activities were normalized on the basis of Renilla luciferase activities.

2.11. Knockdown of Mcl-1 by RNAi

The small interfering RNAs (siRNAs) were designed and synthesized by Genepharma Company (Suzhou, China) and used to knock down the expression of Mcl-1 in LMH cells. The sequences of siRNA against Mcl-1 in LMH cells included RNAi#1 and RNAi#2 and are shown in Table 4. LMH cells were seeded onto 12-well plates and cultured for 24 h before transfection with siRNA or controls using Lipofectamine 3000. Double transfections were performed at 24 h intervals. Twenty-four hours after the second transfection, cells were harvested for further analysis.

Table 4. Sheet of siRNAs sequences.

The Name of siRNA	Sense Primers (5′–3′)	Antisense Primers (5′–3′)
RNAi#1	CUCAUCUCAUUUGGUGCCUTT	AGGCACCAAAUGAGAUGAGTT
RNAi#2	GCCUACAUGAUCCGAAAGUTT	ACUUUCGGAUCAUGUAGGCTT
negative siRNA control	UUCUCCGAACGUGUCACGUTT	ACGUGACACGUUCGGAGAATT

2.12. Measurement of Cytochrome C Release

The isolation of mitochondria and cytosol was performed using a cell mitochondrion isolation kit (Beyotime Biotechnology, Beijing, China). Briefly, LMH cells were transfected with miRNAs for 24 h, followed by mock infection or infection with FAdV-4 at an MOI of 1 for 24 and 48 h. Untreated cells or cells receiving miRNAs were incubated in 100 μL of ice-cold mitochondrion lysis buffer on ice for 15 min, and cell suspensions were homogenized on ice with a Dounce grinder. The homogenates were centrifuged at 600× *g* for 10 min at 4 °C, and the supernatant was obtained and then centrifuged again at 12,000× *g* for 20 min at 4 °C. The supernatant was examined for cytochrome *c* release by Western Blot using anti-cytochrome *c* antibodies.

2.13. Caspase-3 Activity Assays

LMH cells were seeded on 6-well plates before being transfected with gga-miR-30c-5p mimics, miRNA controls, gga-miR-30c-5p inhibitors, or miRNA inhibitor controls. Twenty-four hours after transfection, cells were mock-infected or infected with FAdV-4 at an MOI of 1. Twenty-four or forty-eight hours after infection, cell lysates were prepared and examined for caspase-3 activities using caspase-3 activity assay kits per the manufacturer's instructions. The samples were measured at 405 nm with a microplate reader (Tecan: Sunrise) using the fluorescent substrate DEVD-pNA (synthetic caspase-3 substrate).

2.14. Cell Viability Assay

The LMH cells were seeded on 96-well culture plates and transfected with Mcl-1-RNAi#1 or control-RNAi. At different time points (24, 48, and 72 h) post-transfection, 10 μL of CCK-8 solution (Beyotime Biotechnology, China) was added to each well, followed by incubation at 37 °C for 1 h, and the absorbance of the solution was finally determined at 450 nm using a microplate spectrophotometer. The results are representative of three independent experiments.

2.15. Statistical Analysis

The statistical analysis was performed using GraphPad Prism version 9.0. The significance of the differences between FAdV-4 infection and mock controls and between gga-miR-30c-5p mimics/inhibitors and miRNA controls in gene expression, apoptosis, cytochrome *c* release, caspase activity, cell viability and viral growth was determined by a Mann–Whitney test or analysis of variance (ANOVA) accordingly.

3. Results

3.1. Infection of LMH Cells with FAdV-4 Reduced gga-miR-30c-5p Expression

To explore the roles of miRNA in the host response against FAdV-4 infection, we performed deep sequencing to analyze the miRNA expressions in LMH cells infected with the FAdV-4 HuBWH strain at an MOI of 1 for 24 h. As shown in Figure 1A, 67 miRNAs were encoded by the FAdV-4 DNA genome, 552 miRNAs were found changed in LMH cells after FAdV-4 infection, while 132 were unchanged in LMH cells. To explore the role of miRNAs in the host response to FAdV-4 infection, we selected 12 miRNAs with altered expression levels and examined their effects on host cell apoptosis and virus replication. Among them, gga-miR-30c-5p was apparently decreased after FAdV-4 infection (Figure 1B), attracting our attention. It was reported that gga-miR-30c-5p inhibited glioma proliferation and invasion via targeting Bcl-2 [32] and regulated cisplatin-induced apoptosis of renal tubular epithelial cells via targeting Bnip3L and Hspa5 [33]. Thus, we set out to examine the role of gga-miR-30c-5p in the host response to FAdV-4 infection. We infected LMH cells with different doses of FAdV-4 and examined the expression of gga-miR-30c-5p at 24 h post-infection. As shown in Figure 1C, the expression of gga-miR-30c-5p markedly decreased in a dose-dependent manner 24 h post-FAdV-4 infection, indicating that gga-miR-30c-5p might play a role in host cell response to FAdV-4 infection.

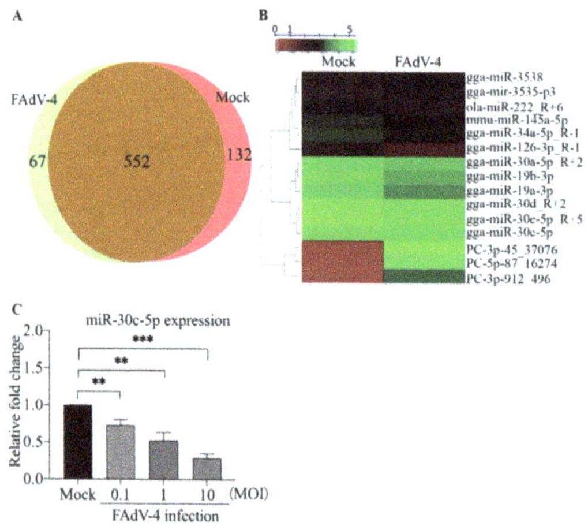

Figure 1. Infection of LMH cells with FAdV-4 strain HuBWH decreases gga-miR-30c-5p expression. (**A**) Venn diagram represents the numbers of miRNAs coded by FAdV-4 and changed miRNAs in LMH cells 24 h after infection with FAdV-4 at an MOI of 1. (**B**) The heat map illustrating the expression profiles of 12 miRNAs in LMH cells with FAdV-4 infection and 3 miRNAs encoded by FAdV-4. The red color stands for the decreased expressions of miRNAs while the green for increased expressions of miRNAs. (**C**) LMH cells were mock-infected or infected with FAdV-4 strain HuBWH at an MOI of 0.1, 1, or 10. Twenty-four hours after FAdV-4 infection, the expression levels of gga-miR-30c-5p were examined by qRT-PCR. The expression of U6 was used as an internal control. The relative levels of gga-miR-30c-5p expression were calculated as follows: gga-miR-30c-5p expression in FAdV-4-infected cells/that of mock controls. Data are representative of three independent experiments and presented as means ± SD. (*** stands for $p < 0.001$; ** for $p < 0.01$).

3.2. Gga-miR-30c-5p Enhances Apoptosis in LMH Cells with or without FAdV-4 Infection

As we previously found that FAdV-4-induced apoptosis in LMH cells [20], it was intriguing to examine the effect of gga-miR-30c-5p on FAdV-4-induced apoptosis. We transfected LMH cells with gga-miR-30c-5p mimics or inhibitors and examined apoptosis in cells 24 and 48 h post-transfection using flow cytometry. As shown in Figure 2A,B, the cells receiving gga-miR-30c-5p displayed marked apoptosis compared to that of miRNA controls ($p < 0.01$), while the inhibition of endogenous gga-miR-30c-5p with its inhibitors significantly inhibited apoptosis ($p < 0.05$), indicating that gga-miR-30c-5p is involved in spontaneous cell death in LMH cells. To explore the role of gga-miR-30c-5p in FAdV-4-induced apoptosis, we transfected LMH cells with gga-miR-30c-5p mimics, gga-miR-30c-5p inhibitors, or miRNA controls, and examined apoptosis 24 and 48 h post-FAdV-4 infection. As shown in Figure 2C,D, the overexpression of gga-miR-30c-5p markedly enhanced apoptosis in FAdV-4-infected LMH cells ($p < 0.01$). On the contrary, the inhibition of endogenous gga-miR-30c-5p significantly reduced apoptosis in FAdV-4-infected cells ($p < 0.05$). These data clearly indicate that gga-miR-30c-5p serves as a pro-apoptotic factor, enhancing apoptosis in host cells with or without FAdV-4 infection.

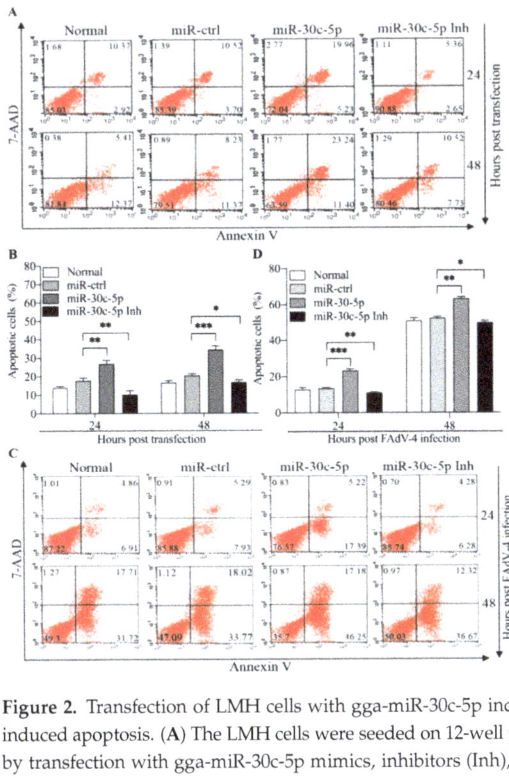

Figure 2. Transfection of LMH cells with gga-miR-30c-5p induces apoptosis and enhances FAdV-4-induced apoptosis. (**A**) The LMH cells were seeded on 12-well plates and cultured overnight, followed by transfection with gga-miR-30c-5p mimics, inhibitors (Inh), or miRNA controls. At different time points (24 and 48 h) post-transfection, cells were harvested, stained with Annexin V-PE and 7-AAD, and examined for apoptosis by flow cytometry. (**B**) The percentage of apoptotic cells in each group shown in panel A was graphed and statistically analyzed. (**C**) The LMH cells were seeded on 12-well plates and cultured overnight, followed by transfection with gga-miR-30c-5p mimics, gga-miR-30c-5p inhibitors or miRNA controls. Twenty-four hours after transfection, cells were infected with FAdV-4 at an MOI of 1. At different time points (24 and 48 h) post-FAdV-4 infection, the cells were examined for apoptosis by flow cytometry as described above. (**D**) The percentage of apoptotic cells in each group shown in panel (**C**) was graphed and statistically analyzed. Data are representative of three independent experiments and presented as means ± SD. (*** stands for $p < 0.001$; ** for $p < 0.01$; * for $p < 0.05$).

3.3. Gga-miR-30c-5p Promotes FAdV-4 Replication in LMH Cells

Since gga-miR-30c-5p promoted FAdV-4-induced apoptosis in host cells, we attempted to examine the effect of gga-miR-30c-5p on FAdV-4 replication. We transfected LMH cells with different doses of gga-miR-30c-5p and examined the expression of Hexon and PX, structural proteins of FAdV-4, in gga-miR-30c-5p-transfected cells post FAdV-4 infection by Western Blot using anti-Hexon and anti-PX monoclonal antibodies. As shown in Figure 3A–C, both Hexon and PX expressions increased in FAdV-4-infected cells with gga-miR-30c-5p transfection in a dose-dependent manner ($p < 0.05$), suggesting that gga-miR-30c-5p promotes FAdV-4 replication. Furthermore, we examined the viral titers in gga-miR-30c-5p-transfected LMH cells at different time points (12, 24, 48, and 72 h) post-FAdV-4 infection using $TCID_{50}$ assay. As a result, the transfection of LMH cells with gga-miR-30c-5p markedly promoted FAdV-4 replication compared to that of control cells ($p < 0.05$) (Figure 3D). These results indicate that gga-miR-30c-5p promoted FAdV-4 replication in host cells. To consolidate the above results, we inhibited the expression of endogenous gga-miR-30c-5p in LMH cells by specific miRNA inhibitors and examined the viral titers in these cells. As shown in Figure 4A, the inhibition of endogenous gga-miR-30c-5p markedly reduced viral titers in FAdV-4-infected cells ($p < 0.05$). Consistently, the expressions of both Hexon and PX proteins significantly decreased in gga-miR-30c-5p inhibitor-treated cells post FAdV-4 infection compared with that of controls as examined by Western Blot assay ($p < 0.05$) (Figure 4B–D). These data clearly show that gga-miR-30c-5p promotes FAdV-4 replication in host cells.

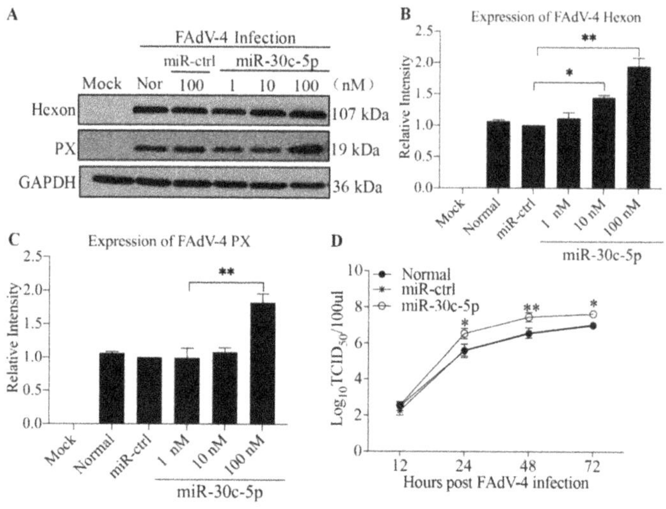

Figure 3. gga-miR-30c-5p facilitates FAdV-4 replication in LMH cells. (**A–C**) The LMH cells were transfected with gga-miR-30c-5p mimics or controls at different concentrations. Twenty-four hours after transfection, the cells were infected with FAdV-4 at an MOI of 1. Twenty-four hours after FAdV-4 infection, cell lysates were prepared and subjected to Western Blot analysis using anti-Hexon and anti-PX antibodies. Endogenous GAPDH expression was examined as an internal control. The band densities of Hexon and PX shown in panel (**A**) were quantitated by densitometry, as shown in panels (**B**,**C**). The relative levels of Hexon and PX were calculated as follows: band density of Hexon or PX/that of GAPDH. (**D**) The LMH cells were transfected with gga-miR-30c-5p mimics or miRNA controls. Twenty-four hours after transfection, the cells were infected with FAdV-4 at an MOI of 1. At different time points (12, 24, 48, and 72 h) after FAdV-4 infection, the viral loads in the cell cultures were determined by $TCID_{50}$ assays. The significance of the differences between gga-miR-30c-5p-transfected cells and controls in terms of viral growth was determined by ANOVA. Data are representative of three independent experiments and presented as means ± SD. (** stands for $p < 0.01$; * for $p < 0.05$).

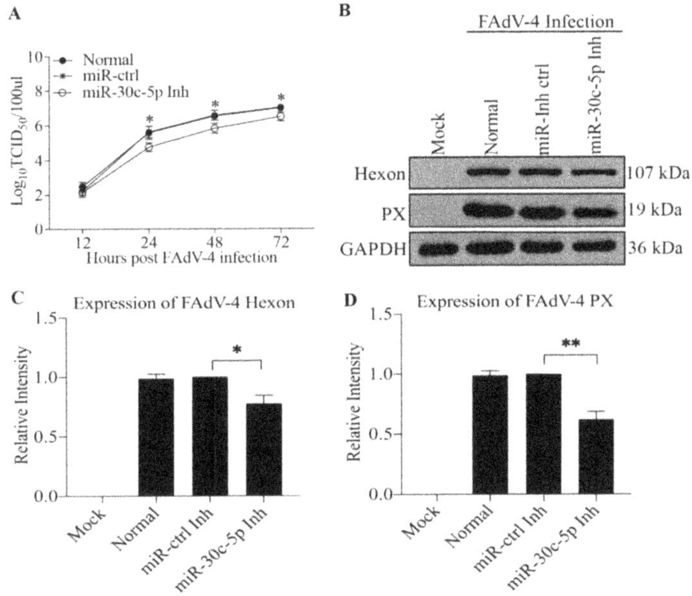

Figure 4. Inhibition of endogenous gga-miR-30c-5p suppresses FAdV-4 replication. (**A**) Knockdown of endogenous gga-miR-30c-5p in LMH cells suppressed FAdV-4 replication. The LMH cells were transfected with gga-miR-30c-5p inhibitor (Inh) or miR-Inh-control (ctrl). Twenty-four hours after transfection, the cells were infected with FAdV-4 at an MOI of 1. At different time points (12, 24, 48, and 72 h) after FAdV-4 infection, the viral loads in the cell cultures were determined by $TCID_{50}$ assays. (**B–D**) The LMH cells were transfected with gga-miR-30c-5p Inh or miR-Inh-ctrl. Twenty-four hours after transfection, the cells were infected with FAdV-4 at an MOI of 1. Twenty-four hours after FAdV-4 infection, cell lysates were prepared and subjected to Western Blot analysis using anti-Hexon and anti-PX antibodies. Endogenous GAPDH expression was examined as an internal control. The band densities of Hexon and PX shown in panel (**B**) were quantitated by densitometry, as shown in panels (**C**,**D**). The relative levels of Hexon and PX were calculated as follows: band density of Hexon or PX/that of GAPDH. The significance of the differences between gga-miR-30c-5p Inh-transfected cells and miR-Inh-ctrl in terms of viral growth was determined by ANOVA. Data are representative of three independent experiments and presented as means ± SD. (** stands for $p < 0.01$; * for $p < 0.05$).

3.4. The Mcl-1 Gene Is a Target of gga-miR-30c-5p in LMH Cells

Since gga-miR-30c-5p is involved in apoptosis, it is intriguing to explore the underlying molecular mechanism. Using the RNA22 (version 2) databases, we identified Mcl-1 as a target of gga-miR-30c-5p in host cells. The region of the Mcl-1 CDS at bp 874 to 895 contains the target site for gga-miR-30c-5p (Figure 5A). Mcl-1, an anti-apoptotic oncoprotein, is a member of the Bcl-2 family [34]. It was reported that human miR-30c-5p caused apoptosis via targeting apoptosis-related genes, such as Bcl-2 [32], and that some miRNAs induced apoptosis via targeting Mcl-1 [35,36]. Therefore, we hypothesized that gga-miR-30c-5p might regulate apoptosis by targeting Mcl-1 in FAdV-4-infected cells. To test this hypothesis, we constructed a firefly luciferase reporter gene plasmid (pGL3-Mcl-1-WT, where WT is wild type) containing the predicted target site in Mcl-1 and another construct (pGL3-Mcl-1-Mut) with mutations in the targeted regions as controls, transfected LMH cells with these reporter gene plasmids and miRNAs and performed luciferase reporter gene assays. As shown in Figure 5B, the transfection of cells with gga-miR-30c-5p together with pGL3-Mcl-1-WT significantly reduced luciferase activity of the reporter gene compared with that of controls ($p < 0.01$), but this reduction could be completely abolished by transfection with pGL3-Mcl-1-Mut, indicating that gga-miR-30c-5p inhibited Mcl-1 expression by targeting its specific sequence in the Mcl-1 gene CDS. On the contrary, the inhibition of endogenous gga-miR-30c-5p by miRNA inhibitors promoted luciferase activity of the reporter gene ($p < 0.05$), suggesting that Mcl-1 expression is enhanced in cells with a reduced level of gga-miR-30c-5p. Furthermore, we examined the effect of gga-miR-30c-5p on Mcl-1 mRNA expression by qRT-PCR. Consistently, our results show that the transfection of LMH cells with gga-miR-30c-5p decreased Mcl-1 mRNA expression compared with that of miRNA controls ($p < 0.01$), while the knockdown of endogenous gga-miR-30c-5p increased the mRNA expression of Mcl-1 ($p < 0.001$) (Figure 5C). These data indicate that gga-miR-30c-5p reduced Mcl-1 expression at an mRNA level. To consolidate these findings, we next examined the effect of gga-miR-30c-5p on Mcl-1 protein expression by Western Blot assay. As a result, the overexpression of gga-miR-30c-5p in LMH cells markedly suppressed Mcl-1 expression ($p < 0.001$) (Figure 5D,E), while the knockdown of endogenous gga-miR-30c-5p by its inhibitors enhanced Mcl-1 expression compared with that of miRNA inhibitor controls ($p < 0.01$) (Figure 5F,G). These data clearly show that gga-miR-30c-5p inhibits Mcl-1 expression by directly targeting Mcl-1 gene expression in LMH cells.

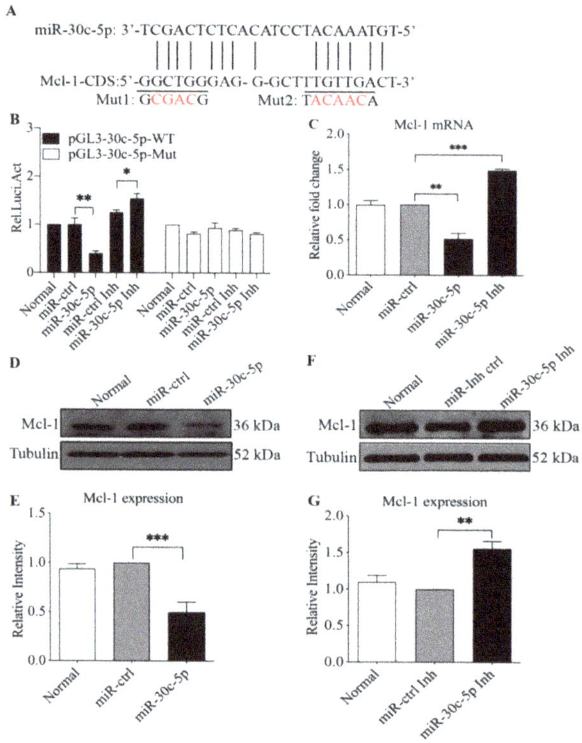

Figure 5. The Mcl-1 gene is a cellular target of gga-miR-30c-5p. (**A**) Diagram of the predicted target site for gga-miR-30c-5p in Mcl-1 gene. The target sequences of gga-miR-30c-5p are underlined and were mutated as indicated. (**B**) The transfection of LMH cells with gga-miR-30c-5p reduced the expression of Mcl-1 but not its mutant. The LMH cells were co-transfected with gga-miR-30c-5p mimics (80 nM), inhibitors (Inh) (200 nM), or miRNA controls, together with luciferase reporter gene vectors (200 ng) and pRL-TK (20 ng). Forty-eight hours post-transfection, the cells were lysed, and a luciferase reporter gene assay was performed to measure Mcl-1 expression. The relative levels of luciferase activity (Rel Luc Act) were calculated as follows: luciferase activity of cells transfected with the reporter gene plasmids together with gga-miR-30c-5p mimics or inhibitors/that of cells co-transfected with the reporter gene plasmids and miRNA controls. (**C**) Effect of gga-miR-30c-5p on

Mcl-1 expression. LMH cells were transfected with gga-miR-30c-5p mimics or miRNA controls at 80 nM or gga-miR-30c-5p Inh at 200 nM. Twenty-four hours after transfection, the mRNA expression of Mcl-1 was examined by qRT-PCR. GAPDH was used as an internal control. The relative levels of gene expression were calculated as follows: mRNA expression of Mcl-1 in cells transfected with gga-miR-30c-5p mimics or miRNA inhibitor or miRNA control/that of Mcl-1 in normal cells. (**D**,**E**) The transfection of LMH cells with gga-miR-30c-5p reduced the expression of Mcl-1 protein. The LMH cells were transfected with gga-miR-30c-5p mimics or miRNA controls at 80 nM. Forty-eight hours after transfection, cell lysates were prepared and subjected to Western Blot analysis using anti-Mcl-1 antibodies, and the band densities of Mcl-1 in panel (**D**) were quantitated by densitometry (**E**). Endogenous GAPDH expression was examined as an internal control. The relative levels of Mcl-1 protein were calculated as follows: band density of Mcl-1/that of GAPDH in the same sample. (**F**,**G**) The knockdown of endogenous gga-miR-30c-5p enhanced the expression of Mcl-1 protein. LMH cells were transfected with gga-miR-30c-5p Inh or miR-Inh-ctrl as described in (**C**). Forty-eight hours after transfection, the expression of Mcl-1 protein was examined as described in (**D**), the band densities of Mcl-1 shown in panel (**F**) were quantitated by densitometry (**G**), and the relative levels of Mcl-1 expression were calculated as described in (**E**). Data are representative of three independent experiments and are presented as means ± SD. (*** stands for $p < 0.001$; ** for $p < 0.01$; * for $p < 0.05$).

3.5. Mcl-1 Suppressed gga-miR-30c-5p-Induced Apoptosis in LMH Cells

The facts that gga-miR-30c-5p promoted apoptosis in LMH cells and that gga-miR-30c-5p inhibited Mcl-1 expression suggest that gga-miR-30c-5p might promote apoptosis by targeting Mcl-1 and that the overexpression of Mcl-1 in host cells would, therefore, block gga-miR-30c-5p-induced apoptosis while the knockdown of Mcl-1 would promote apoptosis as gga-miR-30c-5p did. To test this hypothesis, we transfected LMH cells with gga-miR-30c-5p alone or together with pRK5-Flag-Mcl-1 and examined the effect of over-expressed Mcl-1 on gga-miR-30c-5p-induced apoptosis by flow cytometry. As shown in Figure 6A,B, the overexpression of Mcl-1 markedly blocked gga-miR-30c-5p-induced apoptosis in LMH ($p < 0.01$), suggesting that Mcl-1 is an anti-apoptotic molecule, suppressing miR-30c-5p-induced apoptosis. Furthermore, we made two Mcl-1 RNAi#1 and RNAi#2 constructs and found that both constructs could effectively lower the cellular level of Mcl-1 (Figure 6C). As shown in Figure 6D, transfection with Mcl-1 RNAi#1 in cells decreased the viability and proliferation of cells compared to that of control RNAi, indicating that the knockdown of endogenous Mcl-1 suppressed the viability and proliferation of LMH cells. We then examined apoptosis in LMH cells receiving the Mcl-1 RNAi#1 or control RNAi. As a result, the knockdown of Mcl-1 by RNAi markedly promoted apoptosis in LMH cells as gga-miR-30c-5p did ($p < 0.001$) (Figure 6E,F), indicating that Mcl-1 serves as an anti-apoptotic factor in host cells and that gga-miR-30c-5p induces apoptosis in cells by targeting Mcl-1.

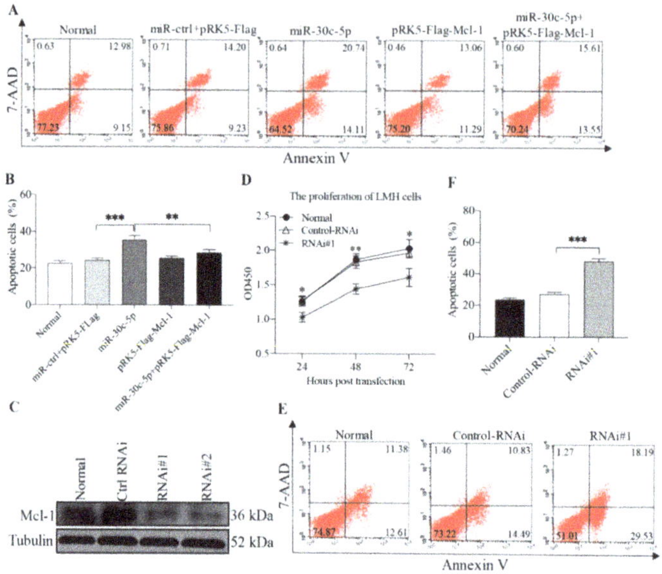

Figure 6. Overexpression of Mcl-1 blocked gga-miR-30c-5p-induced apoptosis in LMH cells, and the knockdown of Mcl-1 promotes apoptosis in LMH cells. (**A**) The LMH cells were co-transfected with gga-miR-30c-5p mimics (80 nM) or miRNA controls (80 nM), pRK5-Flag-Mcl-1(1 μg), or pRK5-Flag (1 μg). Forty-eight hours after transfection, the cells were collected, stained with Annexin V-PE and 7-AAD, and examined for apoptosis by flow cytometry. (**B**) The percentage of apoptotic cells in each group in panel (**A**) was graphed and statistically analyzed. (**C**) Effects of Mcl-1 RNAi on the expression of endogenous Mcl-1. The LMH cells were transfected with siRNA (RNAi#1 and RNAi#2) or controls. Twenty-four hours after the second transfection, cell lysates were prepared and examined by Western Blot using anti-Mcl-1 and anti-Tubulin antibodies. Endogenous Tubulin expression was examined as an internal control. (**D**) The knockdown of Mcl-1 by RNAi suppressed the viability and proliferation of cells. The LMH cells were seeded on a 96-well culture plate and transfected with Mcl-1-RNAi#1 or control-RNAi at a final concentration of 60 nM. At different time points (24, 48, and 72 h) post-transfection, 10 μL of CCK-8 solution was added to each well, followed by incubation at 37 °C for 1 h, and the absorbance of the solution was finally determined at 450 nm using a microplate spectrophotometer. (**E**,**F**) The knockdown of Mcl-1 promotes LMH cells apoptosis. The LMH cells were double transfected with siRNA (RNAi#1) or RNAi controls at an interval of 24 h. Twenty-four hours after the second transfection, the cells were collected, stained with Annexin V-PE and 7-AAD, and examined for apoptosis by flow cytometry. The percentage of apoptotic cells in each group in panel (**D**) was graphed and statistically analyzed (**E**). Data are representative of three independent experiments and presented as means ± SD. (*** stands for $p < 0.001$; ** for $p < 0.01$; * for $p < 0.05$).

3.6. Mcl-1 Is Involved in FAdV-4-Induced Apoptosis and Suppressed FAdV-4 Replication

Since gga-miR-30c-5p promoted FAdV-4-induced apoptosis and viral replication by targeting Mcl-1, it is likely that Mcl-1 is involved in FAdV-4-induced apoptosis and affects viral replication. Thus, we transfected the LMH cells with pRK5-Flag-Mcl-1 or pRK5-Flag as a control and examined apoptosis and viral replication in LMH cells post FAdV-4 infection. As a result, the overexpression of Mcl-1 in FAdV-4-infected cells markedly reduced FAdV-4-induced apoptosis in host cells ($p < 0.01$) (Figure 7A,B), accompanied by reduced expression of viral proteins Hexon and PX in FAdV-4-infected cells ($p < 0.01$) (Figure 7C–E), indicating that Mcl-1 suppressed FAdV-4-mediated apoptosis, suppressing viral infection. To confirm the results, we transfected LMH cells with pRK5-Flag-Mcl-1 or pRK5-Flag and examined the viral titers in these cell cultures 24-h-post FAdV-4 infection using $TCID_{50}$ assay. Consistently, the transfection of LMH cells with pRK5-Flag-Mcl-1 significantly inhibited FAdV-4 replication compared to that of controls ($p < 0.01$) (Figure 7F). These data indicate that Mcl-1 acts as an anti-apoptotic factor in FAdV-4-infected cells and suppresses FAdV-4 replication.

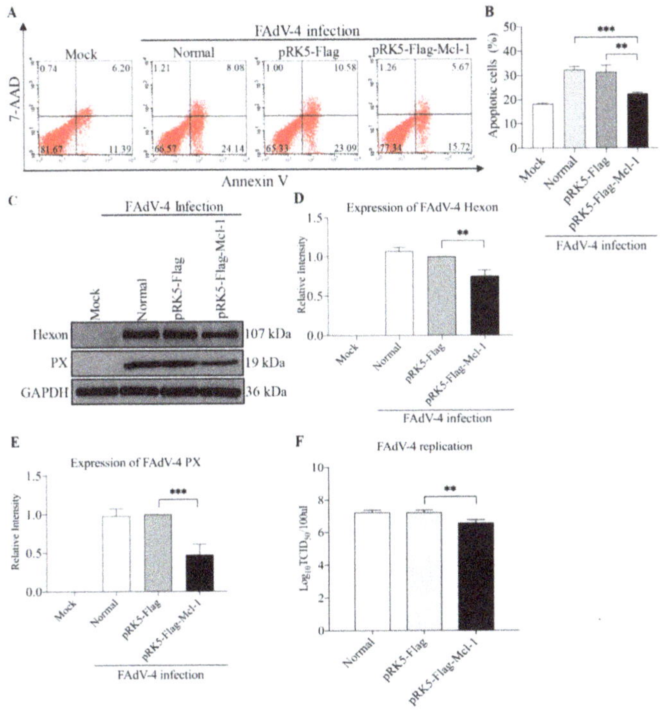

Figure 7. The overexpression of Mcl-1 inhibited FAdV-4-induced apoptosis and suppressed FAdV-4 replication. (**A**,**B**) The overexpression of Mcl-1 inhibited FAdV-4-induced apoptosis. The LMH cells were transfected with pRK5-Flag-Mcl-1 or controls. Twenty-four hours after transfection, the cells were infected with FAdV-4 at an MOI of 1. Forty-eight hours after FAdV-4 infection, the cells were collected, stained with Annexin V-PE and 7-AAD, and examined for apoptosis by flow cytometry (**A**). The percentage of apoptotic cells in each group in panel (**A**) was graphed and statistically analyzed (**B**). (**C**–**E**) The overexpression of Mcl-1 suppressed FAdV-4 replication. LMH cells were treated as described above. Twenty-four hours after transfection, the cells were infected with FAdV-4 at an MOI of 1. Forty-eight hours after FAdV-4 infection, cell lysates were prepared and subjected to Western Blot analysis using anti-Hexon and anti-PX antibodies. Endogenous GAPDH expression was examined as an internal control. The band densities of Hexon and PX in (**C**) were quantitated by densitometry as shown in (**D**,**E**). (**F**) The overexpression of Mcl-1 reduced viral titers in FAdV-4-infected cells. LMH cells were transfected with pRK5-Flag-Mcl-1 vectors or pRK5-Flag as controls, followed by infection with FAdV-4 as described above. Forty-eight hours after FAdV-4 infection, the viral loads in the cell cultures were determined by $TCID_{50}$ assays. Data are representative of three independent experiments and presented as means ± SD. (*** stands for $p < 0.001$; ** for $p < 0.01$).

3.7. Overexpression of Gga-miR-30c-5p Enhanced FAdV-4-Induced Cytochrome C Release and Activation of Caspase-3

As one of the vital anti-apoptotic molecules of the Bcl-2 family, Mcl-1 interferes in a cascade of events leading to the release of cytochrome *c* from mitochondria to promote cell survival [37]. Since gga-miR-30c-5p inhibits the expression of Mcl-1 and Mcl-1-mediated apoptosis is associated with cytochrome *c* release, we hypothesized that gga-miR-30c-5p enhanced apoptosis via facilitating cytochrome *c* release and the activation of caspase-3. To test this hypothesis, we transfected LMH cells with the miRNAs or miRNA controls and examined cytochrome *c* release and caspase-3 activities 24 and 48 h post-FAdV-4 infection. As shown in Figure 8A,B, the transfection of LMH cells with gga-miR-30c-5p significantly increased cytochrome *c* release compared to that of controls ($p < 0.01$). Similarly, the overexpression of gga-miR-30c-5p significantly increased caspase-3 activity ($p < 0.01$), while the inhibition of endogenous gga-miR-30c-5p by inhibitors suppressed the activity of caspase-3 (Figure 8C,D). These data demonstrate that gga-miR-30c-5p enhances apoptosis via facilitating cytochrome *c* release and activation of caspase-3.

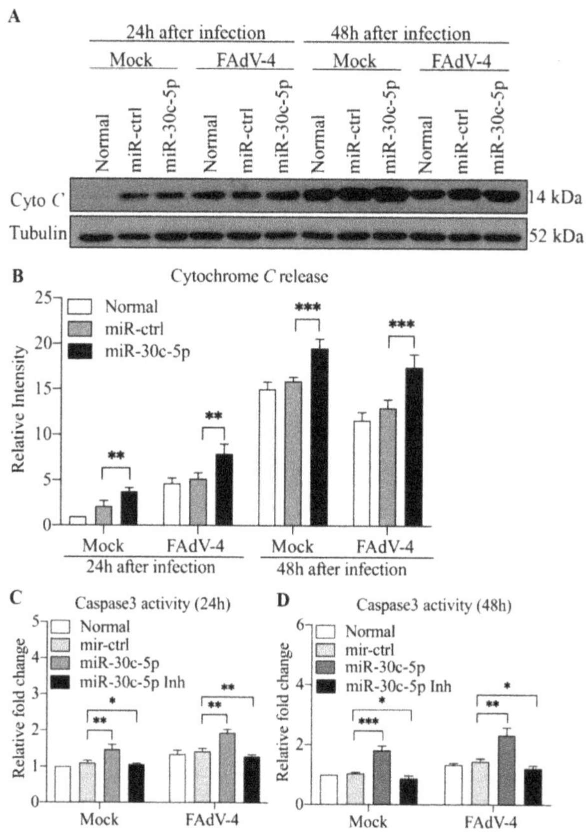

Figure 8. The transfection of cells with gga-miR-30c-5p enhanced FAdV-4-induced cytochrome *c* release and caspase-3 activation. (**A**,**B**) The transfection of cells with gga-miR-30c-5p enhanced the FAdV-4-induced release of cytochrome *c*. The LMH cells were transfected with the indicated miRNAs or miRNA controls. Twenty-four hours after transfection, the cells were mock-infected or infected with FAdV-4 at an MOI of 1. At the indicated time points (24 and 48 h) after FAdV-4 infection, cytosolic proteins were prepared and subjected to Western Blot analysis for the measurement of cytochrome *c* in the cytosol of cells. The band densities of cytochrome *c* in (**A**) were quantitated by densitometry. The relative levels of cytochrome *c* were calculated as follows: band density of cytochrome *c*/that of Tubulin. (**C**,**D**) The transfection of cells with gga-miR-30c-5p enhanced FAdV-4-induced activation of caspase-3. The LMH cells were treated as described above in (**A**). Twenty-four or 48 h after FAdV-4 infection, the activities of caspase-3 were measured at 405 nm with a microplate reader using the substrates DEVD-pNA (synthetic caspase-3 substrate). Data are representative of three independent experiments and presented as means ± SD. (*** stands for $p < 0.001$; ** for $p < 0.01$; * for $p < 0.05$).

Taken together, our data show that the infection of LMH cells with FAdV-4 decreased gga-miR-30c-5p expression. Importantly, gga-miR-30c-5p enhanced FAdV-4-induced apoptosis in LMH cells by directly targeting Mcl-1, facilitating viral replication in host cells. The overexpression of Mcl-1 inhibited gga-miR-30c-5p or FAdV-4-induced apoptosis in LMH cells, reducing FAdV-4 replication, while the knockdown of Mcl-1 by RNAi enhanced apoptosis in LMH cells. Furthermore, gga-miR-30c-5p enhanced FAdV-4-induced apoptosis associated with increased cytochrome *c* release and caspase-3 activation. Thus, gga-miR-30c-5p enhances apoptosis in FAdV-4-infected cells by directly targeting Mcl-1, a cellular anti-apoptotic protein, facilitating FAdV-4 replication in host cells.

4. Discussion

FAdV-4 is one of the most important pathogens for chickens and was first identified as a causative agent responsible for HHS in three to six-week-old broilers in Pakistan in 1987 [38], and subsequently, HHS occurred around the world [8,39–41]. In contrast to mild disease caused by other FAdV serotypes, the outbreaks of HHS in chickens infected by FAdV-4 caused tremendous economic losses to the poultry industry in China in 2015. Thus, effective control of HHS by vaccination is crucial to the income of stakeholders. A complete understanding of the mechanisms underlying FAdV-induced pathogenesis is required for the development of highly efficacious novel vaccines against HHS. However, currently, the pathogenesis of FAdV infection is still largely unknown. In the present study, first, our data show that the infection of LMH cells decreased gga-miR-30c-5p expression. Second, the overexpression of gga-miR-30c-5p significantly enhanced FAdV-induced apoptosis and facilitated viral replication in LMH cells, while the knockdown of gga-miR-30c-5p by specific inhibitors reduced apoptosis and suppressed viral replication in FAdV-infected cells. Thirdly, and importantly, the overexpression of gga-miR-30c-5p in LMH cells induced apoptosis by directly targeting cellular anti-apoptotic factor Mcl-1, while this induction could be blocked by overexpressed Mcl-1. Fourth, the knockdown of Mcl-1 by RNAi induced apoptosis just as the overexpression of gga-miR-30c-5p did. Finally, gga-miR-30c-5p-enhanced apoptosis in FAdV-4-infected cells was associated with increased cytochrome *c* release and caspase-3 activation. Thus, gga-miR-30c-5p enhances FAdV-4-induced apoptosis by directly targeting Mcl-1 and facilitates FAdV-4 replication in host cells, highlighting a critical role of miRNAs in host defense against FAdV-4 infection.

It is well known that miRNAs, small and non-coding RNA molecules, negatively regulate gene expression by binding to mRNAs and thereby promoting degradation of the target mRNA or blocking its translation into protein [42]. Studies about the downregulating expression of miRNAs post-viral infection are relatively rare. In our study, FAdV-4 infection with LMH cells reduced the expression of gga-miR-30c-5p, and the overexpression of gga-miR-30c-5p promoted the replication of FAdV-4. FAdV-4 infection triggered host response against viral infection by downregulating gga-miR-30c-5p expression. It has been reported that VSV decreased miR-27a in different immune cells and found that the IFN/JAK/STAT1 signal pathway induced downregulation of downstream miR-27a by RUNX1, which was inhibited by VSV [43]. Therefore, we will focus on the mechanism of regulating gga-miR-30c-5p by FAdV-4 in the future. miRNAs play important roles in host responses to pathogenic infection, while some viruses have evolved with the capacity of manipulating host miRNAs to evade host response to benefit viral replication [44,45]. For example, gga-miR-181a-5p facilitated FAdV-4 replication via targeting of STING [46], miR-30c promoted type 2 PRRSV infection by targeting the interferon–alpha/beta receptor beta chain [47], gga-miR-142-5p attenuated IRF7 signaling and promoted replication of IBDV by targeting the chMDA5′s 3′ untranslated region [48], and gga-miR-16c-5p induced apoptosis in IBDV-infected cells via targeting Bcl-2, facilitating viral replication [26]. In the present study, we found that gga-miR-30c-5p enhanced FAdV-4-induced apoptosis by directly targeting Mcl-1, promoting FAdV-4 replication in host cells. Thus, it seems that cellular miRNAs could not only serve as components of host response in maintaining homeostasis but also be taken advantage of by pathogens for their own survival. In humans, miR-30c-5p was involved in apoptosis, inflammation, autophagy, and other physiological processes by different targets, including ATG5 [49], eIF2α [50], SIRT1 [51], and FOXO3 [52]. As far as we know, this study is the first report to show that gga-miR-30c-5p is involved in FAdV-4-induced apoptosis in host cells by targeting Mcl-1, an anti-apoptotic member of the Bcl-2 family.

Apoptosis is a strictly controlled physiological process that eliminates unwanted or potentially harmful cells. We previously reported that some miRNAs were involved in apoptosis in host cells, affecting viral replication, for instance, gga-miR-16c-5p enhanced IBDV-induced apoptosis and viral replication by targeting Bcl-2 [26], and gga-miR-29a-3p suppressed avian reovirus-induced apoptosis and viral replication via targeting caspase-3 [53]. In the present study, our results show that gga-miR-30c-5p enhanced apoptosis in FAdV-4-infected cells by targeting Mcl-1, facilitating viral replication. Furthermore, our previous study demonstrated that PX of FAdV-4 acts as a major viral factor inducing apoptosis in FAdV-4-infected LMH cells [20]. It is highly possible that FAdV-4-induced apoptosis is attributable to multiple factors at both protein and RNA levels. Interestingly, we noted that all these viruses (IBDV, Reovirus, and FAdV) are non-enveloped viruses; apoptosis may be one of the important means required for their release and spread. Thus, from an evolutionary point of view, it is very important for FAdV-4 to have the capability of inducing apoptosis in host cells at both the protein and RNA levels.

Mcl-1, a member of the Bcl-2 anti-apoptotic protein subfamily, was first identified in the ML-1 human myeloid leukemia cell line in 1993 [54]. Mcl-1 was expressed differentially in all cells, especially in cancer cells, and its expression was downregulated during apoptosis in many cell types in contrast to the anti-apoptotic Bcl-2 and Bcl-X_L proteins [55,56]. It has been reported that several miRNAs, such as miR-101 [57], miR-519d [58], miR-1469 [35], and miR-26b [59], play a role in apoptosis by targeting Mcl-1. In this study, we found that gga-miR-30c-5p enhanced apoptosis in FAdV-4-infected cells by targeting Mcl-1, facilitating viral replication. Of note, there are several questions to be addressed: for example, (1) what is the mechanism underlying FAdV4-induced gga-miR-30c-5p expression? (2) Is the cellular sensor, such as c-GAS or ALRs (AIM-2-like receptors), involved in the recognition of genomic DNA of FAdV-4 to initiate host response? (3) If so, which transcription factors are involved in these biological processes? More efforts will be required to elucidate the molecular mechanisms of FAdV-4-induced pathogenesis.

In conclusion, our data show that the infection of LMH cells with FAdV-4 altered gga-miR-30c-5p expression. Importantly, gga-miR-30c-5p enhanced apoptosis in FAdV-4-infected cells by directly targeting Mcl-1, facilitating viral replication. Furthermore, gga-miR-30c-5p enhanced FAdV-4-induced apoptosis associated with increased cytochrome *c* release and caspase-3 activation. Thus, gga-miR-30c-5p plays an important role in FAdV-4-induced apoptosis, favoring FAdV-4 survival in host cells, which might be used as a potential target for intervening FAdV-4 infection. These findings contribute to the understanding of the mechanisms underlying FAdV-4-induced apoptosis at an RNA level.

Author Contributions: Conceptualization, S.J.Z.; Data curation, A.H.; Formal analysis, A.H. and S.J.Z.; Funding acquisition, X.L. and S.J.Z.; Investigation, A.H.; Methodology, A.H. and S.J.Z.; Project administration, X.L. and S.J.Z.; Resources, L.Z., J.L., W.L., L.G., H.C. and Y.W.; Supervision, S.J.Z.; Validation, A.H.; Visualization, A.H.; Writing—original draft, A.H.; Writing—review and editing, S.J.Z. All authors have read and agreed to the published version of the manuscript.

Funding: This work was supported by the Beijing Natural Science Foundation (# 6222037) and the Earmarked Fund for Modern Agro-Industry Technology Research System (#CARS-40), China.

Institutional Review Board Statement: Not applicable.

Informed Consent Statement: Not applicable.

Acknowledgments: We thank Jinhua Liu for his kind assistance.

Conflicts of Interest: The authors declare no conflict of interest.

References

1. Zsak, L.; Kisary, J. Characterisation of adenoviruses isolated from geese. *Avian Pathol.* **1984**, *13*, 253–264. [CrossRef] [PubMed]
2. Hess, M. Detection and differentiation of avian adenoviruses: A review. *Avian Pathol.* **2000**, *29*, 195–206. [CrossRef] [PubMed]
3. Mohamed, M.H.A.; El-Sabagh, I.M.; Abdelaziz, A.M.; Al-Ali, A.M.; Alramadan, M.; Lebdah, M.A.; Ibrahim, A.M.; Al-Ankari, A.S. Molecular characterization of fowl aviadenoviruses species D and E associated with inclusion body hepatitis in chickens and falcons indicates possible cross-species transmission. *Avian Pathol.* **2018**, *47*, 384–390. [CrossRef] [PubMed]
4. Xia, J.; Yao, K.C.; Liu, Y.Y.; You, G.J.; Li, S.Y.; Liu, P.; Zhao, Q.; Wen, R.; Wu, Y.P.; Huang, X.B.; et al. Isolation and molecular characterization of prevalent Fowl adenovirus strains in southwestern China during 2015–2016 for the development of a control strategy. *Emerg. Microbes Infect.* **2017**, *6*, e103. [CrossRef]
5. Grafl, B.; Prokofieva, I.; Wernsdorf, P.; Steinborn, R.; Hess, M. Infection with an apathogenic fowl adenovirus serotype-1 strain (CELO) prevents adenoviral gizzard erosion in broilers. *Vet. Microbiol.* **2014**, *172*, 177–185. [CrossRef]
6. Schachner, A.; Matos, M.; Grafl, B.; Hess, M. Fowl adenovirus-induced diseases and strategies for their control—A review on the current global situation. *Avian Pathol.* **2018**, *47*, 111–126. [CrossRef]
7. Li, H.; Wang, J.; Qiu, L.; Han, Z.; Liu, S. Fowl adenovirus species C serotype 4 is attributed to the emergence of hepatitis-hydropericardium syndrome in chickens in China. *Infect. Genet. Evol.* **2016**, *45*, 230–241. [CrossRef]
8. Zhao, J.; Zhong, Q.; Zhao, Y.; Hu, Y.X.; Zhang, G.Z. Pathogenicity and Complete Genome Characterization of Fowl Adenoviruses Isolated from Chickens Associated with Inclusion Body Hepatitis and Hydropericardium Syndrome in China. *PLoS ONE* **2015**, *10*, e0133073. [CrossRef]
9. Sun, J.; Zhang, Y.; Gao, S.; Yang, J.; Tang, Y.; Diao, Y. Pathogenicity of fowl adenovirus serotype 4 (FAdV-4) in chickens. *Infect. Genet. Evol.* **2019**, *75*, 104017. [CrossRef]
10. Wu, N.; Yang, B.; Wen, B.; Li, W.; Guo, J.; Qi, X.; Wang, J. Pathogenicity and Immune Responses in Specific-Pathogen-Free Chickens During Fowl Adenovirus Serotype 4 Infection. *Avian Dis.* **2020**, *64*, 315–323. [CrossRef]
11. Kumar, R.; Chandra, R.; Shukla, S.K.; Agrawal, D.K.; Kumar, M. Hydropericardium syndrome (HPS) in India: A preliminary study on the causative agent and control of the disease by inactivated autogenous vaccine. *Trop. Anim. Health Prod.* **1997**, *29*, 158–164. [CrossRef]
12. Lv, L.; Lu, H.; Wang, K.; Shao, H.; Mei, N.; Ye, J.Q.; Chen, H.J. Emerging of a novel natural recombinant fowl adenovirus in China. *Transbound Emerg. Dis.* **2021**, *68*, 283–288. [CrossRef]
13. Nagy, E.C.J. *Avian Virology-Current Research and Future Trends*; Caister Academic Press: Norfolk, UK, 2019; pp. 283–344.
14. Griffin, B.D.; Nagy, E. Coding potential and transcript analysis of fowl adenovirus 4: Insight into upstream ORFs as common sequence features in adenoviral transcripts. *J. Gen. Virol.* **2011**, *92*, 1260. [CrossRef]
15. Xu, L.; Benson, S.D.; Burnett, R.M. Nanoporous crystals of chicken embryo lethal orphan (CELO) adenovirus major coat protein, hexon. *J. Struct. Biol.* **2007**, *157*, 424–431. [CrossRef]
16. Ganesh, K.; Suryanarayana, V.; Raghavan, R.; Gowda, S. Nucleotide sequence of L1 and part of P1 of hexon gene of fowl adenovirus associated with hydropericardium hepatitis syndrome differs with the corresponding region of other fowl adenoviruses. *Vet. Microbiol.* **2001**, *78*, 1–11. [CrossRef]
17. Zhang, Y.; Liu, A.; Wang, Y.; Cui, H.; Wang, X. A Single Amino Acid at Residue 188 of Hexon Protein is Responsible for the Pathogenicity of the Emerging Novel Fowl Adenovirus 4. *J. Virol.* **2021**, *95*, e0060321. [CrossRef]
18. Gao, J.; Zhao, M.; Duan, X.; Wang, Y.; Cao, H.; Li, X.; Zheng, S.J. Requirement of Cellular Protein CCT7 for the Replication of Fowl Adenovirus Serotype 4 (FAdV-4) in Leghorn Male Hepatocellular Cells Via Interaction with the Viral Hexon Protein. *Viruses* **2019**, *11*, 107. [CrossRef]
19. Lee, T.W.R.; Lawrence, F.J.; Dauksaite, V.; Akusjarvi, G.; Blair, G.E.; Matthews, D.A. Precursor of human adenovirus core polypeptide Mu targets the nucleolus and modulates the expression of E2 proteins. *J. Gen. Virol.* **2004**, *85*, 185–196. [CrossRef]
20. Zhao, M.; Duan, X.; Wang, Y.; Gao, L.; Cao, H.; Li, X.; Zheng, S.J. A Novel Role for PX, a Structural Protein of Fowl Adenovirus Serotype 4 (FAdV4), as an Apoptosis-Inducer in Leghorn Male Hepatocellular Cell. *Viruses* **2020**, *12*, 228. [CrossRef]
21. Haiyilati, A.; Li, X.; Zheng, S.J. Fowl Adenovirus: Pathogenesis and Control. *Int. J. Plant Anim. Environ. Sci.* **2021**, *11*, 566–589. [CrossRef]
22. Wang, L.; Liu, Y.; Yu, Z.; Gong, J.; Deng, Z.; Ren, N.; Zhong, Z.; Cai, H.; Tang, Z.; Cheng, H.; et al. Mir-139-5p inhibits glioma cell proliferation and progression by targeting GABRA1. *J. Transl. Med.* **2021**, *19*, 213. [CrossRef]
23. Kornfeld, S.F.; Cummings, S.E.; Fathi, S.; Bonin, S.R.; Kothary, R. MiRNA-145-5p prevents differentiation of oligodendrocyte progenitor cells by regulating expression of myelin gene regulatory factor. *J. Cell. Physiol.* **2021**, *236*, 997–1012. [CrossRef]
24. Chen, X.; Li, A.; Zhan, Q.; Jing, Z.; Chen, Y.; Chen, J. microRNA-637 promotes apoptosis and suppresses proliferation and autophagy in multiple myeloma cell lines via NUPR1. *FEBS Open Bio.* **2021**, *11*, 519–528. [CrossRef]
25. Taefehshokr, S.; Taefehshokr, N.; Hemmat, N.; Hajazimian, S.; Isazadeh, A.; Dadebighlu, P.; Baradaran, B. The pivotal role of MicroRNAs in glucose metabolism in cancer. *Pathol. Res. Pract.* **2021**, *217*, 153314.
26. Duan, X.; Zhao, M.; Wang, Y.; Li, X.; Cao, H.; Zheng, S.J. Epigenetic Upregulation of Chicken MicroRNA-16-5p Expression in DF-1 Cells following Infection with Infectious Bursal Disease Virus (IBDV) Enhances IBDV-Induced Apoptosis and Viral Replication. *J. Virol.* **2020**, *94*, e01724-19. [CrossRef]

27. Fu, M.; Wang, B.; Chen, X.; He, Z.; Wang, Y.; Li, X.; Cao, H.; Zheng, S.J. MicroRNA gga-miR-130b Suppresses Infectious Bursal Disease Virus Replication via Targeting of the Viral Genome and Cellular Suppressors of Cytokine Signaling 5. *J. Virol.* **2018**, *92*, e01646-17. [CrossRef]
28. Lewis, B.P.; Shih, I.H.; Jones-Rhoades, M.W.; Bartel, D.P.; Burge, C.B. Prediction of mammalian microRNA targets. *Cell* **2003**, *115*, 787–798. [CrossRef]
29. Lewis, B.P.; Burge, C.B.; Bartel, D.P. Conserved seed pairing, often flanked by adenosines, indicates that thousands of human genes are microRNA targets. *Cell* **2005**, *120*, 15–20. [CrossRef]
30. Pasquinelli, A.E. MicroRNAs and their targets: Recognition, regulation and an emerging reciprocal relationship. *Nat. Rev. Genet.* **2012**, *13*, 271–282.
31. Reed, L.J.; Muench, H. A simple method of estimating fifty per cent endpoints. *Am. J. Epidemiol.* **1938**, *27*, 493–497. [CrossRef]
32. Yuan, L.-Q.; Zhang, T.; Xu, L.; Han, H.; Liu, S.-H. miR-30c-5p inhibits glioma proliferation and invasion via targeting Bcl2. *Transl. Cancer Res.* **2021**, *10*, 337–348. [CrossRef] [PubMed]
33. Du, B.; Dai, X.M.; Li, S.; Qi, G.L.; Cao, G.X.; Zhong, Y.; Yin, P.D.; Yang, X.S. MiR-30c regulates cisplatin-induced apoptosis of renal tubular epithelial cells by targeting Bnip3L and Hspa5. *Cell Death Dis.* **2017**, *8*, e2987. [CrossRef] [PubMed]
34. Senichkin, V.V.; Streletskaia, A.Y.; Zhivotovsky, B.; Kopeina, G.S. Molecular Comprehension of Mcl-1: From Gene Structure to Cancer Therapy. *Trends Cell Biol.* **2019**, *29*, 549–562. [CrossRef] [PubMed]
35. Ma, C.H.; Zhang, Y.X.; Tang, L.H.; Yang, X.J.; Cui, W.M.; Han, C.C.; Ji, W.Y. MicroRNA-1469, a p53-responsive microRNA promotes Genistein induced apoptosis by targeting Mcl1 in human laryngeal cancer cells. *BioMed Pharm.* **2018**, *106*, 665–671. [CrossRef]
36. Zou, Y.; Liu, W.; Zhang, J.; Xiang, D. miR-153 regulates apoptosis and autophagy of cardiomyocytes by targeting Mcl-1. *Mol. Med. Rep.* **2016**, *14*, 1033–1039. [CrossRef]
37. Michels, J.; Johnson, P.W.; Packham, G. Mcl-1. *Int. J. Biochem. Cell Biol.* **2005**, *37*, 267–271. [CrossRef]
38. Anjum, A.D.; Sabri, M.A.; Iqbal, Z. Hydropericarditis syndrome in broiler chickens in Pakistan. *Vet. Rec.* **1989**, *124*, 247–248. [CrossRef]
39. Toro, H.; Prusas, C.; Raue, R.; Cerda, L.; Geisse, C.; Gonzalez, C.; Hess, M. Characterization of fowl adenoviruses from outbreaks of inclusion body hepatitis/hydropericardium syndrome in Chile. *Avian Dis.* **1999**, *43*, 262–270. [CrossRef]
40. Abe, T.; Nakamura, K.; Tojo, H.; Mase, M.; Shibahara, T.; Yamaguchi, S.; Yuasa, N. Histology, immunohistochemistry, and ultrastructure of hydropericardium syndrome in adult broiler breeders and broiler chicks. *Avian Dis.* **1998**, *42*, 606–612. [CrossRef]
41. Kim, J.N.; Byun, S.H.; Kim, M.J.; Kim, J.; Sung, H.W.; Mo, I.P. Outbreaks of hydropericardium syndrome and molecular characterization of Korean fowl adenoviral isolates. *Avian Dis.* **2008**, *52*, 526–530. [CrossRef]
42. Huntzinger, E.; Izaurralde, E. Gene silencing by microRNAs: Contributions of translational repression and mRNA decay. *Nat. Rev. Genet.* **2011**, *12*, 99–110. [CrossRef]
43. Zheng, Q.; Hou, J.; Zhou, Y.; Yang, Y.; Cao, X. Type I IFN–Inducible Downregulation of MicroRNA-27a Feedback Inhibits Antiviral Innate Response by Upregulating Siglec1/TRIM27. *J. Immunol.* **2016**, *196*, 1317–1326. [CrossRef]
44. Kumar, A.; Kumar, A.; Ingle, H.; Kumar, S.; Mishra, R.; Verma, M.K.; Biswas, D.; Kumar, N.S.; Mishra, A.; Raut, A.A.; et al. MicroRNA hsa-miR-324-5p Suppresses H5N1 Virus Replication by Targeting the Viral PB1 and Host CUEDC2. *J. Virol.* **2018**, *92*, e01057-18. [CrossRef]
45. Ma, Y.; Wang, C.; Xue, M.; Fu, F.; Zhang, X. Coronavirus TGEV Evades the Type I Interferon Response through IRE1α-Mediated Manipulation of the miR-30a-5p/SOCS1/3 Axis. *J. Virol.* **2018**, *92*, e00728-18. [CrossRef]
46. Yin, D.; Shao, Y.; Yang, K.; Tu, J.; Song, X.; Qi, K.; Pan, X. Fowl adenovirus serotype 4 uses gga-miR-181a-5p expression to facilitate viral replication via targeting of STING. *Vet. Microbiol.* **2021**, *263*, 109276. [CrossRef]
47. Liu, F.; Wang, H.; Du, L.; Wei, Z.; Zhang, Q.; Feng, W.H. MicroRNA-30c targets the interferon-alpha/beta receptor beta chain to promote type 2 PRRSV infection. *J. Gen. Virol.* **2018**, *99*, 1671–1680. [CrossRef]
48. Ouyang, W.; Qian, J.; Pan, Q.X.; Wang, J.Y.; Xia, X.X.; Wang, X.L.; Zhu, Y.M.; Wang, Y.S. gga-miR-142-5p attenuates IRF7 signaling and promotes replication of IBDV by directly targeting the chMDA5's 3′ untranslated region. *Vet. Microbiol.* **2018**, *221*, 74–80. [CrossRef]
49. Zhang, L.; Chen, X.; Chang, M.; Jiao, B. MiR-30c-5p/ATG5 Axis Regulates the Progression of Parkinson's Disease. *Front. Cell. Neurosci.* **2021**, *15*, 644507. [CrossRef]
50. Jin, Y.; Yao, G.; Wang, Y.; Teng, L.; Wang, Y.; Chen, H.; Gao, R.; Lin, W.; Wang, Z.; Chen, J. MiR-30c-5p mediates inflammatory responses and promotes microglia survival by targeting eIF2alpha during Cryptococcus neoformans infection. *Microb. Pathog.* **2020**, *141*, 103959. [CrossRef]
51. Chen, J.; Zhang, M.; Zhang, S.; Wu, J.; Xue, S. Rno-microRNA-30c-5p promotes myocardial ischemia reperfusion injury in rats through activating NF-κB pathway and targeting SIRT1. *BMC Cardiovasc. Disord.* **2020**, *20*, 240. [CrossRef]
52. Peng, L.; Zhong, X.; Li, J.; Liu, H.; Xiang, M.; He, R.; Zhao, Y. MicroRNA-30c-5p inhibits NLRP3 inflammasome-mediated endothelial cell pyroptosis through FOXO3 down-regulation in atherosclerosis. *Biochem. Biophys. Res. Commun.* **2018**, *503*, 2833–2840.
53. Zhou, L.; Li, J.; Haiyilati, A.; Li, X.; Gao, L.; Cao, H.; Wang, Y.; Zheng, S.J. Gga-miR-29a-3p suppresses avian reovirus-induced apoptosis and viral replication via targeting Caspase-3. *Vet. Microbiol.* **2022**, *264*, 109294. [CrossRef]

54. Yang, K.; Zhou, B. MCL1, a gene expressed in programmed myeloid cell differentiation, has sequence similarity to BCL2. *Proc. Natl. Acad. Sci. USA* **1993**, *90*, 3516–3520.
55. Craig, W.R. MCL1 provides a window on the role of the BCL2 family in cell proliferation, differentiation and tumorigenesis. *Leukemia* **2002**, *16*, 444–454.
56. Michels, J.; O'Neill, J.W.; Dallman, C.L.; Mouzakiti, A.; Habens, F.; Brimmell, M.; Zhang, K.Y.; Craig, R.W.; Marcusson, E.G.; Johnson, P.W. Mcl-1 is required for Akata6 B-lymphoma cell survival and is converted to a cell death molecule by efficient caspase-mediated cleavage. *Oncogene* **2004**, *23*, 4818–4827. [CrossRef]
57. Cui, J.; Placzek, W.J. PTBP1 enhances miR-101-guided AGO2 targeting to MCL1 and promotes miR-101-induced apoptosis. *Cell Death Dis.* **2018**, *9*, 552. [CrossRef]
58. Xie, Q.; Shuai, W.; Yue, Z.; Zhang, Z.; Qin, C.; Yang, X. MiR-519d impedes cisplatin-resistance in breast cancer stem cells by down-regulating the expression of MCL-1. *Oncotarget* **2017**, *8*, 22003–22013. [CrossRef]
59. Jiang, C.; Long, J.; Liu, B.; Xie, X.; Kuang, M. Mcl-1 Is a Novel Target of miR-26b That Is Associated with the Apoptosis Induced by TRAIL in HCC Cells. *BioMed Res. Int.* **2015**, *2015*, 572738. [CrossRef] [PubMed]

viruses

MDPI

Article

Investigation of Avian Influenza H5N6 Virus-like Particles as a Broad-Spectrum Vaccine Candidate against H5Nx Viruses

Yu-Hsuan Yang [1], Ching-Hui Tai [2], Dayna Cheng [3], Ya-Fang Wang [2] and Jen-Ren Wang [1,2,3,*]

1 Department of Medical Laboratory Science and Biotechnology, College of Medicine, National Cheng Kung University, Tainan 701, Taiwan; e25487910@gmail.com
2 National Institute of Infectious Diseases and Vaccinology, National Health Research Institutes, Tainan 701, Taiwan; e480216@hotmail.com (C.-H.T.); anavon31@gmail.com (Y.-F.W.)
3 Institute of Basic Medical Sciences, College of Medicine, National Cheng Kung University, Tainan 701, Taiwan; daynac_1201@hotmail.com
* Correspondence: jrwang@mail.ncku.edu.tw

Abstract: Highly pathogenic avian influenza (HPAI) clade 2.3.4.4 viruses have been reported to be the source of infections in several outbreaks in the past decades. In a previous study, we screened out a broad-spectrum virus strain, H5N6-Sichuan subtype, by using a lentiviral pseudovirus system. In this project, we aimed to investigate the potential of H5N6 virus-like particles (VLPs) serving as a broad-spectrum vaccine candidate against H5Nx viruses. We cloned the full-length M1 gene and H5, N6 genes derived from the H5N6-Sichuan into pFASTBac vector and generated the VLPs using the baculovirus-insect cell system. H5N6 VLPs were purified by sucrose gradient centrifugation, and the presence of H5, N6 and M1 proteins was verified by Western blot and SDS-PAGE. The hemagglutination titer of H5N6 VLPs after purification reached 5120 and the particle structure remained as viewed by electron microscopy. The H5N6 VLPs and 293T mammalian cell-expressed H5+N6 proteins were sent for mice immunization. Antisera against the H5+N6 protein showed 80 to 320 neutralizing antibody titers to various H5Nx pseudoviruses. In contrast, H5N6 VLPs not only elicited higher neutralizing antibody titers, ranging from 640 to 1280, but also induced higher IL-2, IL-4, IL-5, IFN-γ and TNF production, thus indicating that H5N6 VLPs may be a potential vaccine candidate for broad-spectrum H5Nx avian influenza vaccines.

Keywords: avian influenza virus; virus-like particles; vaccine; H5N6

Citation: Yang, Y.-H.; Tai, C.-H.; Cheng, D.; Wang, Y.-F.; Wang, J.-R. Investigation of Avian Influenza H5N6 Virus-like Particles as a Broad-Spectrum Vaccine Candidate against H5Nx Viruses. *Viruses* **2022**, *14*, 925. https://doi.org/10.3390/v14050925

Academic Editor: Chi-Young Wang

Received: 7 March 2022
Accepted: 26 April 2022
Published: 28 April 2022

Publisher's Note: MDPI stays neutral with regard to jurisdictional claims in published maps and institutional affiliations.

Copyright: © 2022 by the authors. Licensee MDPI, Basel, Switzerland. This article is an open access article distributed under the terms and conditions of the Creative Commons Attribution (CC BY) license (https://creativecommons.org/licenses/by/4.0/).

1. Introduction

Avian influenza virus (AIV) belongs to the *Orthomyxoviridae* family, *Alphainfluenzavirus* genus, and contains eight negative-sense, single-stranded RNA segments as its genome [1]. During viral replication, reassortant genes help the AIVs to overcome the species barrier to infect humans, thus causing a major threat to public health [2]. The highly pathogenic avian influenza (HPAI) virus A/goose/Guangdong/1/96 H5N1 (Gs/GD) lineage caused global epidemics not only in migratory birds and domestic poultry but also in the human population. Taiwan had previously encountered an outbreak of HPAI clade 2.3.4.4 derived from Gs/GD lineage in poultry [3]. Clade 2.3.4.4 has evolved since 2008 and had several reassortant neuraminidase (NA) genes with N2, N3, N5, N6, and N8 variants. Phylogenetic analysis based on the sequence of hemagglutinin (HA) and M gene reveals that HPAI H5Nx viruses in October 2014 were related to two HPAI H5N8 viruses identified from wild birds in Japan. Two subsequent outbreaks in late 2014 and 2016 resulted from seven H5Nx reassortant genes that spread through Taiwan and eight novel introduced internal genes mostly derived from the gene pool of avian influenza virus circulating in wild birds in Asia [3]. In several outbreaks in past decades, it has been reported that the sources of infection were novel HPAI clade 2.3.4.4 viruses. These newly emerging, highly pathogenic H5Nx avian influenza viruses, including H5N1, H5N2, H5N3, H5N6, and H5N8, have raised great concern worldwide [4–6]. Taiwan also encountered the largest outbreak of avian influenza viruses in 2015, including H5N2, H5N6 and H5N8 [7]. Genetic analysis revealed that the HA and NA gene of H5N2 circulating in Taiwan before 2015 outbreaks were derived from American H5N2 (old Mexican-like lineage), and the other six gene segments were similar to local H6N1 in Taiwan [8,9]. The HA of the new epidemic strain H5N2 in 2015 belonged to clade 2.3.4.4 HPAI; therefore, the old vaccine might not be able to protect against new emerging H5N2, H5N3, H5N8 and H5N6 viruses. Thus, a new strategy of vaccine platform for rapid application is necessary.

There are three major types of influenza vaccine approved by the Food and Drug Administration in the United States, including inactivated vaccine, live-attenuated vaccine and recombinant vaccine. The manufacture of inactivated vaccine or live-attenuated vaccine is derived from a specific strain of AIV that probably confers only strain-specific immunity. In addition, inactivated vaccines and live-attenuated vaccines seldom elicit cross-protective immunity against different subtypes of AIVs. A mismatch between the available vaccines and the circulating AIV strain may contribute to high mortality and wild transmission in the avian and human population.

Virus-like particles (VLPs) are non-infectious nanoparticles made up of assembled viral proteins without viral genetic materials. In comparison to whole inactivated vaccines, VLP vaccines consisting of different combinations of antigen have the advantage to induce a broader immune response [10]. At present, highly conserved ectodomain of M2 protein (M2e) or the stalk of the HA protein have been used as the antigen for a VLP vaccine candidate for influenza viruses [11,12]. Animals vaccinated with M2e VLP have lower morbidity and viral titers, while the viral clearance has not been observed. VLP with stalk HA elicits cross-protective humoral immunity against various strains of AIVs; however, it is less immunogenic [11,13]. Several studies also posed different strategies. For example, Wu et al. fused a conserved M2e protein at the N-terminus of HA, or at the N-terminus of NA in the H5N1 VLP, to protect against heterosubtypic H1N1 and H5N1 [13]. Kang et al. have expressed two HAs derived from a different clade of AIVs to enhance robust immune response [14]; however, expression of multiple genes in one vector may also contribute to assembly instability due to the tandem repetition of similar sequences [15]. In this study, we supposed that the conformation of VLP presenting the HA and NA may be similar to the native virions that induced antibodies and will be able to target the HA, which can effectively inhibit viral replication. VLPs can be produced in several expression systems, including mammalian cells, plants, insect cells, yeast and bacteria. Since the VLPs generated from a baculovirus-infected insect cell line have advantages of high expression yield, good protein folding with glycosylation modification and ease for scale-up production, we aimed to develop a potential VLP-based vaccine by using the baculovirus-insect cell expression system.

In our previous study, we analyzed genetic evolution of HA from H5Nx viruses circulating in Taiwan. The phylogenetic analysis performed by PAUP showed that all the new H5Nx viruses isolated in Taiwan belonged to clade 2.3.4.4, except the old H5N2 circulating prior to 2015. HA and NA gene sequences derived from different avian influenza viruses, including A/Northern pintail/Washington/40964/2014 H5N2 (clade 2.3.4.4b), A/duck/Taiwan/1702004/2017 H5N6 (clade 2.3.4.4e), A/duck/Hyogo/1/2016 H5N6 (clade 2.3.4.4e), A/Sichuan/26221/2014 H5N6 (clade 2.3.4.4a), A/gyrfalcon/Washington/41088-6/2014 (H5N8) (clade 2.3.4.4b) and A/chicken/Taiwan/x37/2016 H5N8 (clade 2.3.4.4b), were used for lentiviral pseudovirus production, as described previously [16]. We further identified an H5N6-Sichuan subtype that exhibited high neutralizing titers against different subtypes [16]. A lentiviral vector-based vaccine is able to directly express the antigen of interest in target cells in vivo, which can effectively induce antigen-specific CD8+ T cell immune response [17]. Although several studies have demonstrated the protective efficacy in preclinical and clinical trials, it was still not approved to be used in humans to date. The main concern is the stability of viral integration in the host genome, which may activate the cellular protooncogene by integrated long terminal repeats (LTRs) or the transcriptional interference and suppression by LTRs [18]. Therefore, we developed a potential vaccine candidate using a virus-like particles expression system that is non-infectious and had been approved to use in humans. In order to further investigate whether H5N6 Sichuan subtype would be a potential vaccine candidate for humans, we aimed to establish a virus-like particle expression system and further evaluate the immune response elicited by VLP and 293T mammalian cell-expressed protein in mice models.

2. Materials and Methods

2.1. Cell Culture

Insect cell lines, Spodoptera frugiperda ovarian cells (Sf21) and Trichoplusia ni ovarian cells (High Five^{TM}) were cultured at 28 °C in an incubator, without CO_2 and water supplement. Sf21 cells were maintained in Grace's Insect medium (Gibco) supplemented with 10% fetal bovine serum (FBS, Hyclone, Thermo Fisher Scientific, Carlsbad, CA, USA) for baculovirus production. For H5N6-Sichuan VLP production, High Five™ cells were cultured in serum-free HyClone™ SFM4 Insect medium (Hyclone) with 1% penicillin streptomycin mixtures (P/S).

293T cells and MDCK cells were cultured at 37 °C in an incubator with 5% CO_2. 293T cells were maintained in DMEM supplemented with 10% FBS, 2% P/S, and 10 μM sodium pyruvate for the production of H5Nx pseudoviruses and 293T expressed H5+N6 proteins. MDCK cells were maintained in Eagle's MEM with 1 μg/mL trypsin, 1mM sodium pyruvate, 100 U/mL penicillin and 0.1 mg/mL streptomycin for titration of pseudoviruses and neutralization assay of VLP or mammalian cell-expressed proteins immunized mice serum against H5Nx pseudoviruses.

2.2. Generation of Virus-like Particle Expression Plasmid

PCR products of NA were cloned into pBac-YCCRG3 plasmid containing an M1 sequence from an H7N9 subtype (pFastBac vector, gift from Dr. Yung-Chih Hu, NHRI) using KpnI and NheI restriction enzyme sites, named as pBac-YCCH7SCN6. By using BssHII and RsrII restriction enzyme sites, the amplified HA genes were cloned into the pBac-YCCH7SCN6, named as pBac3-SC. The inserted HA, NA sequences from H5N6 Sichaun subtype and M1 sequence from H7N9 subtype pBac3-SC plasmid were confirmed by Sanger sequencing. The pBac3-SC plasmid was transformed into a DH 10BacTM competent cell (Thermo Fisher Scientific) for transposition into bacmid. The colonies that contain the desired recombinant bacmid were identified by blue-white screening with 50 μg/mL kanamycin, 10 μg/mL tetracycline, 7 μg/mL gentamycin, 100 μg/mL Bluo-gal and 40 μg/mL IPTG. Sf21 cells were seeded in 6-well plates (8×10^5 cells/mL) and transfected with 2.5 μg/well of pBac3-SC plasmid by TransIT-Insect transfection reagent (Mirus Bio, Madison, WI, USA), and the supernatants were collected at 72 h post-transfection. The P1 viruses were immediately amplified in confluent 6-well plates of Sf21 cells and titrated by plaque assay for further experiments. The viruses were stored at −80 °C and protected from light.

2.3. Production of H5N6 VLPs

The cultures were performed in 250 mL shake flasks. Each flask was inoculated with 2×10^5 cells/mL with a working volume of 100 mL. After three days, when the density of High Five™ cells reached 1.6×10^6 cells/mL, the cells were infected with 1×10^8 to 1×10^9 pfu/mL of recombinant baculovirus. The H5N6 VLPs were harvested at 3 days post-infection by centrifugation at 500× *g* (Kubota TA3780) at 4 °C for 10 min, and the yield of H5N6 VLPs was further determined by hemagglutination assay.

2.4. Sucrose Density Gradient for Vlp Purification

For VLPs concentration and purification, VLPs were first UV inactivated for 30 min. It was filtered with 0.45 μm filters to remove cell debris and precipitated by ultracentrifugation at 360,000× *g* (Beckman L90-K ultracentrifuge, 70Ti rotor) at 4 °C for 4 h with a

20% sucrose cushion and a drop of glycerol at the bottom of the tube. The pellets were resuspended with PBS and loaded on the top of a 20–60% sucrose gradient. After ultracentrifugation at 285,000× *g* for 2.5 h, the visible band at 45% sucrose was collected (Beckman L90-K ultracentrifuge, 40Ti rotor).

2.5. SDS-PAGE and Western Blot Analysis for Protein Composition of VLP

Purified H5N6 VLP samples were quantified by Qubit protein assay. Approximately 2 µg of purified H5N6 VLPs were separated on 12% SDS-PAGE. To confirm the purification efficiency of the sucrose density gradient, the SDS-PAGE gel was stained using silver staining. To detect the presence of target HA, the separated proteins on the SDS-PADE were transferred onto the PVDF membrane, blocked with 5% non-fat milk, and incubated with primary antibody (1:5000, avian flu HA antibody, Mab, Sino Biological, Beijing, China) at 4 °C overnight. The blots were then incubated with secondary antibody (1:5000, goat anti-mouse IgG, HRP-labeled, KPL) and developed using enhanced chemiluminescence (ECL, Bio-Rad, Hercules, CA, USA) substrate. All the images were captured by Amersham Imager 600.

2.6. Transmission Electron Microscopy for Analysis of VLP Morphology

Purified VLPs were absorbed onto a plasma-discharged copper grid. After washing with PBS, 2% phosphotungstic acid was applied for negative staining. The particles were observed under a transmission electron microscope (JEOL JEM-1400), performed by Core Facility Center, National Cheng Kung University, Taiwan.

2.7. Hemagglutination Assay for HA Titer of VLP

Turkey RBCs were washed with ice-cold PBS several times until the PBS remained clear after centrifugation at 500× *g* (Kubota TA3780) for 5 min. A 0.75% of turkey RBC suspension with PBS was then prepared. To measure the hemagglutination titers of VLP, VLPs were first diluted 10-fold, and a 2-fold serial dilution with PBS was performed on a 96-well plate at a total volume of 50 µL/well. Afterwards, 50 µL of 0.75 % turkey RBC suspension was added to each well and incubated at room temperature for 30 min to 1 h [19]. In the presence of VLPs, the HA on the surface of VLPs will bind to the sialic acid on the RBC, showing hemagglutination. The highest dilution titer that had shown hemagglutination was regarded as the hemagglutination titer.

2.8. Expression of H5-HA and N6-NA Protein in 293T Mammalian Cell System

293T cells were seeded in 15 cm dishes (0.75–0.9 × 106 /mL, 20 mL/dish). After 16 h of incubation, plasmids with pTT5 vector (pTT5-SCH5, pTT5-SCN6, Leadgene Biomedical, Inc., Tainan, Taiwan) were transfected with 15 µg of H5 or N6 plasmid DNA per dish, respectively, by Hyfect transfection reagent (Leadgene). Dishes were refreshed with cell medium (FreeStyle 293 expression medium) 24 h later. At 72 h post-transfection, supernatant was collected and centrifuged at 500× *g* to remove the cell debris. The supernatant was further precipitated via ammonium sulfate. The desired HA and NA proteins were visualized by Western blot with anti-His tag mAb (1:5000, Leadgene). The blots were then incubated with secondary antibody (1:5000, goat anti-mouse IgG, HRP-labeled, KPL) and developed using enhanced chemiluminescence (ECL, Bio-Rad, Hercules, CA, USA) substrate. All the images were captured by Amersham Imager 600.

2.9. Mouse Immunization

For H5N6 VLPs and H5+N6 protein antisera, 6- to 8-week-old BALB/c mice (4 mice per group) were initially immunized with the purified H5N6 VLPs and 293T cell expressed H5+N6 proteins with adjuvant (Freund's complete adjuvant), which was performed by Leadgene Biomedical, Inc, Taiwan (IACUC-11006003). Ten mg of VLP in PBS or recombinant proteins emulsified in Complete Freund's Adjuvant (CFA, Sigma) for priming and Incomplete Freund's Adjuvant (IFA, Sigma) for the boost in a total of 100 µL were immunized three times at one-week intervals. Serum samples were collected by submandibular blood collection at three time-points: pre-immunization, day 14 and day 35 post-immunization (antiserum of H5+N6 proteins were not collected at day 14, since these two immunizations were performed at separate times). At day 35, mice were sacrificed and the cardiac chambers punctured for the serum sample collection. The end-point serum samples were used for neutralization assay, and the serums collected at different time-points were used to measure the induction of cytokine levels. Each serum was collected from 4 BALB/c mice and pooled. All the serum samples were aliquoted and stored at −80 °C until further use.

2.10. Enzyme-Linked Immunosorbent Assay (ELISA) for Specific IgG Titer

The 96-well ELISA plates were coated with 50 μL of PBS containing 1 μg/mL of H5N6 VLP or 293T expressed H5+N6 proteins, respectively. The plates were incubated at room temperature for an hour and blocked with 3% skim milk in 0.05% PBS-T. Serum samples were 100-fold and then two-fold serially diluted with blocking buffer. After being washed several times with 0.05% PBS-T, the coated plates were then incubated with 50 μL of serially diluted serum, followed by incubating with HRP-conjugated anti-mouse IgG antibodies (1:5000, Leadgene) at 37 °C for 30 min. Lastly, 50 μL of tetramethyl benzidine dihydrochloride (TMB) substrate was added in each well and stopped with 50 μL/well of 1 N H_2SO_4 two minutes later. The absorbance at 450 nm was measured using an ELISA plate reader (Meter-tech M965+).

2.11. Neutralization Assay for Neutralizing Antibody Titer

To determine the neutralizing antibody titer of H5N6 VLPs and 293T expressed H5+N6 protein immunized mouse serum against H5Nx pseudoviruses, neutralization assay was performed as previously reported [16]. MDCK cells were seeded in 24-well plates with 7.5×10^4 cells/well and incubated at 37 °C for 16 h. Mouse antisera were complement inactivated at 56 °C for 30 min before neutralization assay. H5Nx pseudoviruses (5×10^6–1×10^7 TU/mL) were co-incubated with 2-fold serially diluted mouse serum in a 37 °C incubator for 30 min. MDCK cells were then infected with the mixture and adsorbed for 1.5 h. The serum-free DMEM medium was added after adsorption. The medium was changed to MDCK medium (Eagle's MEM with 1 μg/mL trypsin, 1mM sodium pyruvate, 100 U/mL penicillin and 0.1 mg/mL streptomycin) after 24 h. Infected cells with fluorescence were fixed with 2% paraformaldehyde at 4 days post-infection, and the percentage of fluorescence of infected cells was measured by flow cytometer. The neutralization titers (NT50) were defined as 50% reduction of transduction unit (TU) in both duplications compared with the transduction unit of virus control.

2.12. Multiplex Analysis of Mouse Cytokine Productions

Th1/ Th2-related multiple cytokines were measured by BD cytometric bead array. Five beads with distinct fluorescence intensities had been coated with capture antibodies specific for IL-2, IL-4, IL-5, TNF, and IFN-γ. Antisera against H5N6 VLPs and 293T cell-expressed H5+N6 proteins were collected at different time-points and were diluted three-fold with assay diluent. Standard beads were prepared following the instruction manual. The captured beads were mixed with the serum samples and the PE detection reagents and incubated at room temperature for two hours. After washing with 1 mL of wash buffer for each tube, the tubes were centrifuged at 200× *g* for 5 min. Lastly, the pellet was resuspended with 300 μL of wash buffer, and the fluorescence was analyzed by FACS Calibur flow cytometer.

3. Results

3.1. Establishment of the H5N6 Virus-like Particles Expression System Using Baculovirus-Insect Cell System

3.1.1. Construction of the VLP-Expressing Plasmid

In our previous study, we established a lentiviral pseudovirus system, which is able to rapidly screen out the broad-spectrum virus strain during outbreaks and examine vaccine immunogenicity. Through the neutralization assay of the generated avian influenza H5 subtypes pseudoviruses antisera, we found that H5N6-Sichuan subtype antisera exhibited better cross-protectivity among H5N2, H5N6 and H5N8 subtypes [16]. In order to further investigate whether H5N6-Sichuan strain could be a potential vaccine candidate, we first generated the virus-like particles derived from H5N6-Sichuan strain by using a Bac-to-Bac baculovirus expression system. The full-length HA and NA derived from H5N6-Sichuan were PCR amplified and cloned into pFASTBac vector. The plasmid was then transformed into DH 10BacTM competent cells, which contain a bacmid DNA. Using mini-attTn7 site-specific transposition, the inserted foreign genes will disrupt the LacZα gene, and the recombinant bacmid containing the desired insert can be identified by blue-white selection (Figure 1A).

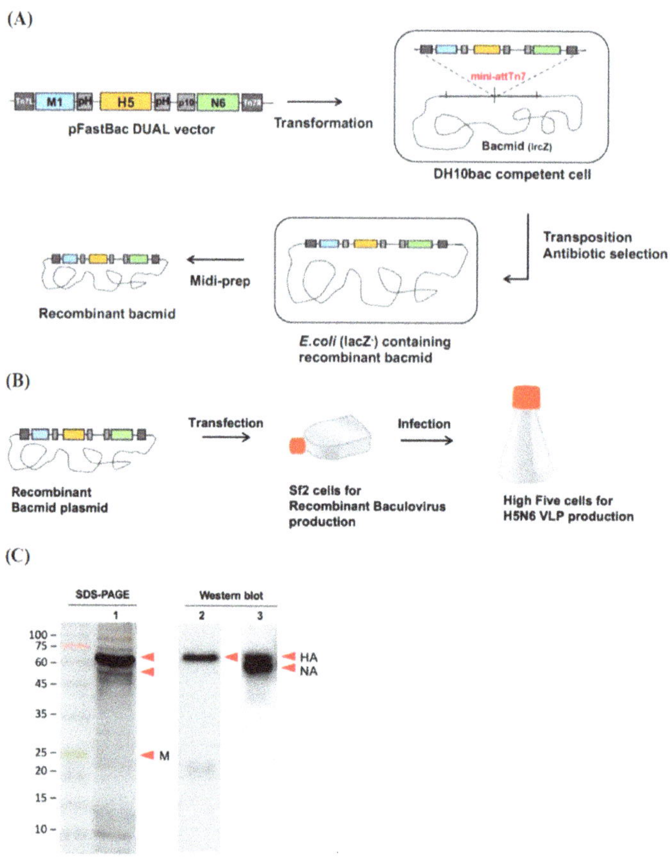

Figure 1. Production of H5N6 VLPs in the baculovirus expression vector system. (**A**) A schematic diagram of H5N6 recombinant bacmid construction. The recombinant bacmid carrying the hemagglutinin (HA), neuraminidase (NA) and matrix protein 1 (M1) genes derived from the H5N6 subtype

was generated by mini attTn7 transposition in DH10bac competent cells. (**B**) Recombinant baculovirus collected from Sf21 cells were used to infect High Five cells for VLP production. (**C**) The expression of the HA, NA and M1 proteins on the VLPs was visualized by SDS-PAGE with silver staining (lane 1) and confirmed by Western blot using avian flu HA antibody (Sino Biological, 11048-MM06) (lane 2) and Lenti-H5N6-SC mouse antisera (lane 3) (indicated by red arrow).

3.1.2. Generation of the H5N6 Virus-like Particles from Sf21 and High Five Cells

After we confirmed the inserted gene in the recombinant bacmid by sequencing, the recombinant bacmid were amplified in LB broth and extracted with Qiagen plasmid midi kit. Previously, Lai et al. had demonstrated that Sf21 cells are suitable for a large amount of baculovirus production, while the VLPs are more easily generated by High Five™ cells [20]. Therefore, we transfected the recombinant bacmid with TransIT transfection reagent into Sf21 cells cultured in an adherent flask to generate the recombinant baculovirus. After three days of culture, the recombinant baculoviruses budded into the supernatant and were immediately propagated by infecting High Five™ cells. High Five™ cells were infected with MOI 1 of recombinant baculovirus from the above culture of Sf21 cells (Figure 1B). SDS-PAGE was first performed to analyze the production of the H5N6 VLPs. The results showed the presence of HA (64.08 kDa) and NA (52.4 kDa) proteins, followed by confirmation by Western blot with specific antibodies of avian flu H5N1 HA antibody and mouse antisera against lenti-H5N6 Sichuan pseudoviruses (Figure 1C).

3.1.3. Precipitation and Purification of H5N6 VLP by Sucrose Density Gradient

In order to investigate whether the H5N6 VLPs induced immunogenicity, purification and concentration of H5N6 VLPs is needed for mice immunization. Previous studies have shown that norovirus capsid VLPs purified by sucrose density gradient obtained high yield and were more stable compared to other methods [21]. At day three post-infection, 250 mL of culture supernatant was clarified and precipitated by low-speed centrifugation with a 20% sucrose cushion and a drop of glycerol at the bottom of the tube. The pellet was resuspended with 1.5 mL of PBS and loaded onto the 20–60% sucrose density gradient. After ultracentrifuge, a visible band at 45% sucrose was observed, which indicated the purified H5N6 VLPs. Fractions from 15%, 25%, 35% and 45% sucrose were collected from the bottom for further analysis of the purification efficiency (Figure 2A,B).

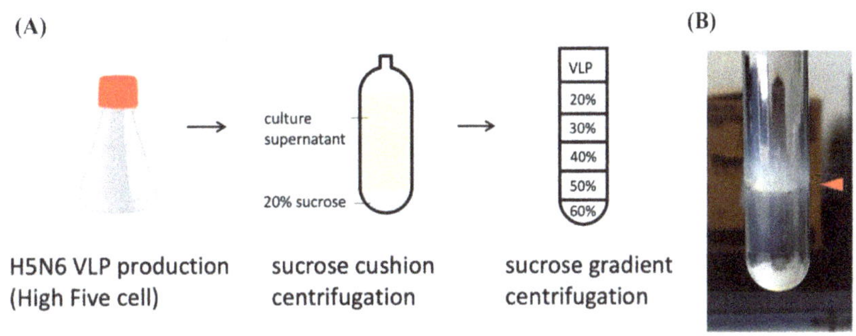

Figure 2. Purification of H5N6 VLPs by sucrose gradient. (**A**) Purification steps for H5N6 VLPs purification. The H5N6 VLPs were precipitated by a 20% sucrose cushion first and further purified by a 20–60% sucrose density gradient. (**B**) A visible band at 45% of sucrose fraction indicated the purified H5N6 VLPs.

3.2. Characterization of H5N6 VLPs Properties

3.2.1. The Protein Composition of the H5N6 VLPs

In order to verify whether the major immune epitope, HA proteins, still remained in the H5N6 VLPs after purification, we detected the proteins by Western blot using avian flu anti-HA antibody. SDS-PAGE with silver stain was also performed to examine the purity of

H5N6 VLPs. The results showed that the fraction of 45% sucrose not only had the majority of the HA but also the NA and M1 proteins; furthermore, other proteins derived from the High Five cells were mostly removed after purification (Figure 3A).

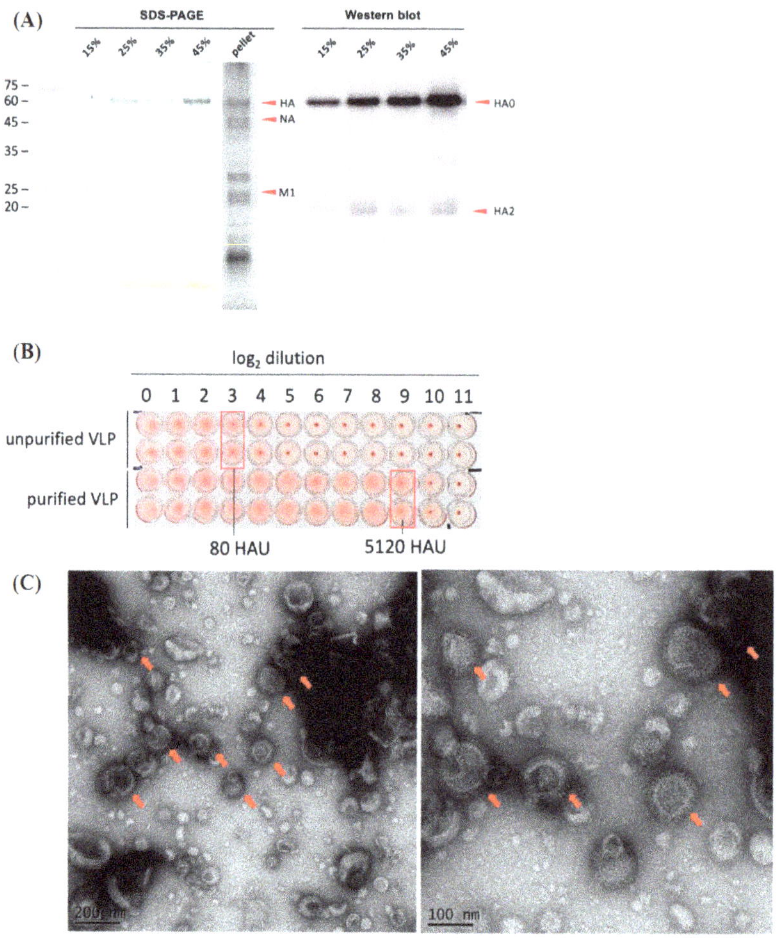

Figure 3. Protein expression, hemagglutination and structure of H5N6 VLP after sucrose gradient purification. (**A**) The expression of the HA, NA and M1 proteins on the VLPs at different percentages of sucrose fraction was visualized by SDS-PAGE with silver staining and confirmed by Western blot using avian flu HA antibody, Mab (Sino Biological, 11048-MM06). The pellet of H5N6 VLPs was precipitated by a 20% sucrose cushion. (**B**) Comparison of hemagglutination activity between unpurified and purified VLPs (at 45% sucrose fraction) by hemagglutination assay. The serum sample started from a 10-fold dilution. (**C**) The H5N6 VLPs after sucrose gradient purification were imaged using transmission electron microscopy by negative staining. Scale bars, Left: 200 nm, Right: 100 nm.

3.2.2. The Hemagglutination Property of the H5N6 VLPs Remained

Hemagglutinin glycoprotein was reported as the receptor-binding protein of the influenza virus. In order to examine whether sucrose gradient purification affects the hemagglutination property, hemagglutination assay was performed. Approximately 250 mL of VLPs in the culture supernatant were precipitated and resuspended with 4 mL of PBS. Then, 4 mL of concentrated VLP was purified and the final volume of purified VLP was about 1 mL. The result indicates that the hemagglutination property of H5N6 VLPs purified from sucrose gradient ultracentrifugation can reach 5120 HAU/ 50 µL. In contrast, the HA titer of H5N6 VLPs before concentration and purification was only 80 HAU/ 50 µL (Figure 3B).

3.2.3. The Morphology of the H5N6 VLPs Remained Structurally Intact after Purification

To examine the formation of the H5N6 VLP, we directly visualized the particles using negative stain by transmission electron microscopy. The results showed that H5N6 VLPs were successfully prepared with morphology that resembled the native virions without genetic materials. The particles were homogenous and approximately 110 nm in size (Figure 3C). These data suggested that VLPs derived from H5N6-Sichuan strain can be successfully generated through a baculovirus expression system. After being concentrated and purified by a sucrose cushion and density gradient, H5N6 VLPs maintained a functional hemagglutination property and were structurally intact.

3.3. Generation of 293T Mammalian Cell Expressed H5+N6 Protein

To investigate whether the virus-like structure of VLPs is helpful to induce superior immunogenicity, we also used 293T mammalian cells to express H5 HA and N6 NA genes derived from the H5N6 Sichuan strain for comparison. The clones of HA and NA constructed in pTT5 expression vector were performed by Leadgene Biomedical, Inc. The plasmids were separately transfected into a different dish of 293T cells, the supernatant containing the desired H5 HA and N6 NA proteins was collected at day 4 post-transfection and the proteins were precipitated by ammonium sulfate (Figure 4A). The H5 and N6 proteins were observed by Western blotting with anti-His tag antibody and pooled for mice immunization (Figure 4B).

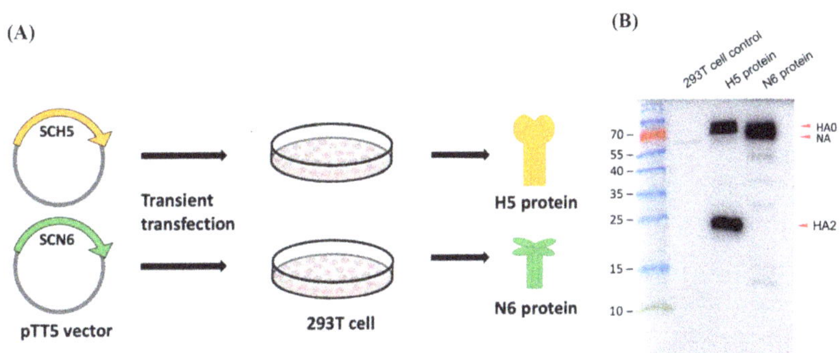

Figure 4. Expression of HA and NA proteins derived from H5N6 Sichuan strain in 293T mammalian cells. (**A**) Schematic diagram of mammalian cell expressed H5-HA and N6-NA proteins. The HA and NA genes were cloned into a pTT5 protein expression vector and transfected into 293T cells. After 4 days of culture, the H5 and N6 proteins were collected from the culture supernatant. (**B**) The presence of H5 (62 kDa) and N6 (56.4 kDa) proteins was verified by anti-His tag antibody in Western blot.

3.4. Investigation of the Immune Response Elicited by the H5N6 VLPs and 293T Cell-Expressed H5+N6 Protein

3.4.1. Specific Total IgG Antibodies Were Produced in Immunized Mice

Groups of BALB/c mice were immunized by intraperitoneal route (i.p) with purified H5N6 VLPs or 293T cell expressed H5+N6 proteins. Serum samples were collected at pre-immunization and day 35 post-immunization (Figure 5A). The level of total IgG was measured by ELISA. The results showed that pre-immunized antisera had no specific IgG antibody against H5N6 VLPs and 293T cell-expressed H5+N6 proteins (Figure 5B,C). At day 35 post-immunization, IgG antibody end-point titers against their immunized antigens were both 100×2^{10} in H5N6 VLPs and 293T cell-expressed H5+N6 proteins antisera, suggesting that mice immunized with H5N6 VLPs and 293T cell-expressed H5+N6 proteins successfully generated the IgG antibodies against the antigens (Figure 5B,C).

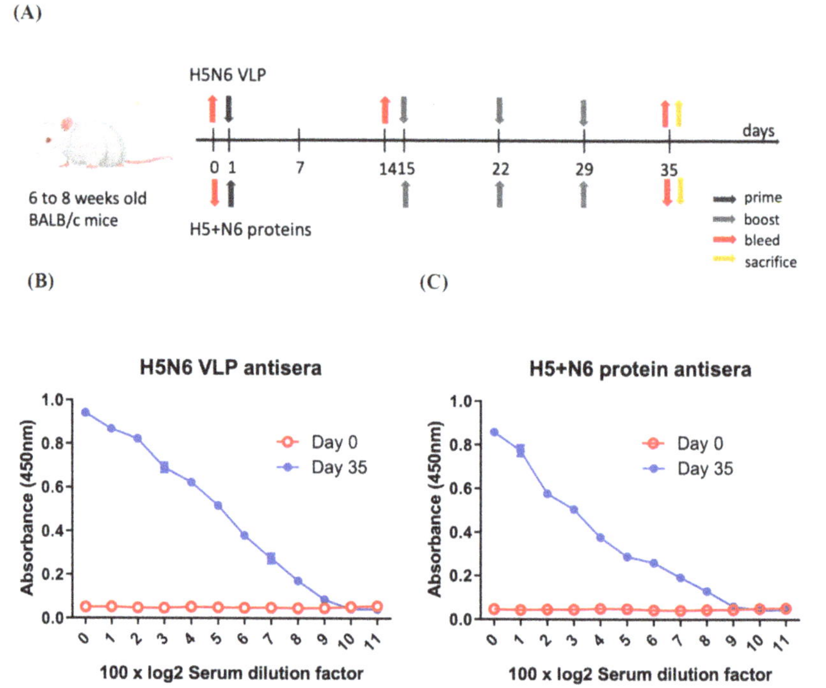

Figure 5. Specific IgG antibody responses in mice immunized with H5N6 VLP or 293T cell-expressed H5+N6 protein. (**A**) Experimental design for mice immunization (n = 4 per group). The mice were immunized and boosted at one-week intervals after 14 days post-immunization (d.p.i). Serum samples were collected by submandibular blood collection at pre-immunization (Day 0), day 14 and day 35 post-immunization (Day 35). The total specific IgG level at Day 0 and Day 35 in H5N6 VLP antiserum (**B**) and H5+N6 protein antiserum (**C**) against their immunized antigens were analyzed by ELISA.

3.4.2. Neutralization Antibodies Were Higher in VLP Immunized Mice

To compare the immunogenicity induced by H5N6 VLPs and 293T cell-expressed H5+N6 protein in mice, after three rounds of immunization with each immunogen formulated with adjuvant, mice were sacrificed and serum samples were collected by submandibular blood collection. Neutralization titers of antisera were distinguished by over 50% reduction of virus transduction in duplicate. The results demonstrated that H5+N6 protein antisera only elicited partial cross-reactive antibodies, ranging from 80 to 320 against different subtypes of avian influenza pseudoviruses, including H5N2, H5N6 and H5N8. In contrast, H5N6 VLPs antisera exhibited higher neutralizing antibody titers, between 640 to 1280 (Table 1). These findings indicated that H5N6 VLP can induce higher neutralizing antibodies than 293T cell-expressed H5+N6 proteins, suggesting that the structure of VLP is helpful in displaying the immune epitope on the surface, which might result in better cross-reactive neutralizing antibodies.

Table 1. Neutralizing antibody titers of H5N6 VLP-induced antisera and 293T cell expressed H5+N6 protein-induced antisera against various H5Nx pseudoviruses.

Antisera	Pseudoviruses Tested (Neutralizing Antibody Titer)					
	Lenti-H5N2 Washington	Lenti-H5N6 TW17	Lenti-H5N6 Hyogo	Lenti-H5N6 Sichuan	Lenti-H5N8 Washington	Lenti-H5N8 TWX37
H5N6 VLP	640	640	640	1280	1280	640
H5+N6 proteins	80	80	80	320	320	80

3.4.3. Cytokine Production

Th1 and Th2 cells play an important role in adaptive immunity. After immunization, Th1 cells secrete IFN-γ, TNF and IL-2 to mediate cytotoxic T cell and opsonizing antibodies. Th2 cells stimulate the humoral immune response by secreted IL-4 and IL-5, which can regulate B cell proliferation and induce antibody production. We measured the levels of multiple cytokines to investigate the 293T cell-expressed H5+N6 protein or H5N6 VLPs induced immune response. H5+N6 protein and H5N6 VLPs antisera were collected at preimmunization, day 14 (H5N6 VLP only) and day 35 post-immunization. The results demonstrated that H5N6 VLPs can induce a higher production of Th1- and Th2-related cytokines (Figure 6). These data suggest that H5N6 VLPs have the advantage of particle formation that can display HA protein on the surface, which stimulated higher levels of cytokine production to eventually help to generate more neutralizing antibodies.

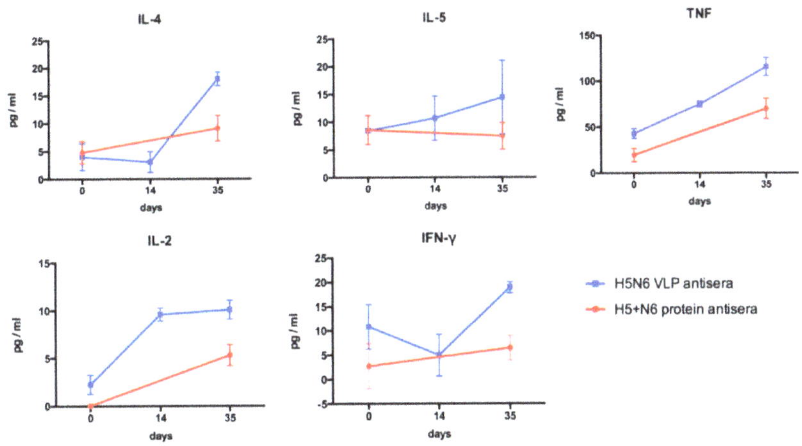

Figure 6. Cytokine profiles of mice immunized with H5N6 VLPs or 293T cell-expressed H5+N6 proteins. Antisera of vaccinated mice were collected at different timepoints: pre-immunization, day 14 and day 35 post-immunization The cytokines (IL-2, IL-4, IL-5, IFN-γ, and TNF) in antisera were measured by cytometric bead array (CBA) multiplex assay. Serum samples collected at each time-point were measured in duplicates.

4. Discussion

Highly pathogenic avian influenza epidemics pose serious threats to socioeconomic and animal health. In the past decades, multiple outbreaks of H5 HPAI viruses have occurred, especially in Asian countries, including H5N1 in China and Vietnam, H5N2 in North America and Taiwan, H5N6 in China, Laos and Vietnam, and H5N8 in China, Europe, Japan, North America, South Korea and Taiwan [22]. Recently, reinvasion of H5N8 subtype viruses in Japan resulted in the cull of 1.9 million birds in October 2020 [23]. In order to develop an effective vaccine with robust protectivity for epidemic prevention, we established a VLP expression system and generated VLPs derived from a broad-spectrum virus strain, H5N6-Sichuan subtype, as a vaccine candidate. We compared the immune response elicited by H5N6 VLPs and 293T mammalian cell-expressed H5+N6 proteins. The results showed that Th1- and Th2-related cytokines were induced by both antigens, while the level of cytokines was higher in H5N6 VLPs antisera, which resulted in a higher titer of cross-reactive neutralizing antibodies. Combined with the lentiviral–pseudovirus system, we can efficiently produce a large amount of VLPs carrying HA from a broad-spectrum virus strain during outbreaks, which may be helpful to the poultry industry and public health in the human population.

Evaluation of humoral and cellular immunity is an essential aspect to examine the effectiveness of vaccination. CD4+ T cells are activated via the secretion of Th2-related cytokines (IL-4, IL-5) to stimulate B cells for the production of antibodies to prevent viral infection. CD8+ T cells, stimulated by Th1-related cytokines, are responsible for killing the infected cells. ELISA was used to detect all the binding antibodies in the antisera. In our results, H5N6 VLPs induced a higher production of Th1- and Th2-related cytokines that lead to higher titers of neutralizing antibody in equal amounts of specific IgG level, measured by ELISA, in H5N6 VLPs and 293T cell-expressed H5+N6 protein antisera. We suggested that H5N6 VLP, with the advantage of displaying the immune epitope on the particle, exhibited a higher titer of neutralizing antibody. A large portion of specific IgG antibodies detected in 293T cell-expressed H5+N6 protein antisera may be the non-neutralizing antibody. Although non-neutralizing antibody is not able to prevent viral infection, it may still contribute to antiviral protection through other antibody-mediated effects, such as antibody-dependent cellular cytotoxicity (ADCC), antibody-dependent cellular phagocytosis (ADCP) or antibody-mediated complement-dependent cytotoxicity (CDC) [24].

In our findings, antisera against H5N6 VLP exhibited a neutralizing antibody titer of 1280 against both Lenti-H5N6-Sichuan and Lenti-H5N8-Washington subtypes, which can be correlated to their close proximity on the antigenic cartography of avian influenza virus H5Nx pseudoviruses [16]. In addition to generating broad-spectrum VLPs with single HA as vaccine candidate, scientists had also proposed some other strategies for VLP-based vaccine development. Kang et al. exploited multi-clade VLPs carrying two HAs derived from two clades, clade 2.3.2.1c and clade 2.3.4.4c, within a single vector, and found that multi-clade VLPs not only show high productivity but also induced effective antibodies [14]. In contrast to variable surface antigens of HA and NA, the ion channel protein, Matrix protein 2 ectodomain (M2e), represents an alternative antigen target for AVI vaccine, since M2e protein has a relatively conserved sequence and plays a critical role in virus assembly and budding during replication. Mice intranasally immunized with a pandemic live attenuated vaccine supplemented with M2e-VLPs were prevented from weight loss and had low viral titers when challenged with heterosubtypic virus, indicating that the combination of split vaccine with M2e-VLPs provides cross protection by eliciting IgG, IgA antibodies as well as M2e-induced T cell response [25]. Apart from targeting the conserved epitope, Carter et al. have reported a new method, computationally optimized broadly reactive antigens (COBRA), in order to overcome the deficiency of weak protectivity induced by M2e or NP antigens when challenged with high-dose viruses. They provided a consensus HA sequence using multiple rounds of consensus building to generate a candidate antigen and expressed the COBRA HA proteins on VLP for broadly reactive, functional antibody responses [26].

Although vaccination is helpful to control the spread of AIV, even highly immunogenic AIV strains were not able to elicit sufficient immune response without appropriate adjuvants. Previous studies had evaluated the immune response between different adjuvants for AIV vaccine in poultry [27,28]. In this study, we vaccinated the mice model with addition of complete Freund's adjuvant (CFA) and incomplete Freund's adjuvant (IFA) to gain a strong and long-lasting immune response. Both of them are a water-in-oil emulsions mixture prepared at a ratio of 85% mineral oil and 15% emulsifier, as described by Stone et al. [29]. The difference between IFA and CFA is that CFA contains additional heat-killed mycobacteria (Mycobacterium tuberculosis), which dramatically enhanced the immune responses by trafficking the macrophage and other immune cells to the inoculation site. In general, CFA was used with initial injection for the potentiation of T helper cells that leads to the production of immunoglobulins and effector T cells. IFA is for subsequent immunization following a strong Th2 immune response through the formation of a depot at the injection site and the stimulation of antibody producing plasma cells. The cytokine responses to H5N6-VLPs and H5N6 protein were increased to a certain degree when induced by CFA and IFA adjuvants. Since the immune response to CFA, IFA and commercial oil-based adjuvants such as Montanide ISA 70VG, 71VG, 760VG, GEL01, etc., were in a large variation [28], the use of CFA and IFA might not be able to exactly mimic the clinical situations. We used CFA and IFA for the purpose of evaluating immune response between H5N6-VLPs and H5N6 protein, while ISA 71 VG and Montanide ISA VG70 oil adjuvant might be a better choice to compare the results with others [14,30].

Taiwan has not approved the use of AIV vaccine to date. The main concerns are that a higher virus mutation rate may result in gene reassortment triggered by inactivated and live-attenuated vaccine, and there may be asymptomatic transmission between vaccinated animals. The distinct host range of AIVs also makes it difficult to raise the population immunity. Although there is still a wide spectrum of opinions about the issue of whether vaccination can actually control the spread of virus, Walker et al. have used the Bayesian Markov Chain Monte Carlo (MCMC) data augmentation technique to evaluate the impact of the vaccination in northern Vietnam. The results demonstrated a significant decrease in the transmissibility of infection following vaccination and the wave of outbreaks were less intense [31]. They suggested that vaccination is helpful to reduce the transmission of infection. In conclusion, based on our results, we suggest that H5N6-VLPs might be a potential vaccine candidate, since they induced potent humoral and cellular immune response to produce high levels of cross-reactive neutralizing antibodies and cytokines. The strategy of generating a broad-spectrum VLP with single HA as a vaccine candidate may be feasible during urgent AIV outbreaks.

Author Contributions: Y.-H.Y., C.-H.T. and J.-R.W. designed the experiments. Y.-H.Y. and C.-H.T. performed the experiments. Y.-H.Y., C.-H.T. and J.-R.W. analyzed the data. Y.-H.Y., D.C. and J.-R.W. wrote the manuscript. Y.-F.W. and J.-R.W. supervised the project and were in charge of conceptualization, funding acquisition and resources. All authors have read and agreed to the published version of the manuscript.

Funding: This study was financially supported by grants from the National Health Research Institutes (09AI-IVGP09-038, IV-109-PP-13, and IV-110-PP-115).

Institutional Review Board Statement: Not applicable.

Informed Consent Statement: Not applicable.

Data Availability Statement: The data that support the findings of this study are available within the article.

Acknowledgments: The authors would like to thank Yung-Chih Hu and Yu-Chieh Cheng of NHRI for providing the plasmids and their advice for VLP generation.

Conflicts of Interest: The authors declare no conflict of interest.

References

1. Spackman, E. A Brief Introduction to Avian Influenza Virus. In *Animal Influenza Virus: Methods and Protocols*; Spackman, E., Ed.; Springer: New York, NY, USA, 2020; pp. 83–92. [CrossRef]
2. Suzuki, Y.; Ito, T.; Suzuki, T.; Holland, R.E.; Chambers, T.M.; Kiso, M.; Ishida, H.; Kawaoka, Y. Sialic Acid Species as a Determinant of the Host Range of Influenza A Viruses. *J. Virol.* **2000**, *74*, 11825–11831. [CrossRef] [PubMed]
3. Huang, C.W.; Chen, L.H.; Lee, D.H.; Liu, Y.P.; Li, W.C.; Lee, M.S.; Chen, Y.P.; Lee, F.; Chiou, C.J.; Lin, Y.J. Evolutionary history of H5 highly pathogenic avian influenza viruses (clade 2.3.4.4c) circulating in Taiwan during 2015–2018. *Infect. Genet. Evol.* **2021**, *92*, 104885. [CrossRef] [PubMed]

4. Hill, S.C.; Lee, Y.J.; Song, B.M.; Kang, H.M.; Lee, E.K.; Hanna, A.; Gilbert, M.; Brown, I.H.; Pybus, O.G. Wild waterfowl migration and domestic duck density shape the epidemiology of highly pathogenic H_5N_8 influenza in the Republic of Korea. *Infect. Genet. Evol.* **2015**, *34*, 267–277. [CrossRef] [PubMed]
5. Lee, D.H.; Torchetti, M.K.; Winker, K.; Ip, H.S.; Song, C.S.; Swayne, D.E. Intercontinental Spread of Asian-Origin H5N8 to North America through Beringia by Migratory Birds. *J. Virol.* **2015**, *89*, 6521–6524. [CrossRef]
6. Yang, L.; Zhu, W.; Li, X.; Bo, H.; Zhang, Y.; Zou, S.; Gao, R.; Dong, J.; Zhao, X.; Chen, W.; et al. Genesis and Dissemination of Highly Pathogenic H5N6 Avian Influenza Viruses. *J. Virol.* **2017**, *91*, e02199-16. [CrossRef]
7. Chang, C.F.; King, C.C.; Wan, C.H.; Chang, Y.C.; Chan, T.C.; David Lee, C.C.; Chou, P.H.; Li, Z.R.; Li, Y.T.; Tseng, T.J.; et al. Lessons from the Largest Epidemic of Avian Influenza Viruses in Taiwan, 2015. *Avian Dis.* **2016**, *60*, 156–171. [CrossRef]
8. Cheng, M.C.; Soda, K.; Lee, M.S.; Lee, S.H.; Sakoda, Y.; Kida, H.; Wang, C.H. Isolation and characterization of potentially pathogenic H5N2 influenza virus from a chicken in Taiwan in 2008. *Avian Dis.* **2010**, *54*, 885–893. [CrossRef]
9. Lee, C.C.; Zhu, H.; Huang, P.Y.; Peng, L.; Chang, Y.C.; Yip, C.H.; Li, Y.T.; Cheung, C.L.; Compans, R.; Yang, C.; et al. Emergence and evolution of avian H5N2 influenza viruses in chickens in Taiwan. *J. Virol.* **2014**, *88*, 5677–5686. [CrossRef]
10. Zepeda-Cervantes, J.; Ramírez-Jarquín, J.O.; Vaca, L. Interaction Between Virus-like Particles (VLPs) and Pattern Recognition Receptors (PRRs) From Dendritic Cells (DCs): Toward Better Engineering of VLPs. *Front. Immunol.* **2020**, *11*, 1100. [CrossRef]
11. Yong, C.Y.; Yeap, S.K.; Ho, K.L.; Omar, A.R.; Tan, W.S. Potential recombinant vaccine against influenza A virus based on M2e displayed on nodaviral capsid nanoparticles. *Int. J. Nanomed.* **2015**, *10*, 2751–2763. [CrossRef]
12. Krammer, F. The Quest for a Universal Flu Vaccine: Headless HA 2.0. *Cell Host Microbe* **2015**, *18*, 395–397. [CrossRef] [PubMed]
13. Wu, C.-Y.; Yeh, Y.-C.; Chan, J.-T.; Yang, Y.-C.; Yang, J.-R.; Liu, M.-T.; Wu, H.-S.; Hsiao, P.-W. A VLP Vaccine Induces Broad-Spectrum Cross-Protective Antibody Immunity against H5N1 and H1N1 Subtypes of Influenza A Virus. *PLoS ONE* **2012**, *7*, e42363. [CrossRef] [PubMed]
14. Kang, Y.-M.; Cho, H.-K.; Kim, J.H.; Lee, S.J.; Park, S.-J.; Kim, D.-Y.; Kim, S.Y.; Park, J.-W.; Lee, M.-H.; Kim, M.-C.; et al. Single dose of multi-clade virus-like particle vaccine protects chickens against clade 2.3.2.1 and clade 2.3.4.4 highly pathogenic avian influenza viruses. *Sci. Rep.* **2021**, *11*, 13786. [CrossRef] [PubMed]
15. Sequeira, D.P.; Correia, R.; Carrondo, M.J.T.; Roldão, A.; Teixeira, A.P.; Alves, P.M. Combining stable insect cell lines with baculovirus-mediated expression for multi-HA influenza VLP production. *Vaccine* **2018**, *36*, 3112–3123. [CrossRef]
16. Huang, S.W.; Tai, C.H.; Hsu, Y.M.; Cheng, D.; Hung, S.J.; Chai, K.M.; Wang, Y.-F.; Wang, J.R. Assessing the application of a pseudovirus system for emerging SARS-CoV-2 and re-emerging avian influenza virus H5 subtypes in vaccine development. *Biomed. J.* **2020**, *43*, 375–387. [CrossRef]
17. Esslinger, C.; Chapatte, L.; Finke, D.; Miconnet, I.; Guillaume, P.; Lévy, F.; MacDonald, H.R. In vivo administration of a lentiviral vaccine targets DCs and induces efficient $CD8^+$ T cell responses. *J. Clin. Investig.* **2003**, *111*, 1673–1681. [CrossRef]
18. Morante, V.; Borghi, M.; Farina, I.; Michelini, Z.; Grasso, F.; Gallinaro, A.; Cecchetti, S.; Di Virgilio, A.; Canitano, A.; Pirillo, M.F.; et al. Integrase-Defective Lentiviral Vector Is an Efficient Vaccine Platform for Cancer Immunotherapy. *Viruses* **2021**, *13*, 355. [CrossRef]
19. Kaufmann, L.; Syedbasha, M.; Vogt, D.; Hollenstein, Y.; Hartmann, J.; Linnik, J.E.; Egli, A. An Optimized Hemagglutination Inhibition (HI) Assay to Quantify Influenza-specific Antibody Titers. *J. Vis. Exp.* **2017**, *130*, e55833. [CrossRef]
20. Lai, C.C.; Cheng, Y.C.; Chen, P.W.; Lin, T.H.; Tzeng, T.T.; Lu, C.C.; Lee, M.S.; Hu, A.Y. Process development for pandemic influenza VLP vaccine production using a baculovirus expression system. *J. Biol. Eng.* **2019**, *13*, 78. [CrossRef]
21. Huhti, L.; Blazevic, V.; Nurminen, K.; Koho, T.; Hytönen, V.P.; Vesikari, T. A comparison of methods for purification and concentration of norovirus GII-4 capsid virus-like particles. *Arch. Virol.* **2010**, *155*, 1855–1858. [CrossRef]
22. Lai, S.; Qin, Y.; Cowling, B.J.; Ren, X.; Wardrop, N.A.; Gilbert, M.; Tsang, T.K.; Wu, P.; Feng, L.; Jiang, H.; et al. Global epidemiology of avian influenza A H5N1 virus infection in humans, 1997–2015: A systematic review of individual case data. *Lancet Infect. Dis.* **2016**, *16*, e108–e118. [CrossRef]
23. Isoda, N.; Twabela, A.T.; Bazarragchaa, E.; Ogasawara, K.; Hayashi, H.; Wang, Z.J.; Kobayashi, D.; Watanabe, Y.; Saito, K.; Kida, H.; et al. Re-Invasion of H5N8 High Pathogenicity Avian Influenza Virus Clade 2.3.4.4b in Hokkaido, Japan, 2020. *Viruses* **2020**, *12*, 1439. [CrossRef] [PubMed]
24. Ko, Y.A.; Yu, Y.H.; Wu, Y.F.; Tseng, Y.C.; Chen, C.L.; Goh, K.S.; Liao, H.Y.; Chen, T.H.; Cheng, T.R.; Yang, A.S.; et al. A non-neutralizing antibody broadly protects against influenza virus infection by engaging effector cells. *PLoS Pathog.* **2021**, *17*, e1009724. [CrossRef] [PubMed]
25. Kim, M.C.; Kim, K.H.; Lee, J.W.; Lee, Y.N.; Choi, H.J.; Jung, Y.J.; Kim, Y.J.; Compans, R.W.; Prausnitz, M.R.; Kang, S.M. Co-Delivery of M2e Virus-like Particles with Influenza Split Vaccine to the Skin Using Microneedles Enhances the Efficacy of Cross Protection. *Pharmaceutics* **2019**, *11*, 188. [CrossRef]
26. Carter, D.M.; Darby, C.A.; Lefoley, B.C.; Crevar, C.J.; Alefantis, T.; Oomen, R.; Anderson, S.F.; Strugnell, T.; Cortés-Garcia, G.; Vogel, T.U.; et al. Design and Characterization of a Computationally Optimized Broadly Reactive Hemagglutinin Vaccine for H1N1 Influenza Viruses. *J. Virol.* **2016**, *90*, 4720–4734. [CrossRef]
27. Lone, N.A.; Spackman, E.; Kapczynski, D. Immunologic evaluation of 10 different adjuvants for use in vaccines for chickens against highly pathogenic avian influenza virus. *Vaccine* **2017**, *35*, 3401–3408. [CrossRef]
28. Liu, C.G.; Liu, M.; Liu, F.; Liu, D.F.; Zhang, Y.; Pan, W.Q.; Chen, H.; Wan, C.H.; Sun, E.C.; Li, H.T.; et al. Evaluation of several adjuvants in avian influenza vaccine to chickens and ducks. *Virol. J.* **2011**, *8*, 321. [CrossRef]

29. Stone, H.D.; Brugh, M.; Hopkins, S.R.; Yoder, H.W.; Beard, C.W. Preparation of Inactivated Oil-Emulsion Vaccines with Avian Viral or Mycoplasma Antigens. *Avian Dis.* **1978**, *22*, 666–674. [CrossRef]
30. Zhu, W.Z.; Wen, Y.C.; Lin, S.Y.; Chen, T.C.; Chen, H.W. Anti-Influenza Protective Efficacy of a H6 Virus-like Particle in Chickens. *Vaccines* **2020**, *8*, 465. [CrossRef]
31. Walker, P.G.; Cauchemez, S.; Métras, R.; Dung, D.H.; Pfeiffer, D.; Ghani, A.C. A Bayesian approach to quantifying the effects of mass poultry vaccination upon the spatial and temporal dynamics of H5N1 in Northern Vietnam. *PLoS Comput. Biol.* **2010**, *6*, e1000683. [CrossRef]

viruses MDPI

Article

Identification and Functional Analyses of Host Proteins Interacting with the p17 Protein of Avian Reovirus

Chengcheng Zhang [1,2,3], Xinyi Liu [1,2,3], Fuxi Zhao [1,2,3], Qingqing Zhang [1,2,3], Wei Zuo [1,2,3], Mengjiao Guo [1,2,3], Xiaorong Zhang [1,2,3] and Yantao Wu [1,2,3,*]

1 College of Veterinary Medicine, Yangzhou University, Yangzhou 225009, China; zcc@yzu.edu.cn (C.Z.); qtzdlxy99215@163.com (X.L.); ZFX951023@163.com (F.Z.); zqq1600676595@163.com (Q.Z.); zw15103447259@163.com (W.Z.); guomj@yzu.edu.cn (M.G.); zxr@yzu.edu.cn (X.Z.)
2 Jiangsu Co-Innovation Center for the Prevention and Control of Important Animal Infectious Disease and Zoonoses, Yangzhou 225009, China
3 Comparative Medicine Research Institute, Yangzhou University, Yangzhou 225009, China
* Correspondence: ytwu@yzu.edu.cn

Abstract: Avian reovirus (ARV) causes viral arthritis, chronic respiratory diseases, retarded growth and malabsorption syndrome. However, the precise molecular mechanism remains unclear. Here, we report the host cellular proteins that interact with ARV p17 by yeast two-hybrid screening. In this study, the p17 gene was cloned into pGBKT7 to obtain the bait plasmid pGBKT7-p17. After several rounds of screening of a chicken cDNA library, 43 positive clones were identified as possible host factors that interacted with p17. A BLAST search of the sequences was performed on the NCBI website, which ultimately revealed 19 interacting proteins. Gene ontology enrichment and Kyoto Encyclopedia of Genes and Genome analyses indicated that the acquired proteins were involved in multicellular organismal processes, metabolic processes, and biological regulation. When the subcellular localization of the host protein and ARV p17 protein was investigated, we observed colocalization of p17-GFP with IGF2BP1-RED and PQBP1-RED in the transfected cells but not with FGF1-RED. The direct interaction of ARV p17 protein with IGF2BP1 and PQBP1 was confirmed by coimmunoprecipitation and GST pulldown assays. We used RT-qPCR to assess the expression variation during ARV infection. The results showed that IGF2BP1, PAPSS2, RPL5, NEDD4L, PRPS2 and IFI16 were significantly upregulated, whereas the expression of FGF1, CDH2 and PQBP1 was markedly decreased in DF-1 cells infected with ARV. Finally, we demonstrated that IGF2BP1 had a positive effect on ARV replication, while PQBP1 had the opposite effect. Our findings provide valuable information for better insights into ARV's pathogenesis and the role of the p17 protein in this process.

Keywords: avian reovirus; p17 protein; yeast two hybrid; protein interaction

Citation: Zhang, C.; Liu, X.; Zhao, F.; Zhang, Q.; Zuo, W.; Guo, M.; Zhang, X.; Wu, Y. Identification and Functional Analyses of Host Proteins Interacting with the p17 Protein of Avian Reovirus. *Viruses* **2022**, *14*, 892. https://doi.org/10.3390/v14050892

Academic Editor: Chi-Young Wang

Received: 12 April 2022
Accepted: 22 April 2022
Published: 25 April 2022

Publisher's Note: MDPI stays neutral with regard to jurisdictional claims in published maps and institutional affiliations.

Copyright: © 2022 by the authors. Licensee MDPI, Basel, Switzerland. This article is an open access article distributed under the terms and conditions of the Creative Commons Attribution (CC BY) license (https://creativecommons.org/licenses/by/4.0/).

1. Introduction

Avian reovirus (ARV) belongs to the Orthoreovirus genus of Reoviridae, which are important pathogens that cause diseases including arthritis, malabsorption syndrome, immunosuppression and other chronic respiratory diseases in chickens [1–3]. ARVs cause considerable economic losses in the poultry industry [4].

As an icosahedral nonenveloped virus, the ARV genome consists of 10 double-stranded RNA segments that express 10 structural proteins and 4 nonstructural proteins [5], but the functions of many of these proteins are largely unknown. The S1 genome segment of avian reovirus contains three open reading frames and has been related to immunosuppression through interactions with host proteins that interfere with the innate immune response [6], one of which is the nonstructural protein p17 [7].

The nonstructural p17 protein of ARVs contains 146 amino acids (aa) and has been reported to play a critical role in virus–host interactions [8,9]. The p17 protein, which is known as a nucleocytoplasmic shuttling protein with a unique sequence [10,11] that specifically shuttles between the nucleus and the cytoplasm to regulate signaling pathways, including autophagy, gene transcription, and DNA binding, promotes viral replication and the cell cycle and interacts with several cellular proteins [11–14]. Several studies have reported that ARV p17 can inhibit cell growth and cause cell cycle retardation by initiating the p53 pathway in different cancer cell lines and interacting with cyclin-dependent kinases and different cyclins [11,15]. In contrast, the direct interaction of p17 with CDK1 leads to its inactivation and to suppression of the serine/threonine-protein kinase Plk1, which is important for a G2/M transition with oncogenic potential [16,17]. Moreover, the ARV p17 protein acts as a positive regulator to stabilize PTEN and enhance the Rak-PTEN interaction to prevent degradation of PTEN [18]. Recently, ARV p17 was reported to possess antiangiogenic activity by increasing the transcription and release of dipeptidyl peptidase 4, which is well known as a tumor suppressor molecule [19].

In this study, we identified the host proteins that interact with p17 from a cDNA library prepared from a chicken liver infected by the ARV GX/2010/1, and all the interactions between 19 host factors and p17 were reconfirmed by Y2H assays. We further examined the subcellular localization of some selected factors and assessed the expression of those proteins after ARV infection. Our study provides valuable information for better insights into ARV's pathogenesis and role of the p17 protein in this process.

2. Materials and Methods

2.1. Cells and Virus

DF-1 (an immortalized chicken embryo fibroblast cell line) and Vero (an African green monkey kidney cell line) cells were obtained from ATCC (USA) and cultured in Dulbecco's modified Eagle's medium (DMEM) (Life Technologies Corp., Grand Island, NY, USA) supplemented with 10% (v/v) fetal bovine serum (FBS), penicillin and streptomycin (50 IU/mL and 50 μg/mL, respectively, Sigma-Aldrich, Burlington, MA, USA) in a humidified atmosphere of 5% CO_2. All cells were cultured until they reached 70–85% confluence before use. ARV strain GX/2010/1 (accession numbers KJ476699-KJ476708) isolated by our lab and propagated in Vero or DF-1 cells was used in the current research. After three freeze–thaw cycles, the supernatant was collected and stored at −70 °C. The virus titer of ARV was examined by plaque assay. Cells were infected with ARV at a multiplicity of infection (MOI) of 2.

2.2. Reagents

All restriction enzymes were purchased from NEB (USA). pEGFP-C1 and pDsRed-N1 vectors were obtained from Clontech (USA). TurboFect was purchased from Thermo Scientific. 40,6-Diamino-2-phenylindole (DAPI) was purchased from Beytime Company (Nanjing, China).

2.3. Cell Culture and Transfection

Vero cells were seeded in 6-well plates. Twenty-four hours later, when the cells had reached 70% confluence, the cells were transfected with the plasmids using the transfection reagent TurboFect according to the manufacturer's protocol.

2.4. Yeast Two-Hybrid Screen

The yeast two-hybrid screen was performed according to the manufacturer's protocol (Clontech, Cat. No. 630489). The pGBKT7-p17 plasmid expressing the bait protein ARV p17 was transfected into competent Saccharomyces cerevisiae AH109 cells. A chicken cDNA library prepared from AVR-infected liver tissues using the pGADT7 plasmid for the fusion of proteins to GAL4-AD was introduced by transformation into competent Saccharomyces cerevisiae Y187 cells. Screening the interacting prey proteins by yeast mating was performed, as previously described [20]. Yeast cells transfected with pGADT7-T and pGBKT7-p53 or pGBKT7-Lam were used as positive and negative controls, respectively.

Positive clones were transferred to quadruple dropout medium lacking tryptophan, leucine, histidine and adenine (SD/-Ade/-His/-Leu/-Trp, Clontech, CA, USA) and assessed by a PCR assay. The primers used are shown in Table 1. The results were sequenced and then subjected to a BLAST search against the NCBI database.

Table 1. Sequence of PCR primers.

Gene Name	Primer Sequence (5′-3′)	Note	Accession Number
AD-F	TAATACGACTCACTATAGGGCT	Amplification of cDNA inserted in AD plasmid	
AD-R	GTGAACTTGCGGGGTTTTTCAGTATCTACGATT		
p17-GFP-F	GCGAATTCTATGCAATGGCTCCGCCATACG	Amplification of ARV p17 gene with GFP label	
p17-GFP-R	GCGGATCCCTCATGGATCGGCGTCAAATCG		
p17-Flag-F	ATTGGATCCTATGCAATGGCTCCGCCATACG	Amplification of ARV p17 gene with Flag tag	
p17-Flag-R	ATTCTCGAGTCACTTATCGTCGTCATCCTTGTAATCCTCATGGATCGGCGTCAAATCG		
p17-GST-F	GCGAATTCTATGCAATGGCTCCGCCATACG	Amplification of ARV p17 gene with GST label	
p17-GST-R	GCCTCGAGCTCATGGATCGGCGTCAAATCG		
PQBP1-RED-F	CGGAATTCATATGCCGCTGCCCGTTG	Amplification of PQPB1 gene with RED label	AJ973596.1
PQBP1-RED-R	CGGGATCCACCTGCTGCTTGGTT		
PQBP1-Myc-F	ATTGGATCCATGCCGCTGCCCGTTG	Amplification of PQBP1 gene with Myc tag	
PQBP1-Myc-R	ATTCTCGAGTCACAGATCCTCTTCAGAGATGAGTTTCTGCTCACCTGCTGCTTGGTT		
IGF2BP1-RED-F	CGGAATTCCCGTTGCTGTCGGG	Amplification of IGF2BP1 gene with RED label	NM_205071.1
IGF2BP1-RED-R	CGGGATCCGTTCTTAGCCCCAT		
IGF2BP1-Myc-F	ATTGGATCCATGCCGTTGCTGTCGGG	Amplification of IGF2BP1 gene with Myc tag	
IGF2BP1-Myc-R	ATTCTCGAGTCACAGATCCTCTTCAGAGATGAGTTTCTGCTCGTTCTTAGCCCCAT		
FGF1-RED-F	CGGAATTCATGGCCGAGGGGG	Amplification of FGF1 gene with RED label	KY747397
FGF1-RED-F	CGGGATCCGGCACGCTTGGATC		
FGF1-Myc-F	ATTGGATCCATGGCCGAGGGGG	Amplification of FGF1 gene with Myc tag	
FGF1-Myc-R	ATTCTCGAGTCACAGATCCTCTTCAGAGATGAGTTTCTGCTCGGCACGCTTGGATC		
IGF2BP1-F	AAGGCACAAGGCAGGATT	Detect the expression of IGF2BP1 by RT qPCR	NM_205071.1
IGF2BP1-R	GCAGCTCATTGACGGTTTT		
FGF1-F	AAAAGCACGCAGACAAGAAC	Detect the expression of FGF1 by RT qPCR	KY747397
FGF1-R	CATTGGAACACCAGGAAGG		
PAPSS2-F	TTGATGCAGGACACTCGC	Detect the expression of PAPSS2 by RT qPCR	XM_040674794
PAPSS2-R	CAATGGTTGACTTGGGAT		
RPL5-F	TTTCCCTGGTTATGACTC	Detect the expression of RPL5 by RT qPCR	XM_031614822.1
RPL5-R	AGCATCTTCATCCTCCTC		
DHRS3-F	ATGCCTGTTGAGGTCTGC	Detect the expression of DHRS3 by RT qPCR	NM_001277910.3
DHRS3-R	AAAGTTTGGTGGAGTGGA		
NEDD4L-F	TGCGGATAGCACCCAATG	Detect the expression of NEDD4L by RT qPCR	XM_031609223.1
NEDD4L-R	TCCTTTCCTCCCAACCAG		
CDH2-F	ATCCTACTGGACGGTTCG	Detect the expression of CDH2 by RT qPCR	NM_001001615
CDH2-R	TTGGCTAATGGCACTTGA		
PRPS2-F	CGTGGAGCGTTGGAGTCGT	Detect the expression of PRPS2 by RT qPCR	NM_001006264.1
PRPS2-R	GCGGGTTCTGCATACAGGTTAT		
PQBP1-F	TGCCGAGGACTATGACGA	Detect the expression of PQBP1 by RT qPCR	AJ973596.1
PQBP1-R	AGCTTCTTGGCCGATTTG		
IFI16-F	CTGGAACGAAAGGGAG	Detect the expression of IFI16 by RT qPCR	NM_001131692.1
IFI16-R	TTGGGTGGAGCTGAT		
ARV-F	CGTATCATTCACCCGCGATT	Detect the replicaiton of ARV by RT qPCR	
ARV-R	TGTTCGCTGTACCATCACCT		
GAPDH-F	GGTGGTGCTAAGCGTGTTA	Detect the expression of GAPDH by RT qPCR	
GAPDH-R	CCCTCCACAATGCCAA		
Si-PQBP1	UGGCCAAGAGAGGCAUCCUCAAACA	SiRNA target of PQBP1	AJ973596.1
Si-IGF2BP1	CAGUGGGCCAUGAAAGCCAUCGAAA	SiRNA target of IGF2BP1	NM_205071.1
Si-FGF1	CGGAAGAUGUGGGCGAGGUCUAUAU	SiRNA target of FGF1	KY747397
Si-Negative	UUCUCCGAACGUGUCACGUUU	Negative control	

The underline sequences represent the restriction enzyme cutting site.

2.5. Functional Classification and Pathway Analysis

Functional classification analysis was performed using gene ontology and the UniProt database. Pathway analysis was performed mainly using the Kyoto Encyclopedia of Genes and Genomes (KEGG) database [21]. A protein interaction network was drawn based on the knowledge of the screened proteins. Based on the correlation between the proteins in the STRING 9.0 database [22], the host protein interaction network was constructed using Cytoscape v3.9.0 software.

2.6. Confocal Laser Scanning Microscopy Assay

For the subcellular localization assays, Vero cells were seeded on glass coverslips in 24-well plates and cultured overnight before transfection with pEGFP-p17 and pDsRed-PQBP1/IGF2BP1/FGF1. Thirty-six hours after transfection, the cells were fixed with 4% paraformaldehyde, and the nuclei were stained with Hoechst 33342 (10 ng/mL; Beyotime, Beijing, China). The images were observed using a laser confocal scanning microscope (LSM510 META; Zeiss, Germany).

2.7. Reciprocal Coimmunoprecipitation (co-IP) Assay

To further confirm whether the ARV p17 protein interacts with PQBP1, IGF2BP1 or FGF1, co-*IP* assays were carried out. After cotransfection of p17-Flag with recombinant vectors expressing PQBP1-Myc, IGF2BP1-Myc or FGF1-Myc for 48 h in DF-1 cells, immunoprecipitation was performed using an anti-c-Myc or anti-Flag agarose affinity gel (Thermo Fisher Scientific, 23620, Waltham, MA, USA) according to the manufacturer's protocol. Briefly, cells were scraped from the culture plate and lysed with a mixture of RIPA buffer (Beyotime, Beijing, China) containing PMSF. After centrifugation for 20 min at 4 °C, the supernatant was collected, mixed with 10 μL of agarose slurry and incubated overnight at 4 °C. The protein agarose mixture was washed 3 times with TBST and resuspended in 2× nonreducing sample buffer. Finally, the liquid was heated for 5 min, with 2 μL of 2-ME added to the samples during denaturation, followed by analysis by Western blotting with specific antibodies.

2.8. GST Pulldown Assays

The procedure for GST pulldown was described in a previous study [15]. Briefly, GST or GST-p17 protein was expressed by induction in Escherichia coli BL21 (DE3) (Invitrogen, Carlsbad, CA, USA). Purified GST or GST-p17 protein (0.8 μg) was coupled to glutathione agarose (Thermo Scientific, 21516, Waltham, MA, USA), followed by washing with 1:1 wash solution (TBS (25 mM Tris-HCl, 0.15 M NaCl, pH 7.2), pull-down lysis buffer (with 10 μg/mL protease inhibitor mixture)) and incubated for 4 h at 4 °C with recombinant Myc-tagged host protein harvested from transfected Vero cells. The eluted proteins were denatured and examined by Western blot analysis with the corresponding antibodies.

2.9. Transfection with siRNA

PQBP1-, IGF2BP1- and FGF1-specific siRNA oligonucleotides and scrambled siRNA (negative control) were synthesized by GenePharma (Shanghai, China). The sequences used are shown in Table 1. DF-1 cells grown to 70% confluence were transfected with siRNA using TransIntro EL Transfection Reagent (TransGen Biotech, Beijing, China) in a 6-well plate. At 24 h post-transfection, one group of cells was collected, and the mRNA levels of specific proteins were assessed by qRT-PCR, while the remaining cells were infected with ARV at an MOI of 2; the protein levels of ARV p17 were analyzed by Western blotting after another 24 h.

2.10. Western Blot Assays

Vero or DF-1 cells were washed with PBS, collected, and lysed with radioimmunoprecipitation buffer containing the protease inhibitor PMSF (Santa Cruz, CA, USA). The lysates were centrifuged at 12,000× *g* for 15 min at 4 °C. The concentration of solubilized protein in the supernatant was evaluated with the Bio-Rad Protein Assay (Bio-Rad Laboratories, Hercules, CA USA). The protein samples were mixed with 5× SDS–PAGE loading buffer and then boiled for 10 min at 100 °C. After electrophoresis at 100 V, proteins were transferred to a PVDF membrane by a semidry transfer instrument. Expression of the individual proteins was determined using the corresponding specific primary antibody and visualized by HRP-labeled secondary antibodies (diluted at 1:5000). The results were detected using a luminescent imager (Tanon 6600, Shanghai, China) after incubation of the membrane with enhanced chemiluminescence reagent (ECL plus) (Beyotime, P0018, Beijing, China). The intensity of the target protein was analyzed by ImageJ software.

2.11. *Quantitative RT-PCR (qRT-PCR)*

Total RNA was extracted from harvested DF-1 cells with or without ARV infection using an RNeasy Mini kit (Qiagen, Valencia, CA, USA). RNA quality was evaluated using RNA Nano Chips on an Agilent Bioanalyzer 2100 (Agilent Technologies, Santa Clara, CA, USA). Briefly, first strand cDNA was synthesized using 1 μg total RNA. Fivefold diluted cDNA products were used as templates for qRT-PCR by using a SYBR Green master mix (Takara, Dalian, China). The chicken β-actin gene was used as an internal reference to normalize the transcriptional value. All the primers used are listed in Table 1. The relative gene transcriptional levels were calculated using the $2^{-\Delta\Delta CT}$ method.

2.12. *Overexpression of Target Protein Mediated by Recombinant Plasmids*

The PQBP1-Myc, IGF2BP1-Myc and FGF1-Myc recombinant vectors encoding PQBP1, IGF2BP1 and FGF1, respectively, were used to study the effect of target protein overexpression on ARV replication. The recombinant plasmid was transfected into DF-1 cells using the TransIntro EL Transfection Reagent. After 24 h, the cells were infected with ARV. At 48 h post-infection, the viral genome copies were assessed by qRT-PCR.

2.13. *MTT Assay*

The cell viability after silencing PQBP1, IGF2BP1 and FGF1 was tested by an MTT method according to the manufacturer's instructions. Briefly, DF-1 cells were cultured in 96-well culture dishes. Different siRNAs were transfected when the cell confluence reached approximately 70%. After cultivation at 37 °C in an incubator with 5% CO_2 for 24 h, the cell culture medium was removed, and 50 μL MTT (2 μg/mL) was added to each well. After incubation at 37 °C for 4 h, MTT was removed, and 200 μL DMSO was added. The plates were placed in an electronic oscillator for 10 min. Cell viability was quantified using a Multiskan FC Microplate Photometer (Thermo Scientific, Waltham, MA, USA).

2.14. *Statistical Analysis*

All experiments were conducted with at least three independent replicates, and all data analyses are expressed as the mean ± S.D. Statistical comparisons were made using the Student's t test. Values of $p < 0.05$ were considered statistically significant.

3. Results

3.1. *Identification of Cellular Proteins That Interact with the ARV p17 Protein through Screening of a Chicken cDNA Library*

The cDNA library was constructed by using chicken liver cells infected with ARV GX2010/1 and then transformed into competent Y187 yeast cells. The pGBKT7-p17 plasmid was transformed into competent yeast Y2H Gold cells, which were then mixed with 1 ml of the Y187 cells that contained cDNA library and cultured at 30 °C and 50 rpm for 20–24 h. The AH109 cells containing the negative or positive control plasmid indicated that the cotransformation process was successful (Figure 1A,B). Then, the cells were spread onto quadruple dropout medium plates and cultivated at 30 °C until colonies appeared. The results are shown in Figure 1C. For the screened transformants, 65 clones grew on the quadruple dropout medium plates. After 3 repeated inoculations on the quadruple dropout medium plates, 43 blue colonies were considered positive candidate proteins (Figure 1D).

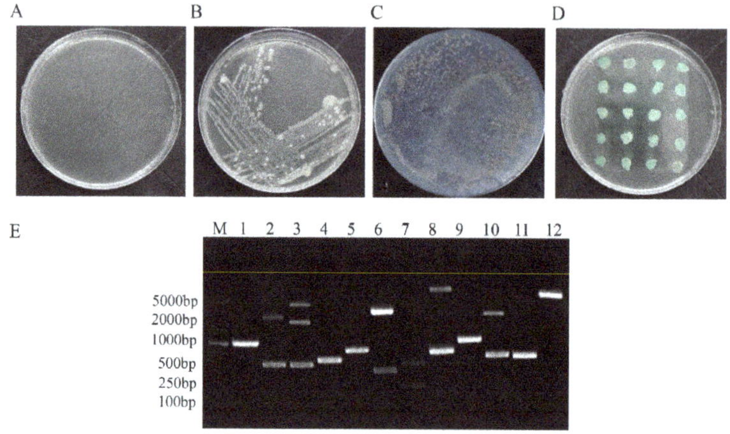

Figure 1. Screening for host proteins that interacted with ARV p17 in a chicken liver cDNA library by yeast two-hybrid assay. (**A**,**B**) show the negative and positive controls of the transformed competent cells grown on the SD/-His/-Leu/-Trp plate. (**C**) One of the transformed yeast strains cultured on SD/-His/-Leu/-Trp plates. (**D**) One of the SD/-His/-Leu/-Trp/-Ade/X-α-Gal plates on which the blue colonies indicate the positive proteins. (**E**) Identification of the positive clones. Plasmids were extracted from the blue colonies, which was used to amplify cDNA by PCR in the colony.

3.2. *Positive Protein Confirmation and Sequence Analysis*

As shown in Figure 1D, out of 43 isolated yeast colonies, 35 were reconfirmed as positive after a second cotransformation with bait plasmid into the Y2H competent cells. The positive samples were sequentially tested by PCR for the library plasmids, and the screened cDNAs from the library were primarily between 500 and 2000 bp in length (Figure 1E). The sequences were then aligned with the basic local alignment search tool (BLAST) on the NCBI website against chicken nonrefSeq databases (Table 2). The BLAST results indicated that there were 19 interacting proteins: ribose-phosphate pyrophosphokinase 2 (PRPS2), gamma-interferon-inducible protein 16 (IFI16), nucleolar GTP-binding protein 1 (GTPBP4), polyglutamine-binding protein 1 (PQBP1), insulin-like growth factor 2 mRNA-binding protein 1 (IGF2BP1), neutral cholesterol ester hydrolase 1 (NCEH1), fibroblast growth factor (FGF1), 3′-phosphoadenosine-5′-phosphosulfate synthase (PAPSS2), cadherin-2 (CDH2), Phasianus colchicus ribosomal protein L5 (RPL5), Coturnix japonica ribosomal protein L7 (RPL7), Gallus gallus discs large MAGUK scaffold protein 1 (DLG1), Gallus gallus dehydrogenase/reductase 3 (DHRS3), Phasianus colchicus NEDD4 like E3 ubiquitin protein ligase (NEDD4L), Gallus gallus FERM domain containing 8 (FRMD8), Gallus gallus zinc finger protein 598 (ZNF598), hypothetical protein and 2 predicted proteins.

Table 2. The positive proteins used for yeast two-hybrid analysis.

No.	GenBank	Protein Name	Description	ORF (bp)
1	NM_001006264.1	Ribose-phosphate pyrophosphokinase 2 (PRPS2)	Activated by magnesium and inorganic phosphate. Competitively or non-competitively inhibited by ADP, or GDP.	1231
2	NM_001131692.1	Gamma-interferon-inducible protein 16 (IFI16)	A predominantly nuclear protein involved in transcriptional regulation, also functions as an innate immune response DNA sensor and induces the IL-1β and antiviral type-1 interferon-β (IFN-β) cytokines.	3012

Table 2. *Cont.*

No.	GenBank	Protein Name	Description	ORF (bp)
3	NM_001006354.1	Nucleolar GTP-binding protein 1 (GTPBP4)	Nucleolar GTP-binding protein 1 is a protein that in humans is encoded by the GTPBP4 gene.	1983
4	AJ973596.1	Polyglutamine-binding protein 1 (PQBP1)	A protein predominantly expressed in lymphoid and myeloid cells was initially identified in the pre-mRNA splicing by interacting with splicing-related factors.	1133
5	NM_205071.1	Insulin-like growth factor 2 mRNA-binding protein 1 (IGF2BP1)	It serves as a post-transcriptional fine-tuner regulating the expression of some essential mRNA targets required for the control of tumor cell proliferation and growth, invasion, and chemo-resistance, and metastasis in various types of human cancers.	2021
6	XM_040679547	Neutral cholesterol ester hydrolase 1 (NCEH1)	Hydrolysis of intracellular CE is the rate-limiting step in the cholesterol efflux from macrophage foam cells. As the hydrolysis of CE takes place at neutral pH, the enzymes catalyzing it have been collectively called neutral CE hydrolases (NCEHs).	5520
7	KY747397	Fibroblast growth factor (FGF1)	Fibroblast growth factors (FGFs) that signal through FGF receptors (FGFRs) regulate a broad spectrum of biological functions, including cellular proliferation, survival, migration, and differentiation.	556
8	XM_040674794	3′-phosphoadenosine-5′-phosphosulfate synthase (PAPSS2)	Blood clotting, bone development 3′-Adenosine phosphate-5′-Sulphate biosynthesis	6730
9	NM_001001615	Cadherin-2 (CDH2)	This protein plays a role in the establishment of left-right asymmetry, development of the nervous system and the formation of cartilage and bone.	3203
10	XM_031614822.1	Phasianus colchicus ribosomal protein L5 (RPL5)	The encoded protein binds 5S rRNA to form a stable complex called the 5S ribonucleoprotein particle (RNP), which is necessary for the transport of non-ribosome-associated cytoplasmic 5S rRNA to the nucleolus for assembly into ribosomes.	1034
11	XM_015855784.2	Coturnix japonica ribosomal protein L7 (RPL7)	The protein has been shown to be an autoantigen in patients with systemic autoimmune diseases. As is typical for genes encoding ribosomal proteins, there are multiple processed pseudogenes of this gene dispersed through the genome.	921
12	XM_025153616.1	Gallus gallus discs large MAGUK scaffold protein 1 (DLG1)	This gene encodes a multi-domain scaffolding protein that is required for normal development. This protein may have a role in septate junction formation, signal transduction, cell proliferation, synaptogenesis and lymphocyte activation.	4586
13	NM_001277910.3	Gallus gallus dehydrogenase/reductase 3 (DHRS3)	DHRS3 (dehydrogenase/reductase 3) is a protein coding gene. Diseases associated with DHRS3 include amphetamine abuse and neuroblastoma. Among its related pathways are vitamin A and carotenoid metabolism and metabolism of fat-soluble vitamins.	4798

Table 2. *Cont.*

No.	GenBank	Protein Name	Description	ORF (bp)
14	XM_031609223.1	Phasianus colchicus NEDD4 like E3 ubiquitin protein ligase (NEDD4L)	NEDD4Lis a protein coding gene. Diseases associated with NEDD4L include periventricular nodular heterotopia 7 and periventricular nodular heterotopia.	7445
15	XM_040694238.1	Gallus gallus FERM domain containing 8 (FRMD8)	It has confirmed the pathophysiological significance of FRMD8 in iPSC-derived human macrophages and mouse tissues, thus demonstrating its role in the regulated release of multiple cytokine and growth factor signals.	9815
16	XM_015294442	Gallus gallus zinc finger protein 598 (ZNF598)	It plays an important role in regulating gene expression, cell differentiation, embryo development and other life processes.	2797
17	AJ720578.1	Gallus gallus mRNA for hypothetical protein	Hypothetical protein	838
18	XR_004306526.1	Coturnix japonica uncharacterized LOC116653138	Predicted protein	2760
19	XM_025145628.1	Gallus gallus NADH-ubiquinone oxidoreductase chain 1-like LOC112530942	Predicted protein	1821

3.3. Functional Classification and Pathway Analysis

After identifying the 19 host proteins that interacted with the ARV p17 protein, GO and KEGG databases were used to identify the main functional groups of the acquired genes. The 19 identified proteins were divided into 41 functional groups, which belonged to biological processes (BP), cellular components (CC), and molecular functions (MF). Multicellular organismal processes, metabolic processes, and biological regulation were the majority of the BP terms. The main subcategories in the MF category were catalytic activity and translation regulators. The five main CC subcategories were cell, organelle, cell parts, protein-containing complex and bindings (Figure 2A). In addition, 26 KEGG pathways were obtained for the identified proteins. The pathways of metabolism, genetic information processing, environmental information processing and diseases were associated with the highest numbers of identified proteins (Figure 2B).

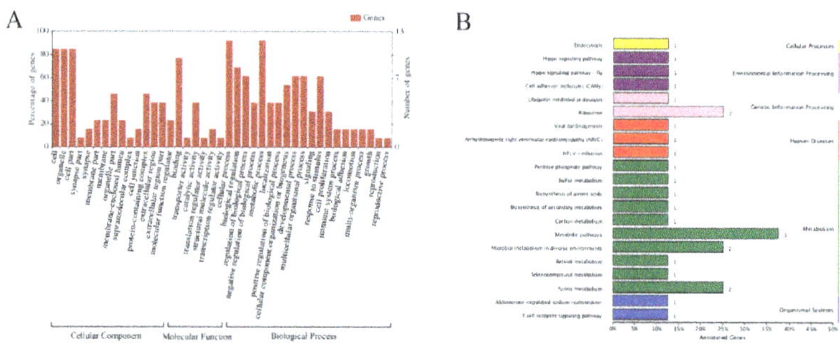

Figure 2. Functional classification and pathway analysis of the screened proteins. (**A**) The classification of the screened proteins was performed according to GO biological processes. All GO categories with a *p* value were chosen. The vertical axis is the percentage of genes, and the horizontal axis is the GO functional category. (**B**) Kyoto Encyclopedia of Genes and Genomes (KEGG) pathway enrichment analyses for the 19 host proteins interacting with ARV p17.

3.4. Construction of the ARV p17-Cellular Protein Interaction Network

In the p17-cellular protein interaction network (Figure 3), GTPBP4, IFI16, and PRPS2 were the most remarkable node proteins, whereas PQPB1, IGF2BP1, FGF1, CDH2, DLG1, NEDD4L and ZNF598 were less remarkable node proteins.

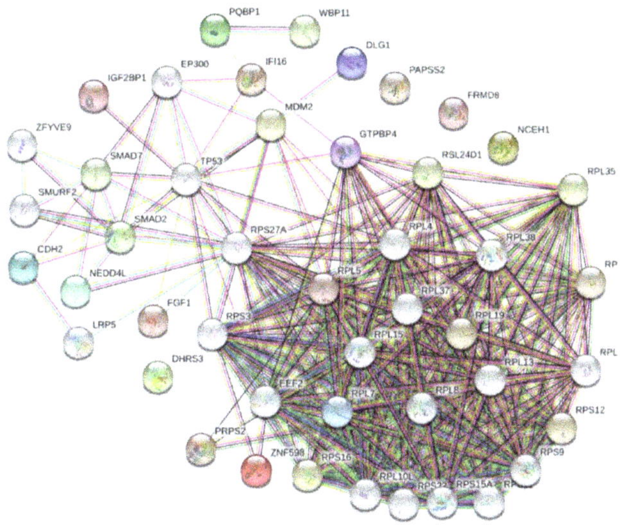

Figure 3. The network of cellular proteins. The network of host proteins interacting with ARV p17 was constructed based on the STRING 9.0 database. GTPBP4, RPL5, RPL7, IFI16, and PRPS2 were the most remarkable node proteins in the interaction network of the cellular factors.

3.5. ARV p17 Protein Colocalizes with PQBP1/IGF2BP1 in Host Cells

To determine the subcellular colocalizations of the identified ARV p17-interacting host factors, we performed a confocal microscopy assay by cotransfecting three proteins as fusions to the N-terminus of RED (PQBP1-RED/IGF2BP1-RED/FGF1-RED) with p17-GFP in Vero cells (Figure 4). When cells were transfected with both plasmids, we observed the p17-GFP, IGF2BP1-RED, PQBP1-RED and FGF1 all location both in the cytoplasm and the nucleus in the transfected cells. However, the colocalization phenomenon exists between p17-GFP with IGF2BP1-RED and PQBP1, but not with FGF1-RED (Figure 4).

3.6. ARV p17 Protein Interacts with PQBP1/IGF2BP1 In Vivo and In Vitro

Because the ARV p17 protein colocalized with PQBP1/IGF2BP1, coimmunoprecipitation and GST pulldown assays were performed to further verify the interaction between p17 and PQBP1/IGF2BP1. The p17-Flag plasmid was cotransfected with those expressing PQBP1-Myc, IGF2BP1-Myc and FGF1-Myc in Vero cells for 24 h, and the cell lysates were coimmunoprecipitated with anti-Flag or anti-Myc affinity gel. After the co-IP process, the proteins in the complexes were assessed by Western blotting. As shown in Figure 5, the ARV p17 protein could interact with IGF2BP1 and PQBP1 but not with FGF1.

The GST pull-down assay was performed to confirm the interaction between p17 and PQBP1/IGF2BP1 in vitro. Recombinant full-length p17-GST fusion protein and GST protein were mixed with the proteins PQBP1-Myc or IGF2BP1-Myc expressed in Vero cells and incubated at 4 °C overnight. As shown in Figure 5, PQBP1-Myc and IGF2BP1-Myc were detected in a complex with p17-GST. These results illustrate that p17 can interact with PQBP1/IGF2BP1 in vitro.

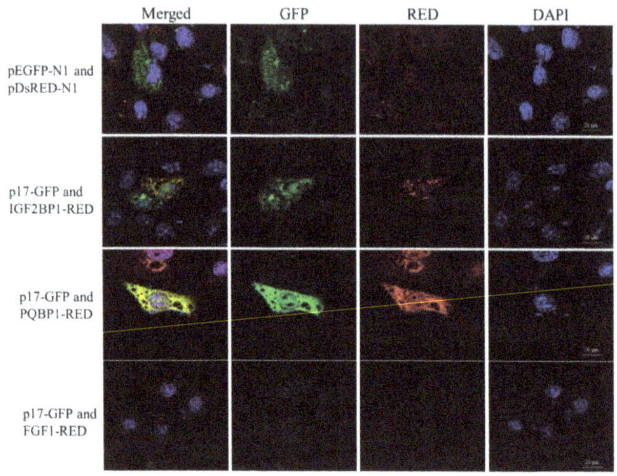

Figure 4. The colocalization of ARV p17 with the screened host proteins. Vero cells were cotransfected with p17-GFP and IGF2BP1, PQBP2 or FGF1-RED and then analyzed by laser confocal microscopy after 36 h. All cells were stained with Hoechst 33342.

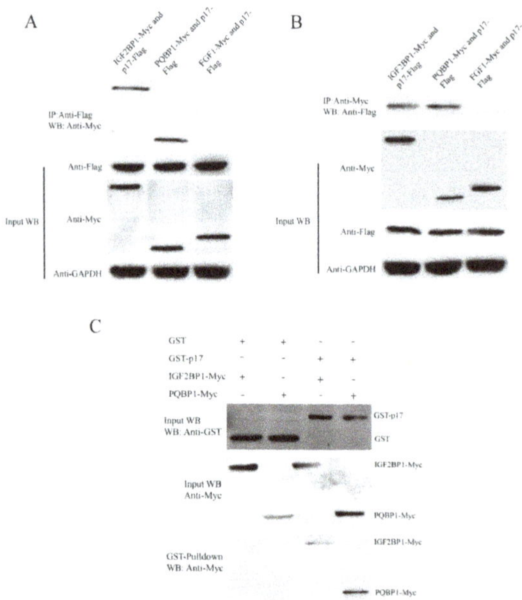

Figure 5. The ARV p17 protein interacts with PQBP1/IGF2BP1. (**A**) A coimmunoprecipitation assay demonstrated that p17-Flag bound to IGF2BP1-Myc and PQBP1-Myc but not to FGF1-Myc in cotransfected cells. Vero cells were transfected with p17-Flag and IGF2BP1-Myc or PQBP1-Myc plasmids for 36 h and then harvested. Cell lysates were immunoprecipitated with an antibody against Flag, followed by Western blotting analysis. (**B**) Reciprocal co-IP experiments showed that the anti-Myc antibody precipitated p17-Flag. (**C**) GST pulldown assay. Glutathione beads conjugated to GST or the GST-p17 fusion protein were incubated with recombinant IGF2BP1-Myc or PQBP1-Myc protein. After washing, proteins were eluted from the beads. The IGF2BP1-Myc and PQBP1-Myc proteins were assessed by immunoblotting with an anti-Myc mAb. GST and GST-LC3 protein expression was confirmed by immunoblotting with a rabbit anti-GST pAb.

3.7. The mRNA Expression of Interacting Proteins after ARV Infection

To study whether the identified host factors play a role in viral replication, the mRNA expression levels of 10 interacting proteins were assessed by using qRT-PCR analysis in ARV-infected DF-1 cells. The results showed that the transcriptional level of IGF2BP1, PAPSS2, RPL5, NEDD4L, PRPS2 and IFI16 were significantly upregulated after ARV infection for 12 h, whereas FGF1, CDH2 and PQBP1 were significantly downregulated (Figure 6).

3.8. The Selected Proteins Have Different Effects on Viral Replication

To further study the effect of selected proteins on ARV replication, IGF2BP1, PQBP1 and FGF1 were successfully overexpressed or knocked down in DF-1 cells, as shown in Figure 7B,D. Then, the transfected cells were infected with ARV for 36 h. The qRT-PCR and WB results showed that IGF2BP1 had a positive effect, while PQBP1 had an inhibitory effect on ARV proliferation (Figure 7C,E–G). Under the same conditions, FGF1 showed no significant effect on ARV replication.

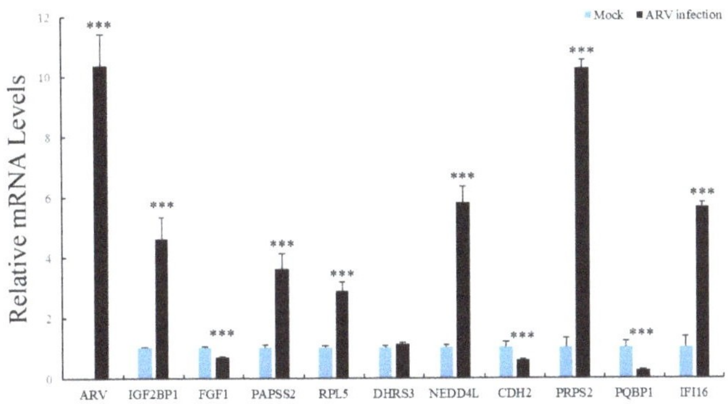

Figure 6. Detection of the mRNA expression of the identified protein interacting with ARV p17 after ARV infection for 12 h by qRT-PCR. Data were pooled across experiments and analyzed using *t* tests. Bars indicate the grand means ± standard deviation (SD). ***, $p < 0.001$.

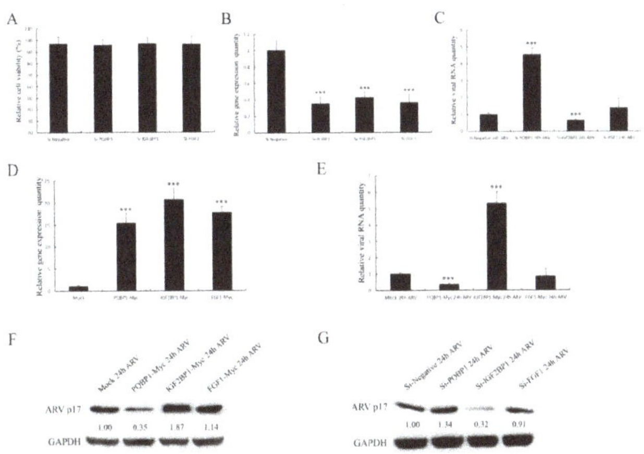

Figure 7. The effect of the selected proteins on ARV replication. (**A**) An MTT assay assessed cell viability. (**B**) Knockdown of IGF2BP1, PQBP1 and FGF1 mediated by siRNA. (**C**) The replication level of ARV assessed by qRT–PCR. (**D**) Overexpression of IGF2BP1, PQBP1 and FGF1 mediated by plasmids. (**E**) The replication level of ARV assessed by qRT–PCR. (**F**,**G**) The replication level of ARV assessed by Western blotting. Data in the graph represent the mean ± S.D. (error bars) calculated from three independent experiments. ***, $p < 0.001$. Signals for all blots were quantified using ImageJ software.

4. Discussion

ARV causes viral arthritis, chronic respiratory diseases, retarded growth and malabsorption syndrome, leading to considerable losses to the poultry industry. To successfully infect cells, viruses always tend to manipulate or utilize many host factors through directly interactions. The ARV p17 protein is also a multifunctional protein that regulates signaling pathways, including autophagy, gene transcription, DNA binding and the cell cycle, and interacts with several cellular proteins [11,12], but its molecular mechanism is largely unknown. The identification of interacting host factors will provide insights into the p17 protein-mediated viral process. In the current study, we identified 19 host proteins that can interact with ARV p17. The interactions of p17 with IGF2BP1 and PQBP1 were verified by co-IP and GST pulldown assays (Figure 5). From our research, IGF2BP1 was shown to play a positive role in ARV replication, while PQBP1 was shown to play an opposite role (Figure 7). Through these host factors, we can further study the mechanism of the function of ARV p17 and the pathogenesis of ARV infection.

IFI16 belongs to the IFN-inducible PYHIN-200 gene family, which contains the two signature HIN domains that bind to dsDNA or ssDNA and a PYRIN domain that mediates interactions with proteins [23–25]. Ribose phosphate pyrophosphate kinase 2 (PRPS2) is a key enzyme in the synthesis of 5-phosphate ribose 1-pyrophosphate (PRPP), an important raw material for DNA synthesis, and plays a critical role in promoting cell apoptosis and inhibiting cell proliferation [26]. Gtp-binding protein 4 (GTPBP4), a member of the GTPBPS family, is a novel G protein located in the nucleolus that mainly participates in the synthesis and maturation of 60S subunits, which are closely related to cell proliferation and growth [27]. Polyglutamine binding protein 1 (PQBP1), a nuclear protein expressed mainly in lymphoid and myeloid cells, can bind to proteins containing poly Q regions and participate in the transcription and RNA modification process, which has been shown to lead to the production of type I interferon in retrovirus infection [28]. Insulin-like growth factor 2 mRNA binding protein 1 (IGF2BP1) is a component of messenger ribonucleoprotein particles, a conserved single-stranded RNA-binding protein family that mediates the transcription of β-actin mRNA and related proteins, regulates cell metabolism and promotes cell adhesion and survival [29]. Fibroblast growth factor 1 (FGF1), a member of the growth factor family, plays an important role in embryonic development, vascular growth, wound healing and other processes [30]. 3′-Adenosine-5′-phosphosulfate synthase 2 (PAPSS2) is an important enzyme gene that catalyzes the synthesis of active sulfate donors in vivo, and its activity is important for normal bone development [31]. Cadherin 2 (CDH2) is a class of calcium-dependent cell adhesion molecules that regulate cell–cell adhesion through homozygous binding in fixed epithelial tissues [32].

Studies that include screening host factors interacting with ARV proteins are relatively rare. In our study, we observed the colocalization of p17-GFP with IGF2BP1-RED and PQBP1-RED both in the cytoplasm and the nucleus in the transfected cells. However, FGF1-RED did not colocalize with ARV p17-GFP (Figure 5). The various distributions of the screened host proteins implied that the ARV p17 protein is indeed a multifunctional protein involved in distinct cellular pathways. As shown in the network of cellular proteins, GTPBP4, FGF1 and IFI16 interacted with ARV p17 but were also involved in interactions with other signaling pathway proteins (Figure 4). These notable node proteins tended to be more essential than the other less notable proteins in the regulation and influence of p17-mediated functions. As shown in Figure 5, the ARV p17 protein interacts with PQBP1/IGF2BP1 verified by co-IP and GST-pulldown assays. PQBP1 is essential to induce the nuclear translocation of nuclear factor κB (NFκB), NFκB-dependent transcription of inflammation genes, brain inflammation in vivo, and eventually mouse cognitive impairment. As shown in Figure 7, PQBP1 had an inhibitory effect on ARV proliferation; interestingly, ARV p17 can interact with PQBP1, and those interactions may play a key role in regulating PQBP1's function.

Several host proteins have been shown to specifically interact with structural ARV proteins. In most of these cases, these interactions play a critical role in adjusting virus replication and, moreover, may be involved in host cell fate or viral virulence [33,34]. The identified host proteins that interact with ARV p17 increase our understanding of the molecular mechanism of ARV infection. We believe that further study of these host proteins and their relationship with p17 can provide valuable information for better insights into ARV's pathogenesis and the role of the p17 protein in this process.

Author Contributions: C.Z. and X.L. conceptualized the experiments; Q.Z., W.Z. and M.G. performed the experiments; X.Z. and F.Z. analyzed the results; C.Z. and Y.W. wrote the paper. All authors have read and agreed to the published version of the manuscript.

Funding: This work was financially supported by 31902299 from National Natural Science Foundation of China, the Earmarked Fund for Modern Agroindustry Technology Research System (CARS-40-K16), and 'High-end talent support program' of Yangzhou University and A Project Funded by the Priority Academic Program Development of Jiangsu Higher Education Institutions (PAPD).

Institutional Review Board Statement: Not applicable.

Data Availability Statement: The data presented in this study are available in the article.

Conflicts of Interest: The authors declare that they have no conflict of interest.

References

1. Lu, H.; Tang, Y.; Dunn, P.A.; Wallner-Pendleton, E.A.; Lin, L.; Knoll, E.A. Isolation and molecular characterization of newly emerging avian reovirus variants and novel strains in Pennsylvania, USA, 2011–2014. *Sci. Rep.* **2015**, *5*, 14727. [CrossRef] [PubMed]
2. Zhang, C.; Hu, J.; Wang, X.; Wang, Y.; Guo, M.; Zhang, X.; Wu, Y. Avian reovirus infection activate the cellular unfold protein response and induced apoptosis via ATF6-dependent mechanism. *Virus Res.* **2021**, *297*, 198346. [CrossRef]
3. Wang, S.; Xie, L.; Xie, Z.; Wan, L.; Huang, J.; Deng, X.; Xie, Z.Q.; Luo, S.; Zeng, T.; Zhang, Y.; et al. Dynamic Changes in the Expression of Interferon-Stimulated Genes in Joints of SPF Chickens Infected With Avian Reovirus. *Front. Vet. Sci.* **2021**, *8*, 618124. [CrossRef] [PubMed]
4. Glass, S.E.; Naqi, S.A.; Hall, C.F.; Kerr, K.M. Isolation and characterization of a virus associated with arthritis of chickens. *Avian Dis.* **1973**, *17*, 415–424. [CrossRef]
5. Spandidos, D.A.; Graham, A.F. Physical and chemical characterization of an avian reovirus. *J. Virol.* **1976**, *19*, 968–976. [CrossRef]
6. Boehme, K.W.; Guglielmi, K.M.; Dermody, T.S. Reovirus nonstructural protein sigma1s is required for establishment of viremia and systemic dissemination. *Proc. Natl. Acad. Sci. USA* **2009**, *106*, 19986–19991. [CrossRef] [PubMed]
7. Garzelli, C.; Onodera, T.; Ray, U.R.; Notkins, A.L. The S1 gene from reovirus type 1 is required for immunosuppression. *J. Infect. Dis.* **1985**, *152*, 640–643. [CrossRef] [PubMed]
8. Bodelon, G.; Labrada, L.; Martinez-Costas, J.; Benavente, J. The avian reovirus genome segment S1 is a functionally tricistronic gene that expresses one structural and two nonstructural proteins in infected cells. *Virology* **2001**, *290*, 181–191. [CrossRef]
9. Huang, W.R.; Li, J.Y.; Wu, Y.Y.; Liao, T.L.; Nielsen, B.L.; Liu, H.J. p17-modulated Hsp90/Cdc37 complex governs oncolytic avian reovirus replication by chaperoning p17 that promotes viral protein synthesis and accumulation of viral proteins sigmaC and sigmaA in viral factories. *J. Virol.* **2022**, *96*, e00074-22. [CrossRef] [PubMed]
10. Costas, C.; Martinez-Costas, J.; Bodelon, G.; Benavente, J. The second open reading frame of the avian reovirus S1 gene encodes a transcription-dependent and CRM1-independent nucleocytoplasmic shuttling protein. *J. Virol.* **2005**, *79*, 2141–2150. [CrossRef]
11. Huang, W.R.; Chi, P.I.; Chiu, H.C.; Hsu, J.L.; Nielsen, B.L.; Liao, T.L.; Liu, H.J. Avian reovirus p17 and sigmaA act cooperatively to downregulate Akt by suppressing mTORC2 and CDK2/cyclin A2 and upregulating proteasome PSMB6. *Sci. Rep.* **2017**, *7*, 5226. [CrossRef]
12. Liu, H.J.; Lin, P.Y.; Lee, J.W.; Hsu, H.Y.; Shih, W.L. Retardation of cell growth by avian reovirus p17 through the activation of p53 pathway. *Biochem. Biophys. Res. Commun.* **2005**, *336*, 709–715. [CrossRef] [PubMed]
13. Chi, P.I.; Huang, W.R.; Lai, I.H.; Cheng, C.Y.; Liu, H.J. The p17 nonstructural protein of avian reovirus triggers autophagy enhancing virus replication via activation of PTEN and AMPK as well as PKR/eIF2α signaling pathways. *J. Biol. Chem.* **2013**, *288*, 3571–3584. [CrossRef]
14. Chiu, H.C.; Huang, W.R.; Wang, Y.Y.; Li, J.Y.; Liao, T.L.; Nielsen, B.L.; Liu, H.J. Heterogeneous Nuclear Ribonucleoprotein A1 and Lamin A/C Modulate Nucleocytoplasmic Shuttling of Avian Reovirus p17. *J. Virol.* **2019**, *93*, e00851-19. [CrossRef] [PubMed]
15. Chiu, H.C.; Huang, W.R.; Liao, T.L.; Chi, P.I.; Nielsen, B.L.; Liu, J.H.; Liu, H.J. Mechanistic insights into avian reovirus p17-modulated suppression of cell cycle CDK-cyclin complexes and enhancement of p53 and cyclin H interaction. *J. Biol. Chem.* **2018**, *293*, 12542–12562. [CrossRef] [PubMed]
16. Ding, L.; Cao, J.; Lin, W.; Chen, H.; Xiong, X.; Ao, H.; Yu, M.; Lin, J.; Cui, Q. The Roles of Cyclin-Dependent Kinases in Cell-Cycle Progression and Therapeutic Strategies in Human Breast Cancer. *Int. J. Mol. Sci.* **2020**, *21*, 1960. [CrossRef] [PubMed]

17. Chiu, H.C.; Huang, W.R.; Liao, T.L.; Wu, H.Y.; Munir, M.; Shih, W.L.; Liu, H.J. Suppression of Vimentin Phosphorylation by the Avian Reovirus p17 through Inhibition of CDK1 and Plk1 Impacting the G2/M Phase of the Cell Cycle. *PLoS ONE* **2016**, *11*, e0162356. [CrossRef]
18. Huang, W.R.; Chiu, H.C.; Liao, T.L.; Chuang, K.P.; Shih, W.L.; Liu, H.J. Avian Reovirus Protein p17 Functions as a Nucleoporin Tpr Suppressor Leading to Activation of p53, p21 and PTEN and Inactivation of PI3K/AKT/mTOR and ERK Signaling Pathways. *PLoS ONE* **2015**, *10*, e0133699. [CrossRef] [PubMed]
19. Manocha, E.; Bugatti, A.; Belleri, M.; Zani, A.; Marsico, S.; Caccuri, F.; Presta, M.; Caruso, A. Avian Reovirus P17 Suppresses Angiogenesis by Promoting DPP4 Secretion. *Cells* **2021**, *10*, 259. [CrossRef] [PubMed]
20. Zhang, C.; He, L.; Kang, K.; Chen, H.; Xu, L.; Zhang, Y. Screening of cellular proteins that interact with the classical swine fever virus non-structural protein 5A by yeast two-hybrid analysis. *J. Biosci.* **2014**, *39*, 63–74. [CrossRef]
21. Jamialahmadi, O.; Motamedian, E.; Hashemi-Najafabadi, S. BiKEGG: A COBRA toolbox extension for bridging the BiGG and KEGG databases. *Mol. Biosyst.* **2016**, *12*, 3459–3466. [CrossRef] [PubMed]
22. Szklarczyk, D.; Gable, A.L.; Nastou, K.C.; Lyon, D.; Kirsch, R.; Pyysalo, S.; Doncheva, N.T.; Legeay, M.; Fang, T.; Bork, P.; et al. Correction to the STRING database in 2021: Customizable protein-protein networks, and functional characterization of user-uploaded gene/measurement sets. *Nucleic Acids Res.* **2021**, *49*, 10800. [CrossRef]
23. Jiang, Z.; Wei, F.; Zhang, Y.; Wang, T.; Gao, W.; Yu, S.; Sun, H.; Pu, J.; Sun, Y.; Wang, M.; et al. IFI16 directly senses viral RNA and enhances RIG-I transcription and activation to restrict influenza virus infection. *Nat. Microbiol.* **2021**, *6*, 932–945. [CrossRef]
24. Ka, N.L.; Lim, G.Y.; Hwang, S.; Kim, S.S.; Lee, M.O. IFI16 inhibits DNA repair that potentiates type-I interferon-induced antitumor effects in triple negative breast cancer. *Cell Rep.* **2021**, *37*, 110138. [CrossRef] [PubMed]
25. Johnson, K.E.; Bottero, V.; Flaherty, S.; Dutta, S.; Singh, V.V.; Chandran, B. Correction: IFI16 Restricts HSV-1 Replication by Accumulating on the HSV-1 Genome, Repressing HSV-1 Gene Expression, and Directly or Indirectly Modulating Histone Modifications. *PLoS Pathog.* **2018**, *14*, e1007113. [CrossRef]
26. Jimenez, A.; Santos, M.A.; Revuelta, J.L. Phosphoribosyl pyrophosphate synthetase activity affects growth and riboflavin production in Ashbya gossypii. *BMC Biotechnol.* **2008**, *8*, 67. [CrossRef]
27. Li, L.; Pang, X.; Zhu, Z.; Lu, L.; Yang, J.; Cao, J.; Fei, S. GTPBP4 Promotes Gastric Cancer Progression via Regulating P53 Activity. *Cell. Physiol. Biochem.* **2018**, *45*, 667–676. [CrossRef] [PubMed]
28. Rahman, S.K.; Okazawa, H.; Chen, Y.W. Frameshift PQBP-1 mutants K192S(fs*7) and R153S(fs*41) implicated in X-linked intellectual disability form stable dimers. *J. Struct. Biol.* **2019**, *206*, 305–313. [CrossRef] [PubMed]
29. Zhang, X.; Yao, Y.; Han, J.; Yang, Y.; Chen, Y.; Tang, Z.; Gao, F. Longitudinal epitranscriptome profiling reveals the crucial role of N(6)-methyladenosine methylation in porcine prenatal skeletal muscle development. *J. Genet. Genom.* **2020**, *47*, 466–476. [CrossRef] [PubMed]
30. Bhushan, B.; Apte, U. Liver Regeneration after Acetaminophen Hepatotoxicity: Mechanisms and Therapeutic Opportunities. *Am. J. Pathol.* **2019**, *189*, 719–729. [CrossRef] [PubMed]
31. Ramaswamy, G.; Sohn, P.; Eberhardt, A.; Serra, R. Altered responsiveness to TGF-beta results in reduced Papss2 expression and alterations in the biomechanical properties of mouse articular cartilage. *Arthritis Res. Ther.* **2012**, *14*, R49. [CrossRef] [PubMed]
32. Tsai, T.Y.; Sikora, M.; Xia, P.; Colak-Champollion, T.; Knaut, H.; Heisenberg, C.P.; Megason, S.G. An adhesion code ensures robust pattern formation during tissue morphogenesis. *Science* **2020**, *370*, 113–116. [CrossRef]
33. Zhou, L.; Li, J.; Haiyilati, A.; Li, X.; Gao, L.; Cao, H.; Wang, Y.; Zheng, S.J. Gga-miR-29a-3p suppresses avian reovirus-induced apoptosis and viral replication via targeting Caspase-3. *Vet. Microbiol.* **2022**, *264*, 109294. [CrossRef] [PubMed]
34. Xie, L.; Wang, S.; Xie, Z.; Wang, X.; Wan, L.; Deng, X.; Xie, Z.; Luo, S.; Zeng, T.; Zhang, M.; et al. Gallus NME/NM23 nucleoside diphosphate kinase 2 interacts with viral sigmaA and affects the replication of avian reovirus. *Vet. Microbiol.* **2021**, *252*, 108926. [CrossRef] [PubMed]

viruses

MDPI

Article

Genetic Analysis of the Complete S1 Gene in Japanese Infectious Bronchitis Virus Strains

Masaji Mase [1,2,3,*], Kanae Hiramatsu [4], Satoko Watanabe [1] and Hiroshi Iseki [1]

1 National Institute of Animal Health, National Agriculture and Food Research Organization, 3-1-5 Kannondai, Tsukuba 305-0856, Japan; satochan@affrc.go.jp (S.W.); hiseki@affrc.go.jp (H.I.)
2 United Graduate School of Veterinary Sciences, Gifu University, 1-1 Yanagido, Gifu 501-1193, Japan
3 Graduate School of Life and Environmental Sciences, Osaka Prefecture University, Izumisano 598-8531, Japan
4 Oita Livestock Hygiene Service Center of Oita Prefecture, 442 Onozuru, Oita 870-1153, Japan; hiramatsu-kanae@pref.oita.lg.jp
* Correspondence: masema@affrc.go.jp; Tel.: +81-29-838-7738

Abstract: The complete nucleotide sequence of the S1 glycoprotein gene of the Japanese infectious bronchitis virus (IBV) strains was determined and genetically analyzed. A total of 61 Japanese IBV strains were classified into seven genotypes, namely GI-1, 3, 7, 13, 18, 19, and GVI-1 using the classification scheme that was proposed by Valastro et al, with three exceptions. These genotypes practically corresponded to those defined in Japan, namely Mass, Gray, JP-II, 4/91, JP-I, JP-III, and JP-IV, which have been identified through their partial nucleotide sequences containing hypervariable regions 1 and 2. In addition, three exceptive strains were considered to be derived from recombination within the S1 gene of IBV strains G1-13 and GI-19. By analyzing the amino acid polymorphism of the S1 glycoprotein among Japanese genotypes, a diversity was observed based on the genotype-specific amino acid residue, the proteolytic cleavage motif at the S1/S2 cleavage site, and the position of the potential N-glycosylation sites.

Keywords: genotype; infectious bronchitis virus; phylogeny; S1 gene

Citation: Mase, M.; Hiramatsu, K.; Watanabe, S.; Iseki, H. Genetic Analysis of the Complete S1 Gene in Japanese Infectious Bronchitis Virus Strains. *Viruses* **2022**, *14*, 716. https://doi.org/10.3390/v14040716

Academic Editor: Chi-Young Wang

Received: 9 March 2022
Accepted: 28 March 2022
Published: 29 March 2022
Corrected: 22 September 2022

Publisher's Note: MDPI stays neutral with regard to jurisdictional claims in published maps and institutional affiliations.

Copyright: © 2022 by the authors. Licensee MDPI, Basel, Switzerland. This article is an open access article distributed under the terms and conditions of the Creative Commons Attribution (CC BY) license (https://creativecommons.org/licenses/by/4.0/).

1. Introduction

Avian infectious bronchitis virus (IBV) causes a highly contagious respiratory, and sometimes urogenital, disease that affects egg production and shell quality in layer chickens [1]. Mortality due to IBV infection alone is usually very low, but it can be significant following secondary infections with bacteria such as *Escherichia coli*. Therefore, IBV infection is considered the second most damaging poultry disease worldwide after highly pathogenic influenza [2]. To protect poultry from it, live or inactivated vaccines are used [3]. However, the protection that is conferred by vaccination is incomplete because the high mutation frequency of IBV leads to the emergence of new strains [1].

The causative coronavirus is an enveloped and positive-stranded RNA virus, containing an unsegmented genome of approximately 27.6 kb. IBV has three major virus-encoded structural proteins: the spike (S) glycoprotein, membrane (M) protein, and nucleocapsid (N) protein. The IBV spike is formed by the post-translational cleavage of S1 and S2 polypeptides [4]. Among these spike components, the S1 glycoprotein is associated with viral attachment and is a primary target of neutralizing antibodies in chickens [5,6]. Therefore, an analysis of the S1 gene is important to characterize the isolated IBVs.

Consequently, based on the nucleotide sequence of the S1 region on the spike gene (S1 gene), the genetic grouping of IBV has mainly been conducted [7–9]. As a result, prevalent nucleotide diversity was revealed in this region containing three hypervariable regions (HVRs) (aa 38–67, 91–141, and 274–387). Moreover, using complete nucleotide sequences of the S1 gene, IBV strains recently distributed worldwide between 1937 and 2013 have been classified into 32 phylogenetic lineages (GI-1 to GI-27 and GII to GVI) [10]. In contrast,

IBV strains in Japan have been genotyped using partial nucleotide sequences containing the HVR-1 and 2 regions [11,12]. Based on this classification, they were classified into seven genotypes: the JP-I, J-II, JP-III, JP-IV, Mass, 4/91, and Gray. However, it remains unknown if this classification system that was adopted in Japan corresponds to that which was described in Valastro et al. [10] since the complete S1 gene has been determined only for a limited number of IBV strains in Japan.

Furthermore, the cryo-electron microscopy structure of the M41 strain spike has recently been resolved [13]. Therefore, to understand the infectivity and pathogenicity of viruses, it would be imperative to examine the amino acids that are related to the receptor-biding domain (RBD) of the virus. This study indicated that the S1 subunit comprised of two independent folding domains: the N-terminal domain (NTD) (amino acids 21 to 237) and the C-terminal domain (amino acids 269 to 414), with a proposed receptor-binding site at both domains. Additionally, the glycosylation of spike proteins influences many of the processes and outcomes of an IBV infection, including the antigenicity, infectivity, and receptor-binding [14]. Therefore, comparing N-glycosylation sites among IBV strains is also essential to examine such biological functions. However, little analysis on the structure of such S1 glycoproteins, including the amino acids that are related to the RBD or potential N-glycosylation sites, have been performed on Japanese IBV strains.

This study determined the complete S1 gene nucleotide sequences of 61 IBV Japanese strains. Then, their genotype definitions were compared with those that were reported in the classification by Valastro et al. [10]. Additionally, amino acid polymorphisms in the S1 glycoproteins, such as the RBD, the motif at the S1/S2 cleavage site, and the position of the potential N-glycosylation site with reference strains, were analyzed among Japanese genotypes.

2. Materials and Methods

2.1. Viruses

The IBV strains that were examined in this study are shown in Table 1. The allantoic cavities of embryonated eggs were used for virus propagation. After inoculation, the eggs were incubated at 37 °C for 3–4 days unless the embryo died. The inoculated eggs were then chilled to 4 °C, and the allantoic fluids were harvested.

Table 1. Japanese IBV strains employed in this study.

Strain	Isolation Year	Length (bp)	Clinical Signs	Genotype Based on HVR-1,2	Genotype Based on Complete Sequence of S1 Gene	Cleavage Site
JP/KH/64	1964	1632	Respiratory	JP-I	GI-18	RRSRR
JP/Shizuoka/71	1971	1632	Respiratory	JP-I	GI-18	RRSRR
JP/Chiba/74	1974	1632	Respiratory	JP-I	GI-18	RRSRR
JP/Chiba/77	1977	1632	Egg drop	JP-I	GI-18	RRSRR
JP/Chiba/80	1980	1632	Respiratory	JP-I	GI-18	RRSRR
JP/Mie/92	1992	1629	Nephritis	JP-I	GI-18	HRFRR
JP/Akita/92	1992	1632	Respiratory	JP-I	GI-18	RRFKR
JP/Nagano/95	1995	1629	Nephritis	JP-I	GI-18	RRSKR
JP/Yamanashi/95	1995	1632	Respiratory	JP-I	GI-18	RRSRR
JP/Shizuoka/98	1998	1626	Nephritis	JP-I	GI-18	RRSRR
JP/Chiba/98	1998	1632	Nephritis	JP-I	GI-18	RRFKR
JP/Toyama/2000	2000	1632	Nephritis	JP-I	GI-18	HRFRR
JP/Okayama-1/2000	2000	1632	Nephritis	JP-I	GI-18	RRFKR
JP/Okayama-2/2000	2000	1632	Nephritis	JP-I	GI-18	RRFKR
JP/Ibaraki/2003	2003	1632	Nephritis	JP-I	GI-18	RRSKR
JP/Ehime/2003	2003	1632	Depression, respiratory	JP-I	GI-18	RRFKR
JP/Kagoshima/2008	2008	1629	Depression, respiratory	JP-I	GI-18	RRSRR
JP/Kagoshima-1/2014	2014	1632	Nephritis	JP-I	GI-18	RRFRR
JP/Yamagata/2017	2017	1629	Rise in mortality	JP-I	GI-18	RRSRR
JP/Kumamoto/2019	2019	1632	Nephritis	JP-I	GI-18	RRFRR

Table 1. *Cont.*

Strain	Isolation Year	Length (bp)	Clinical Signs	Genotype Based on HVR-1,2	Genotype Based on Complete Sequence of S1 Gene	Cleavage Site
C-78		1629	Vaccine strain	JP-I	GI-18	RRSRR
GN		1632	Vaccine strain	JP-I	GI-18	RRFKR
S95		1632	Vaccine strain	JP-I	GI-18	RRSKR
JP/Miyazaki/89	1989	1614	Nephritis	JP-II	GI-7	RRFRR
JP/Yamanashi/93	1993	1614	Nephritis	JP-II	GI-7	RRFRR
JP/Osaka/2000	2000	1614	Nephritis	JP-II	GI-7	RRFRR
JP/Kanagawa/2001	2001	1614	Nephritis	JP-II	GI-7	RRSKR
JP/Yamagata/2011	2011	1614	Nephritis	JP-II	GI-7	RRFKR
JP/Nagasaki/2015	2015	1614	Respiratory	JP-II	GI-7	RRFRR
Miyazaki		1614	Vaccine strain	JP-II	GI-7	RRFRR
TM86		1614	Vaccine strain	JP-II	GI-7	RRFRR
JP/Shimane/98	1998	1617	Respiratory	JP-III	GI-19	HRFRR
JP/Aichi/2000	2000	1620	Nephritis	JP-III	GI-19	RRFRR
JP/Fukui/2000	2000	1620	Respiratory	JP-III	GI-19	RRFRR
JP/Shimane/2002	2002	1614	Nephritis	JP-III	GI-19	RRFRR
JP/Okayama-5/2004	2004	1611	Egg drop	JP-III	GI-19	RRFRR
JP/Chiba/2004	2004	1617	Nephritis	JP-III	GI-19	RRFRR
JP/Wakayama-13/2006	2006	1620	Rise in mortality	JP-III	GI-19	RRFRR
JP/Saitama/2007	2007	1617	Egg drop	JP-III	GI-19	RRFRR
JP/Kagoshima-1/2009	2009	1620	Depression, diarrhea	JP-III	GI-19	RRFRR
JP/Kagoshima-2/2009	2009	1620	Egg drop	JP-III	GI-19	RRFRR
JP/Kochi/2013	2013	1617	Rise in mortality	JP-III	Recombinant	RRFRR
JP/Nagasaki/2013	2013	1620	Depression, diarrhea	JP-III	Recombinant	RRFRR
JP/Gifu/2015	2015	1620	Egg drop	JP-III	GI-19	RRFRR
JP/Nagasaki/2016	2016	1617	Respiratory, diarrhea	JP-III	Recombinant	RRFKR
JP/Chiba/2018	2018	1620	Respiratory	JP-III	GI-19	RRFRR
AK01		1620	Vaccine strain	JP-III	GI-19	RRFRR
JP/Ibaraki/168-1/2009	2009	1638	Egg drop	JP-IV	GVI-1	HRRKR
JP/Chiba/2010	2010	1638	Nephritis	JP-IV	GVI-1	HRRKR
JP/Kagoshima-3/2014	2014	1638	Respiratory	JP-IV	GVI-1	HRRKR
JP/Ishida/51	1951	1611	Respiratory	Mass	GI-1	RRFRR
JP/Nerima/53	1953	1611	Respiratory	Mass	GI-1	RRFRR
JP/Kagoshima-2/2014	2014	1611	Nephritis	Mass	GI-1	RRFRR
Nerima		1611	Vaccine strain	Mass	GI-1	RRFRR
Kita-1		1605	Vaccine strain	Mass	GI-1	RRFRR
KU		1611	Vaccine strain	Mass	GI-1	RRFRR
JP/Kagoshima-4/2014	2014	1623	Rise in mortality	Gray	GI-3	RRSRR
ON		1629	Vaccine strain	Gray	GI-3	RRSRR
JP/Wakayama/2003	2003	1617	Depression, diarrhea	4/91	GI-13	RRFRR
JP/Okayama-7/2004	2004	1617	Nephritis	4/91	GI-13	RRSRR
JP/Saitama/2006	2006	1617	Respiratory	4/91	GI-13	RRFRR

2.2. *RNA Extraction, RT-PCR, and Sequencing*

Viral ribonucleic acid (RNA) was extracted from the infected allantoic fluids using a commercial kit (QIAamp Viral RNA Mini Kit; QIAGEN, Valencia, CA, USA) according to the manufacturer's instructions. The primer sequences that were used for polymerase chain reaction (PCR) are shown in Table 2. These primers were designed to amplify the complete S1 gene and were used for detection and sequencing. After reverse transcription with Superscript IIi (Life Technologies, Gaithersburg, MD) using with random 9-mer oligonucleotides at 25 °C for 10 min and 42 °C for 60 min, cDNAs were amplified by PCR. PCR was performed at the following conditions: 35 cycles at 94 °C for 30 s, 50 °C for 30 s, and 72 °C for 30 s.

Table 2. List of RT-PCR primers.

Name		Sequence (5′-3′)	Position [a]	Length (bp)	Reference
Forward	Reverse				
15F		AGGAATGGTAAGTTRCTRGTWAGAG	20343–20367	671	Mase et al., 2004 [11]
	26Rm	GCGCAGTACCRTTRAYAAAATAAGC	21013–20989		Mase et al., 2004 [11]
19F		GCAGTGTTTGTTACGCATTG	20689–20708	1333	in this study
	1R	CATAACTAACATAAGGGCAA	22021–22002		in this study

[a] Position is given for the S1 gene of strain Beadette42 (Acc.No.NC001451).

Each PCR product was purified with Montage (Millipore, Billerica, MA, USA) according to the manufacturer's instructions. The purified PCR products were used as a template for sequencing on an Applied Biosystems 3130xl Genetic Analyzer using dye terminator cycle sequencing chemistry (Big Dye; Applied Biosystems, Foster City, CA, USA). The purified PCR products were sequenced from both directions.

2.3. Genetic Analysis of the S1 Gene

The sequenced fragments were assembled using ATGC-Mac ver.5 (GENETYX CORPORATION, Tokyo, Japan). The derived nucleotide sequences and deduced amino acids were then analyzed using GENETYX-Mac ver. 18.0 (GENETYX CORPORATION). The detection of potential N-glycosylation sites were also predicted using this software. The nucleotide sequences of the complete S1 glycoprotein-coding region were used to construct a phylogenetic tree in MEGA7 [15]. All tools were run with default parameters unless otherwise specified. All of the sequences that were obtained in this study were deposited in the GenBank database (accession numbers LC662545-LC662604).

2.4. Phylogenetic Analysis

As previously described, phylogenetic trees were constructed using the neighbor-joining method [12]. Subsequently, to compare the Japanese genotypes with those that were reported in Valastro et al. [10], phylogenetic trees were constructed using two portions of the S1 gene (approximately from nt 1 to 1630 and from nt 1 to 600) (Figure 1a,b).

2.5. Recombination Analysis

Potential recombination events were further verified in SimPlot (version 3.5.1.) [16,17], after which nucleotide identity was determined using the Kimura (2-parameter) method with a transition–transversion ratio of 2 and a window width/step size of 200 bp and 20 bp, respectively. Subsequently, BootScan analysis was conducted, employing a subprogram embedded in SimPlot, in addition to signals of 70% or more of the observed permuted trees to indicate potential recombination events [16,17].

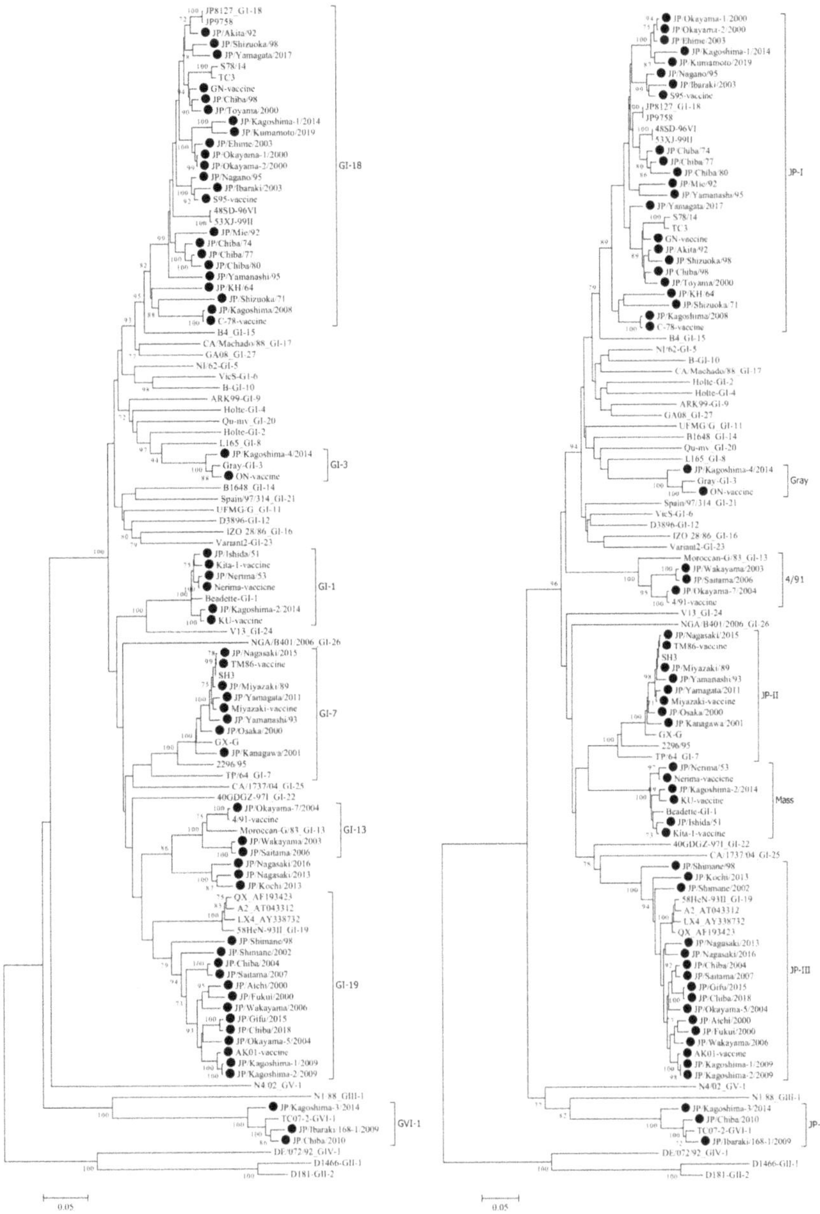

(a) (b)

Figure 1. Phylogenetic trees that are based on the complete (**a**) and partial (**b**) S1 glycoprotein gene of the infectious bronchitis virus (IBV), strain Beaudette (GI-1, GenBank Accession No. NC001451). For (**a**,**b**), nucleotides 20368–21978 (1632 bases) and nucleotides 20368–20988 (621 bases), respectively, were subjected to phylogenetic analysis. Subsequently, both trees were generated using the neighbor-joining method in MEGA 7 [15] with 1000 bootstrap replications. All tools were run with the default parameters unless otherwise specified. Then, horizontal distances were proportionally set to the minimum number of nucleotide differences that were required to join nodes and sequences. The IBV genotypes were defined as described previously [10].

3. Results

3.1. Phylogenetic Analysis

DNA fragments of the expected size were successfully amplified from all IBV samples. The complete nucleotide sequences of the S1 glycoprotein gene were generated from the sequences of two overlapping PCR products. To follow the classification of Valastro et al., the complete length of the S1 gene was analyzed, whereas a partial analysis of the S1 gene was conducted to follow the Japanese classification. To determine the genetic relationships among the Japanese strains and foreign reference ones, a phylogenetic tree was constructed using the nucleotide sequences of the complete S1 gene of the 61 Japanese and 45 reference strains (Figure 1a,b).

The Japanese IBV strains were classified into seven distinct genetic groups (Figure 1a). These groups corresponded to clades GI-1,3,7,13,18,19, and GVI-1, as defined in previous report [10]. Subsequently, as shown in Figure 1b, they were also classified into seven groups by the Japanese definition based on the partial nucleotide sequences, including the HVR-1 and HVR-2 [11,12]. Except for three strains, GI-1,3,7,13,18,19, and GVI-1, corresponded to Mass, Gray, JP-II, 4/91, JP-I, JP-III, and JP-VI, respectively (Table 1). The exceptions were the three IBV strains that were classified as JP-III in Japan based on partial sequences of the S1 gene. Nevertheless, they were not classified as GI-19 when using the complete S1 gene. This result proposes that these strains were possibly recombinant, thus affecting the above classification.

3.2. Recombination Analysis

For three strains (JP/Nagasaki/2013, JP/Kochi/2013, and JP/Nagasaki/2016), in which the classification by Valastro et al. [10], and the Japanese one did not correspond (as shown in Table 1), a Simplot and BootScan analysis was conducted to verify the possibility of genetic recombination. These analyses were conducted to identify the recombinant JP/Nagasaki/2013 and reference JP/Wakayama/2003 (GI-13) and JP/Shimane/98 (GI-19) strains to be used as putative parental strains (Figure 2a,b).

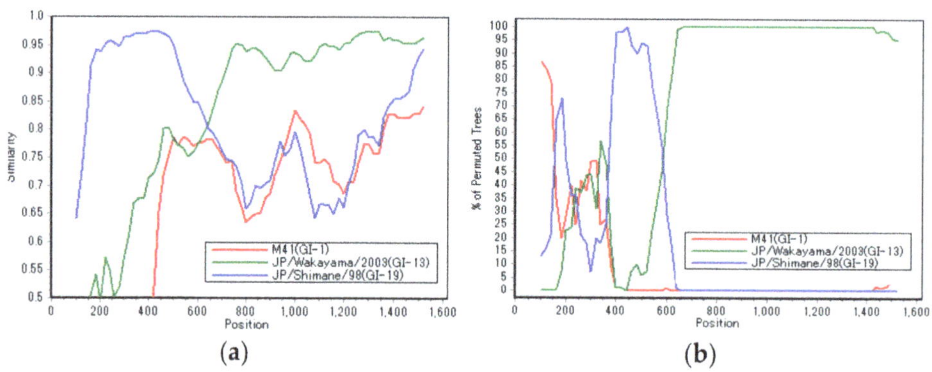

Figure 2. Similarity analysis (**a**) and BootScan analysis (**b**) on the putative recombinant JP/Nagasaki/2013 strain. Reference strains JP/Wakayama/2003 (GI-13, green) and JP/Shimane/98 (GI-19, blue) strains were used as putative parental strains. Additionally, the M41 strain (G1-1, red) was used as an outlier sequence. The *y*-axis indicates the percentage of identity within a 200-bp wide sliding window centered on the plotted position, with a step size of 20 bp between plots.

As shown in Figure 2, the JP/Nagasaki/2013 strain was genetically close to the JP/Shimane/98 strain, belonging to the GI-19 (JP-III) genotype up to around 600 bp on the N-terminal side; however, the remaining on the C-terminal side were close to JP/Wakayama/2003, belonging to the GI-13 (4/91) genotype. Hence, in the S1 gene of these strains (JP/Nagasaki/2013, JP/Kochi/2013, and JP/Nagasaki/2016), approximately 600 bp on the N-terminal side were derived from the GI-19 genotype. However, the remaining region on the C-terminal side was derived from the GI-13 genotype. This result is consistent with that of a previous analysis [10], which showed that most of the recombination breakpoints of field IBV strains occur in the intermediate region between HVR1, HVR2, and HVR3.

3.3. Proteolytic Cleavage Sites in the Spike Glycoprotein

First, to examine the association between the clinical signs and the motif of IBV strains, we determined the amino acids sequences at the cleavage sites. The deduced amino acids that were obtained from the nucleotide sequences at the cleavage site between the S1 and S2 proteins of each strain are shown in Table 1.

The "RRFRR" motif was the most common (30/61) and was found in strains other than the GVI-1 (JP-IV) and GI-3 (Gray) genotypes. However, in the GI-18 (JP-I) genotype, which has the largest number of examined strains, there were only two RRFRR strains, and the RRSRR motif, including various other motifs, was predominant. The second predominant was the "RRSRR" motif, which was more common in the GI-18 (JP-I) genotype, while the "HRRKR" motif was found only in the GVI-1 (JP-IV) genotype.

3.4. Amino Acids Related to the Receptor-Binding Domain (RBD)

It has been reported that the NTD of S1 (aa. 19–272, most N-terminal 253 aa of the mature S1) is both required and sufficient for binding to the respiratory tract [18]. Moreover, amino acids N38, H43, P63, and T69 of M41 S1 were critical to establishing binding within the domain. Therefore, these amino acids' polymorphism (s) were compared among the genotypes. The results that are presented in Table 3 show that the amino acids at positions 38, 43, 63, and 69 were critical for binding to the trachea in terms of M41 spike attachment; however, the diversity at the positions that are mentioned above in these amino acids was also observed in Japanese IBVs.

Table 3. Amino acids based on their relationship to the receptor-binding domain.

Genotype	Amino Acid Position of Amino Acids at the S1 Glycoprotein											
	26	28	29	34	36	37	38	40	42	43	63	69
Mass (GI-1)	Y	Y	Y	F	P	P	D/N	W	L	H/Q	P/S	T/I
Gray (GI-3)	Y	Y	Y	F	P	P	N	W	L	H	S	A
JP-I GI-18)	Y	Y	Y	F/L/Y	P	P/S/G	L/F/P/S/V	W	L/V/I	H	S/H/A	A
JP-II (GI-7)	Y	Y	Y	F	P	P	D/N	W	L	Q	P/L/R	S
JP-III (GI-19)	Y	Y	Y	F	P	P/S	D/N/T/E	W	L	Q	P/S/N/T/A/Q	V
JP-IV (GVI-1)	Y	Y	Y	F	P	P	L/S	W	L	H/Y	G/H	A
4/91 (GI-13)	Y	Y	Y	F	P	G	P/Q	W	L	H/Y	P/S	T

Recently, the motifs "^{26}YxYY^{29}" and "^{34}FxPPxxWxLH^{43}" of high-affinity aminopeptidase N (APN) peptides in chickens have been identified in IBV S1-NTD [19]. APN may be one of the IBV functional receptors. When comparing these amino acid sequences in Japanese IBVs, the motif "^{26}YxYY^{29}" was conserved in all Japanese IBV strains. However, the motif "^{34}FxPPxxWxLH^{43}" was conserved in all the strains belonging to GI-3 and some strains belonging to GI-1, GI-18 and GVI-1 (Table 3). Specifically, in all GI-7 or 19 genotypes strains, the amino acid at position 43 was substituted from H to Q. However, this amino acid was the H that was described above in the M41 strain [19], not the nephropathogenic strain.

Nevertheless, QX-RBD amino acids 110 to 112 (KIP), considered important for binding to the kidney in the nephropathogenic QX strain, and classified into GI-19 [20], were not observed in Japanese IBVs, including the GI-19 (JP-III) genotype (Table 4).

Table 4. Amino acids related to the kidney binding.

Genotype	Corresponding Position of Amino Acids at S1 Glycoprotein		
	110	**111**	**112**
Ck/SWE/0658946/10 (QX)	K	I	P
Mass (GI-1)	M/V	L/V/I	Q
Gray (GI-3)	F/I	L	P
JP-I (GI-18)	L/F/R/S	I	Q/N/A/E
JP-II (GI-7)	F	V	P
JP-III (GI-19)	M/Q/L	I	P/K
JP-IV (GVI-1)	K/I	L	D/K/G
4/91 (GI-13)	M	I	P

3.5. *Potential N-glycosylation Sites*

N-glycosylation occurs at asparagine (N) residues in the amino acid motif asparagine-x-serine/threonine (N-X-S/T) [21]. Therefore, the loss or acquisition of N-glycosylation sites in the spike glycoprotein was predicted, including their positions on representative strains of each genotype (Table 5). Similarities were observed between the positions of the predicted N-glycosylation sites in each S1 glycoprotein of the different Japanese genotype strains, with nine sites shared across all the strains (Table 5). Among these sites, the three N-linked glycosylation sites (at positions 103, 163, and 237 at position in the M41 strain) were considered critically important in previous studies [21], and they were also all conserved in Japanese IBVs. However, their N-glycosylation sites differed among the genotypes. Specifically, although we observed 15 to 17 N-glycosylation sites in the S1 glycoprotein in the GI-1 (Mass) strains, as previously reported [14], 18 or 19 N-glycosylation sites were also present in the GI-19 (JP-III) strains [22]. However, while 19 or 20 N-glycosylation sites in the S1 glycoprotein were observed in the GI-7 (JP-II) strains, only 11 or 12 N-glycosylation sites were present in the GVI-1 (JP-IV) strains. Furthermore, the predicted N-glycosylation site at position 116 (unique to JP-II), and that at position 200, observed in the QX strain (genotype GI-19) [22], were not observed in the GI-19 (JP-III) genotype strains in Japan.

Table 5. Potential N-glycosylation sites of Japanese IBV genotypes.

Strain	Genotype	Amino Acid Position at POTENTIAL N-glycosylation [a]																										Number of Glycosylation Site
M41(AY851295)	Mass (GI-1)	51	-	77	-	103 [b]	-	-	144	163	178	-	-	212	237	247	264	271	276	-	306	-	425	447	-	513	530	17
JP/Ishida/51(LC662594)	Mass (GI-1)	51	-	77	-	103	-	-	144	163	178	-	-	212	237	247	264	271	276	-	306	-	425	447	-	513	530	17
JP/Nerima/53(LC662595)	Mass (GI-1)	51	-	-	-	103	-	-	144	163	178	-	-	212	237	247	264	-	276	-	306	-	425	447	-	513	530	15
Gray(L14069)	GI-3	51	-	75	-	103	-	-	150	169	184	-	-	218	243	253	270	277	282	-	312	-	431	453	-	519	536	17
JP/Kagoshima-4/2014(LC662600)	Gray (GI-3)	-	-	75	-	103	-	-	148	167	182	-	-	216	241	251	268	275	280	-	310	-	429	451	-	517	534	16
QX(MN548289)	QX (GI-19)	52	55	76	92	104	117	141	147	166	181	-	200	215	240	250	267	274	279	282	309	405	428	450	457	516	533	21
JP/Shimane/98(LC662575)	JP-III (GI-19)	51	54	-	-	103	-	140	146	165	180	-	-	214	239	249	266	273	278	281	308	-	427	449	-	515	532	19
JP/Aichi/2000(LC662576)	JP-III (GI-19)	52	55	-	-	104	-	141	147	166	181	-	-	215	240	250	267	274	279	282	309	-	428	450	-	516	-	18
JP/Fukui/2000(LC662577)	JP-III (GI-19)	52	55	76	-	104	-	141	147	166	181	-	-	215	240	250	267	274	279	-	309	405	428	450	-	516	-	19
JP/Shimane/2002(LC662578)	JP-III (GI-19)	50	53	74	-	102	-	139	145	164	179	-	-	213	238	248	265	272	277	280	307	-	426	448	-	514	-	19
JP/Kagoshima-1/2009(LC662583)	JP-III (GI-19)	52	-	76	-	104	-	141	147	166	181	-	-	215	240	250	267	274	279	282	309	-	428	450	-	516	-	18
4/91(KF377577)		-	54	75	-	103	-	-	146	165	180	-	-	214	239	249	266	273	278	281	308	-	427	449	456	515	533	19
JP/Wakayama/2003(LC662602)	4/91 (GI-13)	-	54	75	-	103	-	-	146	165	180	-	-	214	239	249	266	273	278	-	308	-	427	449	456	515	532	18
JP/Okayama-7/2004(LC662603)	4/91 (GI-13)	-	54	75	-	103	-	-	146	165	180	184	-	214	239	249	266	273	278	281	308	-	427	449	456	515	532	20
JP8127(AY296744)	GI-18	51	-	75	-	103	-	-	151	170	185	-	-	219	244	254	271	278	283	-	314	-	433	455	-	521	538	17
JP/KH/64(LC634083)	JP-I (GI-18)	51	-	75	-	103	-	-	151	170	185	-	-	219	244	254	271	278	283	-	313	-	432	454	-	520	-	16
JP/Akita/92(LC662550)	JP-I (GI-18)	51	-	75	-	103	-	-	151	170	185	-	-	219	244	254	271	278	283	-	313	-	432	454	-	520	537	17
JP/Nagano/95(LC662551)	JP-I (GI-18)	51	-	75	91	103	-	-	151	170	185	-	-	219	244	254	271	278	283	-	313	-	431	453	-	519	536	18
JP/Kagoshima-1/2014(LC662561)	JP-I (GI-18)	51	-	75	-	103	-	-	151	170	185	-	-	219	244	254	271	278	283	-	313	-	432	454	461	520	537	18
TP/64(AY606320)	GI-7	51	-	77	-	103	116	139	145	164	179	-	-	213	238	248	265	272	277	-	308	-	427	449	-	515	532	19
JP/Miyazaki/89(LC662567)	JP-II (GI-7)	51	-	77	-	103	116	139	145	164	179	-	-	213	238	248	265	272	277	280	307	-	426	448	-	514	531	20
JP/Nagasaki/2015(LC662572)	JP-II (GI-7)	51	-	-	-	103	116	139	145	164	179	-	-	213	238	248	265	272	277	280	307	-	426	448	-	514	531	19
TC07-2(GQ265948)	GVI-1	-	55	75	-	104	-	-	148	167	182	-	-	216	241	256	-	275	280	-	314	411	-	456	-	522	539	16
JP/Ibaraki/168-1/2009(LC662591)	JP-IV (GVI-1)	-	55	75	-	104	-	-	148	167	182	-	-	216	241	256	-	275	280	-	314	-	-	456	-	522	539	15
JP/Kagoshima-3/2014(LC662593)	JP-IV (GVI-1)	-	55	75	-	104	-	-	148	167	-	-	-	216	241	256	-	275	280	-	314	-	-	456	-	522	539	14

[a] The numbers indicate the amino acid positions of each strain. [b] Bold italics indicate positions that are suggested to be functionally important [19]. -: deletion.

4. Discussion

The S1 glycoprotein determines the antigenicity and tissue tropism of IBVs [23], playing a vital role in inducing neutralizing antibodies and attachment to the host cell receptors [18]. Therefore, the genetic analysis of the complete S1 gene of IBV is critical.

To date, genotyping has been performed on Japanese strains using regions containing HVR-1 and 2 [11,12], but the analysis of the predicted S1 glycoprotein amino acid sequences was insufficient. Here, the complete S1 gene of 61 IBV strains, classified into seven genotypes (including ten vaccine strains which were attenuated strains that were derived from field viruses in Japan), was determined and analyzed.

First, the genotyping using the complete S1 gene was almost consistent with the Japanese typing that was based on partial nucleotide sequences, including HVR-1 and 2, with three exceptions. These strains (JP/Nagasaki/2013, JP/Kochi/2013, and JP/Nagasaki/2016) belonged to genotype JP-III in the classification using Japanese typing. However, they were classified into distinct genotypes from GI-13 and 19 when using the full S1 gene length, whereas most JP-III genotype strains were classified into the GI-19 genotype. Therefore, these three strains that were derived from the GI-19 and GI-13 genotypes were all considered recombinants. This result is consistent with that of a previous analysis [10], which showed that most of the recombination breakpoints of field IBV strains occur in the intermediate region between HVR1, HVR2, and HVR3. Recently, many such field IBV recombinants of this kind have been increasing worldwide [24]. For example, the recent Chinese isolate (CK/CH/SCMY/160315) was identified as a novel recombinant virus that was derived from the H120 (GI-1), 4/91 (GI-13), and LDT3-A (GI-28) live attenuated vaccine strains, and the LJL/08-1 (GI-19) field strain. Besides, in Japan, field strains that are potential recombinants have also been reported through the analysis of S1 and S2 [12]. However, little analysis of Japanese IBV recombinants using the complete genome has been reported. Therefore, it would be necessary to conduct a genetic analysis using the complete genome to clarify the detailed genetic background of domestic epidemic IBV strains.

By analyzing the S1 glycoprotein's predicted amino acid sequence, seven types of S1 protein cleavage recognition motifs were observed among the Japanese IBV strains. However, no association was found between the clinical signs and the motif of the isolates. This finding was consistent with the cleavage recognition motif of the S1 gene being reportedly irrelevant to viral pathogenicity and tissue tropism [25].

In general, IBV infects ciliated epithelial cells in the respiratory tract of chickens. While some IBV strains replicate locally, others can disseminate to various organs, including the kidneys [20]. By conducting studies using the IBV M41 strain, the reference strain of the GI-1 (Mass) genotype, alpha-2,3-linked sialic acids were identified as IBV receptors [23]. Furthermore, it was shown that the amino acids at positions 38, 43, 63, and 69 were critical for binding to the trachea in terms of the M41 spike attachment. By comparing the IBV receptor-related amino acids that were detected using this M41 strain, various substitutions were observed among Japanese IBV genotypes. For example, a recent study on the QX strain, classified into the GI-19 genotype, reported that amino acids 110 to 112 (KIP) of the S1glycoprotein were sufficient to extend its tropism toward the kidney [20]. However, this KIP motif was not found in Japanese IBV strains, some isolated from nephritis, including the GI-19 (JP-III) genotype. Nevertheless, although the two motifs that were found in the recent APN study [19], the amino acid at position 43 of ^{34}FxPPxxWxLH^{43} was substituted with Q instead of H, they were found in genotypes GI-7 (JP-II) and GI-19 (JP-III). Since IBV strains belonging to GI-7 and 19 are known to be the nephropathogenic type in foreign countries [8,16], it would be interesting to examine the effect of this amino acid substitution on the viral pathogenicity or cellular tropism.

Furthermore, studies have reported that the glycosylation of viral structural proteins influences many of the processes and outcomes of an IBV infection, including antigenicity, infectivity, and receptor-binding [14,26]. Therefore, comparing N-glycosylation sites among IBV strains is essential to examine such biological functions. Using bioinformatic techniques, previous analyses of IBV strains identified 17 sites in the S1 sequences of strains belonging to the Mass serotype. Additionally, they observed that the total number of predicted sites was different from that of the QX-like strains [27]. Previous studies have also suggested that some N-linked glycosylation sites were critically important for virus replication and infectivity [21]. Furthermore, these sites were all conserved in Japanese IBVs (Table 5), which is functionally important. For example, N to D/Q mutations in N212 and N276, abolished the infectivity of the recombinant IBV [21]. However, the loss or gain of N-glycosylation sites differed among the genotypes. Interestingly, the JP-II (GI-7) genotype was composed mainly nephritis-derived strains and contained more numbers (20–21) than those that were predicted in the S1 glycoprotein. Therefore, since the mechanism of nephritis that is caused by IBV in chickens remains unknown, it was of particular interest to analyze the virus' nephropathogenicity and the effect of these glycosylations.

In conclusion, we have shown the phylogenetic relationship between Japanese IBV strains and foreign strains in this study. Additionally, by analyzing the amino acid polymorphism of the S1 glycoprotein among Japanese genotypes, a diversity was observed based on the genotype-specific amino acid residue and the position of the potential N-glycosylation sites. This finding is significant since the glycosylation of viral structural proteins has been shown to influence many of the processes and outcomes of IBV infections, including antigenicity, infectivity, and receptor-binding. Thus, to unequivocally determine the contributions of such amino acid residues to the virus' pathogenicity, it is imperative to generate viruses with specific mutations using reverse genetics [28,29]. This study's obtained findings would be useful for elucidating IBV's pathogenicity and improving vaccine development strategies.

Author Contributions: M.M.: designed the study; M.M. and K.H.: performed experiments; M.M., K.H., S.W. and H.I. wrote the manuscript. All authors have read and agreed to the published version of the manuscript.

Funding: This study was supported by a grant from National Agriculture and Food Research Organization.

Data Availability Statement: All data generated and analyzed during this study are included in this article.

Acknowledgments: The authors would like to thank the veterinary officials of each prefecture for their cooperation in collecting the viral samples.

Conflicts of Interest: The authors declare no conflict of interest.

References

1. Jackwood, M.; de Wit, J.J. Infectious bronchitis virus. In *Diseases of Poultry*, 14th ed.; Swayne, D.E., Boulianne, M., Logue, C.M., McDougald, L.R., Nair, V., Suarez, D.L., Eds.; Wiley-Blackwell: Hoboken, NJ, USA, 2020; pp. 167–188.
2. De Wit, J.J.; Cook, J.K.A. Spotlight on avian pathology: Infectious bronchitis virus. *Avian Pathol.* **2019**, *48*, 393–395. [CrossRef] [PubMed]
3. Jordan, B. Vaccination against infectious bronchitis virus: A continuous challenge. *Vet. Microbiol.* **2017**, *206*, 137–143. [CrossRef]
4. Cavanagh, D. Coronavirus avian infectious bronchitis virus. *Vet. Res.* **2007**, *38*, 281–297. [CrossRef] [PubMed]
5. Koch, G.; Hartog, L.; Kant, A.; Van Roozelaar, D.J. Antigenic domains on the peplomer protein of avian infectious bronchitis virus: Correlation with biological functions. *J. Gen. Virol.* **1990**, *71*, 1929–1935. [CrossRef]
6. Kant, A.; Koch, G.; Van Roozelaar, D.J.; Kusters, J.G.; Poelwijk, F.A.J.; Van der Zeijst, B.A.M. Location of antigenic sites defined by neutralizing monoclonal antibodies on the S1 avian infectious bronchitis virus glycopolypeptide. *J. Gen. Virol.* **1992**, *73*, 591–596. [CrossRef] [PubMed]
7. Lee, E.K.; Jeon, W.J.; Lee, Y.J.; Jeong, O.M.; Choi, J.G.; Kwon, J.H.; Choi, K.S. Genetic diversity of avian infectious bronchitis virus isolates in Korea between 2003 and 2006. *Avian Dis.* **2008**, *52*, 332–337. [CrossRef] [PubMed]
8. Ren, M.; Zhang, L.; Hou, Y.; Zhao, Y.; Han, Z.; Sun, J.; Liu, S. Genetic, antigenic, and pathogenic characteristics of infectious bronchitis virus GI-7/TW-II in China. *Avian Dis.* **2020**, *64*, 183–196. [CrossRef] [PubMed]
9. Wang, C.H.; Huang, Y.C. Relationship between serotypes and genotypes based on the hypervariable region of the S1 gene of infectious bronchitis virus. *Arch. Virol.* **2000**, *145*, 291–300. [CrossRef]
10. Valastro, V.; Holmes, E.C.; Britton, P.; Fusaro, A.; Jackwood, M.W.; Cattoli, G.; Monne, I. S1 gene-based phylogeny of infectious bronchitis virus: An attempt to harmonize virus classification. *Infect. Genet. Evol.* **2016**, *39*, 349–364.11. [CrossRef]
11. Mase, M.; Tsukamoto, K.; Imai, K.; Yamaguchi, S. Phylogenetic analysis of avian infectious bronchitis virus strains isolated in Japan. *Arch. Virol.* **2004**, *149*, 2069–2078. [CrossRef]

12. Mase, M.; Gotou, M.; Inoue, D.; Watanabe, S.; Iseki, H. Genotyping of infectious bronchitis viruses isolated in Japan during 2008−2019. *J. Vet. Med Sci.* **2021**, *83*, 522–526. [CrossRef] [PubMed]
13. Shang, J.; Zheng, Y.; Yang, Y.; Liu, C.; Geng, Q.; Luo, C.; Zhang, W.; Li, F. Cryo-EM structure of infectious bronchitis coronavirus spike protein reveals structural and functional evolution of coronavirus spike proteins. *PLoS Pathog.* **2018**, *14*. [CrossRef] [PubMed]
14. Parsons, L.M.; Bouwman, K.M.; Azurmendi, H.; De Vries, R.P.; Cipollo, J.F.; Verheije, M.H. Glycosylation of the viral attachment protein of avian coronavirus is essential for host cell and receptor binding. *J. Biol. Chem.* **2019**, *294*, 7797–7809. [CrossRef] [PubMed]
15. Kumar, S.; Stecher, G.; Tamura, K. MEGA7: Molecular Evolutionary Genetics Analysis Version 7.0 for Bigger Datasets. *Mol. Biol. Evol.* **2016**, *33*, 1870–1874. [CrossRef]
16. Lim, T.H.; Lee, H.J.; Lee, D.H.; Lee, Y.N.; Park, J.K.; Youn, H.N.; Kim, M.S.; Lee, J.B.; Park, S.Y.; Choi, I.S.; et al. An emerging recombinant cluster of nephropathogenic strains of avian infectious bronchitis virus in Korea. *Infect. Genet. Evol.* **2011**, *11*, 678–685. [CrossRef]
17. Pohuang, T.; Chansiripornchai, N.; Tawatsin, A.; Sasipreeyajan, J. Sequence analysis of S1 genes of infectious bronchitis virus isolated in Thailand during 2008-2009: Identification of natural recombination in the field isolates. *Virus Genes* **2011**, *43*, 254–260. [CrossRef]
18. Promkuntod, N.; van Eijndhoven, R.E.W.; de Vrieze, G.; Gröne, A.; Verheije, M.H. Mapping of the receptor-binding domain and amino acids critical for attachment in the spike protein of avian coronavirus infectious bronchitis virus. *Virology* **2014**, *448*, 26–32. [CrossRef]
19. Sun, X.; Li, L.; Pan, L.; Wang, Z.; Chen, H.; Shao, C.; Yu, J.; Ren, Y.; Wang, X.; Huang, X.; et al. Infectious bronchitis virus: Identification of Gallus gallus APN high-affinity ligands with antiviral effects. *Antivir. Res.* **2021**, *186*. [CrossRef]
20. Bouwman, K.M.; Parsons, L.M.; Berends, A.J.; de Vries, R.P.; Cipollo, J.F.; Verheije, M.H. Three Amino Acid Changes in Avian Coronavirus Spike Protein Allow Binding to Kidney Tissue. *J. Virol.* **2020**, *94*, e01363-19. [CrossRef]
21. Zheng, J.; Yamada, Y.; Fung, T.S.; Huang, M.; Chia, R.; Liu, D.X. Identification of N-linked glycosylation sites in the spike protein and their functional impact on the replication and infectivity of coronavirus infectious bronchitis virus in cell culture. *Virology* **2018**, *513*, 65–74. [CrossRef]
22. Stevenson-Leggett, P.; Armstrong, S.; Keep, S.; Britton, P.; Bickerton, E. Analysis of the avian coronavirus spike protein reveals heterogeneity in the glycans present. *J. Gen. Virol.* **2021**, *102*. [CrossRef] [PubMed]
23. Wickramasinghe, I.N.A.; de Vries, R.P.; Gröne, A.; de Haan, C.A.M.; Verheije, M.H. Binding of Avian Coronavirus Spike Proteins to Host Factors Reflects Virus Tropism and Pathogenicity. *J. Virol.* **2011**, *85*, 8903–8912. [CrossRef] [PubMed]
24. Jiang, Y.; Cheng, X.; Zhao, X.; Yu, Y.; Gao, M.; Zhou, S. Recombinant infectious bronchitis coronavirus H120 with the spike protein S1 gene of the nephropathogenic IBYZ strain remains attenuated but induces protective immunity. *Vaccine* **2020**, *38*, 3157–3168. [CrossRef] [PubMed]
25. Jackwood, M.W.; Hilt, D.A.; Callison, S.A.; Lee, C.W.; Plaza, H.; Wade, E. Spike glycoprotein cleavage recognition site analysis of infectious bronchitis virus. *Avian Dis.* **2001**, *45*, 366–372. [CrossRef]
26. Zhang, X.; Deng, T.; Lu, J.; Zhao, P.; Chen, L.; Qian, M.; Guo, Y.; Qiao, H.; Xu, Y.; Wang, Y.; et al. Molecular characterization of variant infectious bronchitis virus in China, 2019: Implications for control programmes. *Transbound. Emerg. Dis.* **2020**, *67*, 1349–1355. [CrossRef]
27. Abro, S.H.; Ullman, K.; Belák, S.; Baule, C. Bioinformatics and evolutionary insight on the spike glycoprotein gene of QX-like and Massachusetts strains of infectious bronchitis virus. *Virol. J.* **2012**, *9*. [CrossRef]
28. Casais, R.; Thiel, V.; Siddell, S.G.; Cavanagh, D.; Britton, P. Reverse genetics system for the avian coronavirus infectious bronchitis virus. *J. Virol.* **2001**, *75*, 12359–12369. [CrossRef]
29. Van Beurden, S.J.; Berends, A.J.; Krämer-Kühl, A.; Spekreijse, D.; Chénard, G.; Philipp, H.C.; Mundt, E.; Rottier, P.J.M.; Verheije, M.H. A reverse genetics system for avian coronavirus infectious bronchitis virus based on targeted RNA recombination. *Virol. J.* **2017**, *14*, 109. [CrossRef]

viruses MDPI

Article

Effect of Insertion and Deletion in the Meq Protein Encoded by Highly Oncogenic Marek's Disease Virus on Transactivation Activity and Virulence

Jumpei Sato [1], Shiro Murata [1,2,*], Zhiyuan Yang [1,3], Benedikt B. Kaufer [4], Sotaro Fujisawa [1], Hikari Seo [1], Naoya Maekawa [2], Tomohiro Okagawa [2], Satoru Konnai [1,2], Nikolaus Osterrieder [4,5], Mark S. Parcells [6] and Kazuhiko Ohashi [1,2]

1 Department of Disease Control, Faculty of Veterinary Medicine, Hokkaido University, Sapporo 060-0818, Japan; s16110010@st.obihiro.ac.jp (J.S.); b20203050402@cau.edu.cn (Z.Y.); s.fujisawa@vetmed.hokudai.ac.jp (S.F.); seo-fuwago@eis.hokudai.ac.jp (H.S.); konnai@vetmed.hokudai.ac.jp (S.K.); okazu@vetmed.hokudai.ac.jp (K.O.)
2 Department of Advanced Pharmaceutics, Faculty of Veterinary Medicine, Hokkaido University, Sapporo 060-0818, Japan; maekawa@vetmed.hokudai.ac.jp (N.M.); okagawa@vetmed.hokudai.ac.jp (T.O.)
3 Institute of Animal Husbandry and Veterinary Medicine, Beijing Academy of Agriculture and Forestry Sciences, Beijing 100097, China
4 Institut für Virologie, Freie Universität Berlin, 14163 Berlin, Germany; benedikt.kaufer@fu-berlin.de (B.B.K.); no.34@fu-berlin.de (N.O.)
5 Department of Infectious Diseases and Public Health, City University of Hong Kong, Kowloon Tong, Hong Kong
6 Department of Animal and Food Sciences, University of Delaware, Newark, DE 19716, USA; parcells@udel.edu
* Correspondence: murata@vetmed.hokudai.ac.jp; Tel.: +81-11-706-5274; Fax: +81-11-706-5217

Citation: Sato, J.; Murata, S.; Yang, Z.; Kaufer, B.B.; Fujisawa, S.; Seo, H.; Maekawa, N.; Okagawa, T.; Konnai, S.; Osterrieder, N.; et al. Effect of Insertion and Deletion in the Meq Protein Encoded by Highly Oncogenic Marek's Disease Virus on Transactivation Activity and Virulence. *Viruses* **2022**, *14*, 382. https://doi.org/10.3390/v14020382

Academic Editor: Chi-Young Wang

Received: 26 January 2022
Accepted: 11 February 2022
Published: 14 February 2022

Publisher's Note: MDPI stays neutral with regard to jurisdictional claims in published maps and institutional affiliations.

Copyright: © 2022 by the authors. Licensee MDPI, Basel, Switzerland. This article is an open access article distributed under the terms and conditions of the Creative Commons Attribution (CC BY) license (https://creativecommons.org/licenses/by/4.0/).

Abstract: Marek's disease virus (MDV) causes malignant lymphoma in chickens (Marek's disease, MD). Although MD is currently controlled by vaccination, MDV strains have continuously increased in virulence over the recent decades. Polymorphisms in Meq, an MDV-encoded oncoprotein that serves as a transcription factor, have been associated with the enhanced virulence of the virus. In addition, insertions and deletions in Meq have been observed in MDV strains of higher virulence, but their contribution to said virulence remains elusive. In this study, we investigated the contribution of an insertion (L-Meq) and a deletion in the Meq gene (S-Meq) to its functions and MDV pathogenicity. Reporter assays revealed that both insertion and deletion enhanced the transactivation potential of Meq. Additionally, we generated RB-1B-based recombinant MDVs (rMDVs) encoding each Meq isoform and analyzed their pathogenic potential. rMDV encoding L-Meq indueced the highest mortality and tumor incidence in infected animals, whereas the rMDV encoding S-Meq exhibited the lowest pathogenicity. Thus, insertion enhanced the transactivation activity of Meq and MDV pathogenicity, whereas deletion reduced pathogenicity despite having increased transactivation activity. These data suggest that other functions of Meq affect MDV virulence. These data improve our understanding of the mechanisms underlying the evolution of MDV virulence.

Keywords: Marek's disease virus; Marek's disease; Meq; CVI988; tumorigenesis; pathogenicity; transactivation activity

1. Introduction

Marek's disease virus (MDV) is an alphaherpesvirus that belongs to the Herpesviridae family (subfamily: *Alphaherpesvirinae*, genus: *Mardivirus*, species: *Gallid alphaherpesvirus 2*). MDV causes Marek's disease (MD), which is characterized by neurological disorders, malignant lymphomas, and immunosuppression in infected chickens. Although MD has previously caused severe economic losses in the poultry industry, the disease is currently controlled by administering attenuated and/or non-pathogenic strains as vaccines [1]. *Meleagrid alphaherpesvirus 1* (turkey herpesvirus, HVT) was initially introduced to the poultry industry, followed by bivalent vaccines consisting of HVT and a strain of non-oncogenic *Gallid alphaherpesvirus 3*, and an attenuated *Gallid alphaherpesvirus 2* strain, Rispens CVI988, was introduced and is widely used as a gold standard vaccine against MD [2]. Introduction of the vaccines minimized the economic losses caused by virulent MDVs; however, the virulence of MDV field strains has continuously increased. Sporadic occurrences of MD are still reported even in vaccinated flocks in some countries [3–8], and the costs to the poultry industry are estimated at approximately USD 2 billion annually [9]. The emergence of MDV strains with increased virulence is considered to be a result of the selection of field strains adapted to vaccine-induced immune pressure [10–12]. Therefore, MDV strains with higher virulence may cause future outbreaks [13].

The *meq* oncogene encodes a 339-amino-acid protein that is expressed during the lytic and latent phases of infection and is essential for transformation of chicken T-cells [14]. Meq consists of a basic region (BR) and a leucine zipper (ZIP), similar to those of the oncoproteins Jun and Fos, at the N-terminus, and a transactivation domain characterized by proline-rich repeats (PRRs) at the C-terminal [14]. Meq can dimerize with itself, c-Jun, JunB, and Fos through its ZIP domain [15]. Microarray analysis identified that the same genes induced by v-Jun transformation, including *JTAP1*, *JAC*, and *HB-EGF*, were also upregulated by the expression of Meq in transformed DF-1 cells [16]. Additionally, Meq upregulated the expression of Bcl-2 and Ski and downregulated the expression of DAP5 and Fas, consistent with its anti-apoptotic properties [16,17]. Furthermore, *meq*-deleted recombinant MDV (rMDV) failed to induce lymphomas in infected chickens, indicating that Meq plays an essential role in the transformation induced by virulent MDV [18]. The distinct diversity in Meq proteins of MDV strains with varying virulence has been highlighted as a factor involved in the increased virulence of MDV [19]. Among them, the polymorphisms in the BR and PRRs in the transactivation domain have been associated with virulence-level, and several amino acid substitutions in these regions affect the transactivation activities of Meq proteins [20–22]. Moreover, in vivo studies using rMDVs generated based on the very virulent strain RB-1B strain revealed that rMDVs encoding Meq from highly virulent MDV strains showed higher virulence than those from low virulent strains [23]. Thus, polymorphisms in the *meq* gene appear to be directly involved in the increased virulence of MDV field strains.

In addition to the polymorphisms in Meq proteins, insertion or deletion in the transactivation domain of Meq have been detected in several MDV strains [8,20,24–28]. A long isoform of Meq containing an insertion (L-Meq) is characterized by a 59/60-amino acid insertion in the transactivation domain [24,27,29], and its insertion causes an increase in the number of PRRs. L-Meq was first identified in CVI988 [29] and has subsequently been detected in low-virulence strains isolated in the USA [19]. Therefore, it was initially believed that the L-Meq isoform contributes to decreased virulence of MDVs encoding this form. However, it was clarified that virulent MDV strains circulating in Australia encode an L-Meq, which also contains polymorphisms in PRRs [24]. Furthermore, recent in vivo experiments using RB-1B-based rMDVs harboring L-Meq or Meq encoded by CVI988, rMDV encoding CVI988-L-Meq exhibited markedly higher pathogenicity and tumorigenicity than did CVI988 [30]. Surprisingly, the rMDV encoding CVI988-L-Meq showed similar or even greater virulence than the rMDV encoding RB-1B-Meq, although RB-1B is an MDV strain with high virulence. Thus, amino acid insertions into the transactivation domain appear to contribute to enhanced virulence.

Since the 2010s, MDV strains encoding the short isoform of Meq (S-Meq) have been detected in some countries [20,25,28,31]. S-Meq contains a 41-amino acid deletion in the transactivation domain, resulting in a decrease in the number of PRRs. In addition, an in vitro study revealed that S-Meq had higher transactivation activity than Meq, suggesting that deletion in the transactivation domain enhances the protein functions of Meq [20]. However, MDV strains encoding S-Meq were observed in chickens without clinical signs in Italy [25] and in diseased but unvaccinated chickens in Japan [20]. Therefore, the contribution of S-Meq to MDV pathogenicity remains unclear. However, because deletion in the transactivation domain was observed in recent field strains and could enhance the transactivation activity of Meq, this deletion may be involved in the evolution towards a greater virulence of these strains [20].

In this study, we systematically investigated the contribution of a common insertion and deletion in the transactivation domain of Meq and evaluated these for their effects on transactivation and MDV pathogenicity. We analyzed the transactivation activities of Meq, L-Meq, and S-Meq, the background sequences of which were identical to those of RB-1B-Meq. In addition, we generated RB-1B-based rMDVs encoding each Meq isoform by replacing the *meq* gene of the RB-1B genome cloned as a bacterial artificial chromosome (BAC) with the L-*meq*/S-*meq* gene that was introduced, and animal experiments were conducted to examine whether insertion/deletion affected MDV pathogenicity.

2. Materials and Methods

2.1. Ethics Statement

All animal experiments were approved by the Institutional Animal Care and Use Committee of Hokkaido University (approval number: 19-0081). All experiments were performed in accordance with the relevant guidelines and regulations of the Faculty of Veterinary Medicine of Hokkaido University, which is fully accredited by the Association for Assessment and Accreditation of Laboratory Animal Care International.

2.2. Cells and Virus

Chicken embryo fibroblasts (CEFs) were prepared from 10-day-old fertilized eggs (Iwamura Hatchery Co., Ltd., Shibata, Japan) as described previously [32]. CEFs were cultured in Eagle's minimal essential medium (Nissui Pharmaceutical Co., Ltd., Tokyo, Japan) supplemented with 10% bovine calf serum (Sigma-Aldrich, St. Louis, MO, USA), 10% tryptose phosphate broth (Difco Laboratories, Detroit, MI, USA), 0.03% L-glutamine, 100 U/mL penicillin, 100 μg/mL streptomycin, and 0.1% $NaHCO_3$. DF-1 cells, a chicken fibroblast cell line, were cultured with 0.5 mL of Dulbecco's modified Eagle's medium (FUJIFILM Wako Pure Chemical Corporation, Osaka, Japan), containing 10% fetal bovine serum, and incubated at 39 °C under 5% CO_2. Viruses were reconstituted by transfecting BAC DNA into CEFs as described previously [33]. Viruses were propagated on CEFs for seven passages, and virus stocks were frozen in Cell Banker 1 (Nippon Zenyaku Kogyo Co., Ltd., Fukushima, Japan) and titrated on CEFs using plaque assays as described previously [34,35].

2.3. Plasmids

The expression plasmids for S-Meq and L-Meq were constructed, and some mutations were introduced by site-directed mutagenesis, according to a previous report [22]. The open reading frames (ORFs) of the S-*meq* and L-*meq* genes derived from the MDV strains Kgw-c2, an MDV strain isolated from unvaccinated chickens with MD symptoms in Japan in the 2010s (accession number: LC385874), and CVI988, an attenuated vaccine strain (accession number: AF493555) were amplified and cloned into the pCI-neo vector (Promega, Madison, WI, USA). The pathotype of Kgw-c2 has not been determined. To match the amino acid sequences with those of Meq from the very virulent RB-1B strain, aside from deletion or insertion in the transactivation domain, we introduced some mutations in the S-*meq* and L-*meq* genes, as shown in Table 1, and their S-Meq and L-Meq were designated as S-Meq (RB-1B) and L-Meq (RB-1B), respectively. In addition, we constructed an expression plasmid for Meq from MDV strain RB-1B (accession number: HM488349.1). For the assay to measure the transactivation activity, a c-Jun expression plasmid was constructed [22], and a reporter plasmid was prepared by inserting the Meq promoter region upstream of the firefly luciferase-coding region in the pGL3-Basic vector (Promega) [22]; a pRL-TK Renilla luciferase plasmid (Promega) was used as the control plasmid.

Table 1. Introduction of the mutation into the S-*meq* and/or L-*meq* genes.

Position in Meq	Substitution		Primers
71	serine–to–alanine	F	5′-GAA TCG TGA CGC CGC TCG GAG AAG ACG-3′
		R	5′-CGT CTT CTC CGA GCG GCG TCA CGA TTC-3′
77, 80	glutamic-acid–to–lysine, tyrosine–to–aspartic-acid	F	5′-CGG AGA AGA CGC AGG AAG CAG ACG GAC TAT GTA GAC AAA C-3′
		R	5′-GTT TGT CTA CAT AGT CCG TCT GCT TCC TGC GTC TTC TCC G-3′
114, 115	cysteine–to–arginine, alanine–to–valine	F	5′-GAG TGC ACG TCC CTG CGT GTA CAG TTG GCT TGT CA-3′
		R	5′-TGA CAA GCC AAC TGT ACA CGC AGG GAC GTG CAC TC-3′
217	alanine–to–proline	F	5′-ATC TGT ACC CCC CCT CCT CCC GAT G-3′
		R	5′-CAT CGG GAG GAG GGG GGG TAC AGA T-3′
326	isoleucine–to–threonine	F	5′-GTT TCC CTC GGA TAC TCA GTC TAC GGT CT-3′
		R	5′-AGA CCG TAG ACT GAG TAT CCG AGG GAA AC-3′

2.4. Dual-Luciferase Reporter Assay

First, DF-1 cells were seeded in 24-well plates at 2.0×10^5 cells per well and incubated for 24 h. The cells in each well were transfected with 300 ng of expression plasmids, Meq, S-Meq (wild-type), S-Meq (RB-1B), L-Meq (wild-type), or L-Meq (RB-1B); 200 ng of the c-Jun expression plasmid; 500 ng of the reporter plasmid; and 5 ng of control pRL-TK using Lipofectamine 2000 (Thermo Fisher Scientific, Waltham, MA, USA) according to the manufacturer's instructions. The transfected cells were lysed 24 h after transfection using $1 \times$ Passive Lysis Buffer (Promega). Luciferase activity was measured using the Dual-Luciferase Reporter Assay System (Promega) and a Luminescencer-JNR AB-2100 (Atto Corp., Tokyo, Japan). The luminescence intensity of firefly luciferase was normalized to that of *Renilla* luciferase, and the results are indicated relative to the value of the luciferase activity in cells transfected with the pCI-neo vector.

2.5. Generation of Recombinant Viruses

To generate recombinant viruses encoding S-Meq (RB-1B), Meq, or L-Meq (RB-1B), we used a BAC clone of the very virulent RB-1B strain that lacks most of the internal repeat long (IRL) region as described previously [35]. The deleted IRL region is rapidly restored during virus reconstitution in the cell culture, resulting in a recombinant virus that harbors the respective *meq* gene in both repeat regions [35]. Therefore, we replaced the *meq* gene in the terminal repeat long (TRL) with the RB-1B-*meq*, S-*meq* (RB-1B), or L-*meq* (RB-1B) gene by two-step Red-mediated mutagenesis as previously described [36,37]. To screen for clones in which each *meq*-isoform was accurately inserted, the obtained plasmids encoding each rMDV genome were digested with BamHI-HF (New England Biolabs Japan Inc., Tokyo, Japan) overnight and subjected to electrophoresis on a 0.8% agarose gel. The insertion of each *meq*-isoform was further confirmed by polymerase chain reaction (PCR) and DNA sequencing, as previously reported [35]. The BAC-based infectious clones were transfected into CEFs using a CalPhos Mammalian Transfection Kit (Takara Bio Inc., Kyoto, Japan) according to the manufacturer's instructions. The pCAGGS-Cre plasmid (Gene Bridges GmbH, Heidelberg, Germany) was co-transfected to remove the BAC sequence from the virus genome. The reconstituted recombinant viruses were referred to as vRB-1B_Meq, vRB-1B_S-Meq, and vRB-1B_L-Meq, respectively, and all viruses were passaged on CEFs. As the IRL region is rapidly restored during viral reconstitution, the restoration of the IRL, in addition to the deletion of the BAC sequence, in each virus was confirmed by PCR [35]. Each virus was titrated using a plaque assay and stored as described previously [34,35].

2.6. In Vitro Replication of the Recombinant Viruses

CEFs were seeded in six-well plates and infected with 50 plaque-forming units (PFU) of recombinant viruses after reaching confluence. The infected cells were collected daily for 5 days. The viral loads in the infected cells were analyzed by quantitative PCR (qPCR) to assess the replication and spread of recombinant viruses in vitro.

2.7. *Confirmation of Meq Expression Levels by RT-PCR*

A reverse transcription (RT)-PCR assay was used to investigate the mRNA expression of each *meq* isoform. In brief, we first seeded CEFs in 24-well plates at 2.0×10^5 cells per well, and after 24 h incubation, the CEFs were infected with 50 PFU of recombinant viruses. At 6 days post-infection (dpi), the infected CEFs were lysed with TRI reagent (Molecular Research Center, Inc., Cincinnati, OH, USA) and the total RNA was extracted according to the manufacturer's instructions. After the total RNA was treated with DNase I (Promega), cDNA was synthesized using PrimeScript Reverse Transcriptase (Takara Bio Inc.). The expression of each *meq* isoform in the cDNA samples was detected by PCR using TaKaRa Ex Taq (Takara Bio Inc.). The primer sets used are listed in Table 2.

Table 2. Primers used for reverse transcription-polymerase chain reaction and quantitative polymerase chain reaction analyses.

Gene	Sequence	Application
meq	F 5′-AGT TGG CTT GTC ATG AGC CAG-3′ R 5′-TGT TCG GGA TCC TCG GTA AGA-3′	RT-PCR
ICP4	F 5′-GCA TCG ACA AGC ACT TAC GG-3′ R 5′-CGA GAG CGT CGT ATT GTT TGG-3′	qPCR
iNOS	F 5′-GAG TGG TTT AAG GAG TTG GAT CTG A-3′ R 5′-TTC CAG ACC TCC CAC CTC AA-3′	qPCR

Abbreviations: qPCR, quantitative polymerase chain reaction; RT-PCR, reverse transcription polymerase chain reaction.

2.8. *In Vivo Characterization of Recombinant Viruses*

2.8.1. Experimental Chickens

Fertilized eggs (Iwamura Hatchery Co., Ltd.) from commercial white leghorn chickens were hatched in the authors' laboratory and the chicks were raised in isolators. To analyze the recombinant viruses, we performed animal experiments as follows:

2.8.2. 1st Animal Experiment

A total of 98 1-day-old chicks (Iwamura Hatchery Co., Ltd.) were randomly divided into four groups and housed separately in different isolators. The chickens were inoculated via the intraabdominal route with 5000 PFU of vRB-1B_Meq (n = 28), vRB-1B_S-Meq (n = 27), vRB-1B_L-Meq (n = 27), or PBS (n = 16) as a negative control.

- In vivo kinetics of recombinant viruses

Four chickens per group were euthanized at 7, 14, 28, and 35 dpi, and the blood, spleens, and feather tips were collected. Peripheral blood mononuclear cells (PBMCs) were isolated by density gradient centrifugation using Percoll solution (GE Healthcare, Chicago, IL, USA). DNA was extracted from the PBMCs, spleens, and feather tips, and the viral loads were analyzed for each sample.

- Pathogenicity of recombinant viruses in unvaccinated chickens

The pathogenicity of the recombinant viruses in infected chickens, vRB-1B_Meq (n = 12), vRB-1B_S-Meq (n = 11), and vRB-1B_L-Meq (n = 12) were compared by monitoring the survival rate. After inoculating the chickens with recombinant viruses, we monitored them for clinical signs of MD daily for 8 weeks, euthanized the infected chickens that showed neurological symptoms and dysphagia during the experimental period, and examined the gross tumor lesions in the internal organs. In addition, we examined the tumor incidence in infected chickens without clinical signs at 56 dpi. We collected the blood, spleens, feather tips, and tumors from euthanized chickens and analyzed the viral loads in each tissue type.

2.8.3. 2nd Animal Experiment: Pathogenicity of Recombinant Viruses in Vaccinated Chickens

A total of 29 1-day-old chicks (Iwamura Hatchery Co., Ltd.) were randomly divided into three groups and housed separately. The chickens were subcutaneously inoculated with 2000 PFU of HVT vaccine (strain FC 126; Kyoritsu Seiyaku, Tokyo, Japan). At 3 days post-vaccination, the chickens were superinfected via the intraabdominal route with 5000 PFU of vRB-1B_Meq ($n = 9$), vRB-1B_S-Meq ($n = 8$), or vRB-1B_L-Meq ($n = 12$), and the survival rate and tumor incidence were monitored for 8 weeks. We collected blood, spleens, feather tips, and tumors from euthanized chickens and analyzed the viral loads in each tissue type. To monitor the viral loads, we also collected blood from the wing veins of four chickens per group at 7, 14, 28, 35, and 49 dpi and monitored the viral loads in the samples from the whole blood of infected chickens.

2.9. *DNA Extraction*

Total cellular DNA was extracted from the whole blood samples of infected chickens using a DNeasy blood and tissue kit (Qiagen, Düsseldorf, Germany) according to the manufacturer's instructions. Total cellular DNA was extracted from the feather tips as previously described [38,39]. Briefly, the two feather tips were cut and immersed overnight at 55 °C in 1 mL of lysis buffer (0.5% sodium dodecyl sulfate, 0.1 M NaCl, 10 mM Tris pH 8.0, 1 mM ethylenediaminetetraacetic acid) containing proteinase K at a final concentration of 200 mg/mL. Total cellular DNA was extracted using phenol-chloroform-isoamyl alcohol (25:24:1), precipitated with ethanol, and treated with RNase A. The total cellular DNA was extracted from the infected CEFs, PBMCs, spleens, and tumor lesions using SepaGene (Sekisui Medical Co., Ltd., Tokyo, Japan) according to the manufacturer's instructions. All samples were subjected to qPCR analysis to determine the viral load.

2.10. *qPCR*

The viral loads in CEFs and chickens infected with the recombinant viruses and exposed to the HVT vaccine were determined by qPCR with primers specific to the *infected cell protein 4* (*ICP4*) gene of MDV and the *HVT070* gene of HVT, respectively. qPCR was performed using TB Green Premix DimerEraser (Takara Bio Inc.) and LightCycler 480 System II (Roche Diagnostics, Mannheim, Germany). The chicken *inducible nitric oxide synthase* (*iNOS*) gene was amplified as a reference gene. The viral loads are indicated as ratios between each target and the *iNOS* gene. The primers used for the qPCR analyses are shown in Table 2.

2.11. *Statistical Analyses*

Statistical analyses were performed using R Statistical Software (version 4.0.3; R Foundation for Statistical Computing, Vienna, Austria). The multi-step growth kinetics were analyzed using the Kruskal–Wallis test. Kaplan–Meier survival curves were generated to analyze the survival rate in infected chickens, and a log-rank test (Mantel–Cox test) was conducted. Tumor incidence was analyzed using Fisher's exact test. The transactivation activities were analyzed using Tukey's multiple comparison test. Statistical significance was set at $p < 0.05$.

3. Results

3.1. *Transactivation Activity of Meq Isoforms*

To determine whether the insertion/deletion in the transactivation domain of Meq affects protein function, we analyzed the transactivation activities of S-Meq from Kgw-c2, an MDV strain isolated in Japan in the 2010s, the Meq from RB-1B (parent), and the L-Meq from CVI988. Because there are some polymorphisms among RB-1B-*meq*, Kgw-c2-S-*meq*, and CVI988-L-*meq* (Figure 1A), we introduced some mutations in wild-type S-*meq* and L-*meq* genes from Kgw-c2 and CVI988, respectively, and manipulated their sequences to the same sequence as RB-1B-*meq*, aside from the insertion or deletion in the transactivation domain (Figure 1A). S-Meq (wild type) demonstrated similar transactivation activity to that of RB-1B-Meq, whereas the transactivation activity of L-Meq (wild type) was significantly lower than that of RB-1B-Meq and S-Meq (wild type) (Figure 1B). The mutation-introduced S-Meq, L-Meq, S-Meq (RB-1B), and L-Meq (RB-1B) showed higher transactivation activities than wild-type S-Meq and L-Meq (Figure 1B), suggesting that these amino acid substitutions were responsible for higher transactivation activity. Moreover, the transactivation activities of S-Meq (RB-1B) and L-Meq (RB-1B) were higher than that of the wild-type RB-1B Meq. Interestingly, the L-Meq (RB-1B) exhibited the highest activity among the mutated L-Meq and S-Meq, and RB-1B-Meq isoforms (Figure 1B). These data indicate that both insertion and deletion in the transactivation domain have the potential to enhance the transactivation activity of Meq.

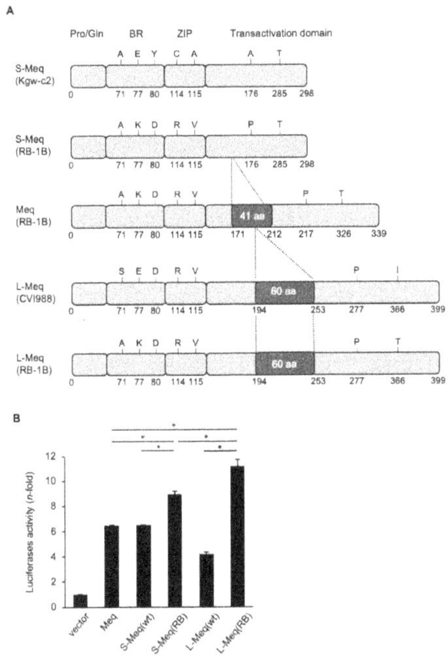

Figure 1. Analysis of the transactivation activity of the three Meq isoforms. (**A**) Schematic representation of the different Meq isoforms. The structures of short-Meq containing the deletion (S-Meq) from Kgw-c2, a Marek's disease virus (MDV) strain, Meq from RB-1B, and long-Meq containing the insertion (L-Meq) from CVI988, and S-Meq (RB-1B) and L-Meq (RB-1B) whose sequences were matched with that of RB-1B-Meq, except for the deletion or insertion in the transactivation domain, are indicated. Meq comprises a proline/glutamine (Pro/Gln) rich domain followed by the basic region (BR) and leucine zipper (ZIP) at the N-terminal region, and the transactivation domain at the C-terminal region. The Meq isoforms include mutations in the BR, ZIP, and transactivation domain. S-Meq contains a 41 amino acid deletion in the transactivation domain and L-Meq is characterized by a 60-amino acid insertion in the transactivation domain. (**B**) Transactivation activity of the Meq isoforms. The transactivation activity of RB-1B-Meq, wild-type S-Meq (Kgw-c2), wild-type L-Meq (CVI988), S-Meq (RB-1B), and L-Meq (RB-1B) was compared on the Meq promoter-driven luciferase activities. DF-1 cells in each well were transfected with 300 ng of expression plasmids, Meq, S-Meq (wild-type), S-Meq (RB-1B), L-Meq (wild-type), or L-Meq (RB-1B); 200 ng of the c-Jun expression plasmid; 500 ng of the reporter plasmid; and 5 ng of control pRL-TK. Luciferase activities were analyzed 24 h post-transfection. Firefly luciferase activity is expressed relative to the mean basal activity in the presence of pCI-neo after normalization to Renilla luciferase activity. Error bars indicate standard deviations. * $p < 0.01$.

3.2. Generation of Recombinant Viruses

To investigate whether insertion/deletion affects MDV pathogenicity, we generated RB-1B-based rMDVs encoding each Meq isoform, as previously reported [30]. We inserted the S-*meq* (RB-1B), RB-1B *meq*, and L-*meq* (RB-1B) genes into the IRL-deleted RB-1B genome cloned as a BAC plasmid by replacing the native *meq* gene in the TRL with them (Figure 2A). The resulting BAC plasmids were screened using restriction fragment length polymorphism analysis (Figures 2B and S1), and the insertion of each *meq*-isoform was confirmed by polymerase chain reaction (PCR) and DNA sequencing (data not shown). The BAC-based infectious clones were transfected into chicken embryo fibroblasts (CEFs), and the restoration of the IRL in the reconstituted viruses, termed vRB-1B_S-Meq, vRB-1B_Meq, and vRB-1B_L-Meq, was confirmed by PCR (data not shown).

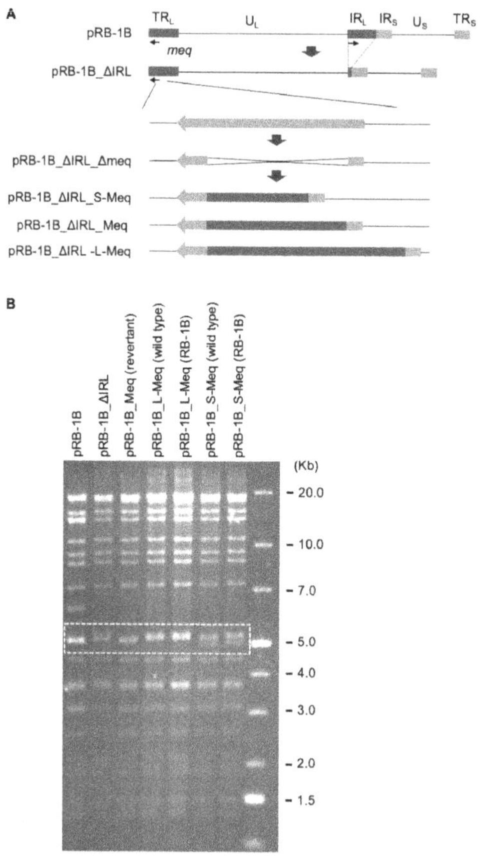

Figure 2. Construction of recombinant Marek's disease viruses. (**A**) Schematic diagrams of the constructs cloned the RB-1B genome used in this study. In the RB-1B genome cloned as the bacterial artificial chromosome (BAC) plasmid (pRB-1B), most of the internal repeat long (IRL) regions were deleted, designated as pRB-1B_ΔIRL, and used for mutagenesis. The *meq* gene in terminal repeat long (TRL) was replaced with the RB-1B-*meq*, short-Meq containing the deletion (S-*meq*) (RB-1B), or long-Meq containing the insertion (L-*meq*) (RB-1B) genes by two-step red-mediated mutagenesis. (**B**) Restriction fragment length polymorphism analysis of the BAC plasmids obtained by mutagenesis. The BAC plasmids were digested with BamHI to determine the accurate insertion of each *meq*-isoform into the RB-1B genome. A dashed box indicates the variation of the recombinant Marek's disease virus genomes with each *meq*-isoform.

3.3. Characterization of rMDVs In Vitro

To determine if insertion and deletion in the transactivation domain affect virus replication in vitro, we analyzed the expression of each *meq*-isoform and viral loads in CEFs infected with vRB-1B_S-Meq, vRB-1B_Meq, and vRB-1B_L-Meq. We confirmed the mRNA expression of each *meq*-isoform in infected CEFs by RT-PCR (Figures 3A and S2), and no significant difference in the growth kinetics among the rMDVs in vitro was observed (Figure 3B). These results suggest that insertion and deletion in the transactivation domain do not affect virus replication of the recombinant viruses in cell culture.

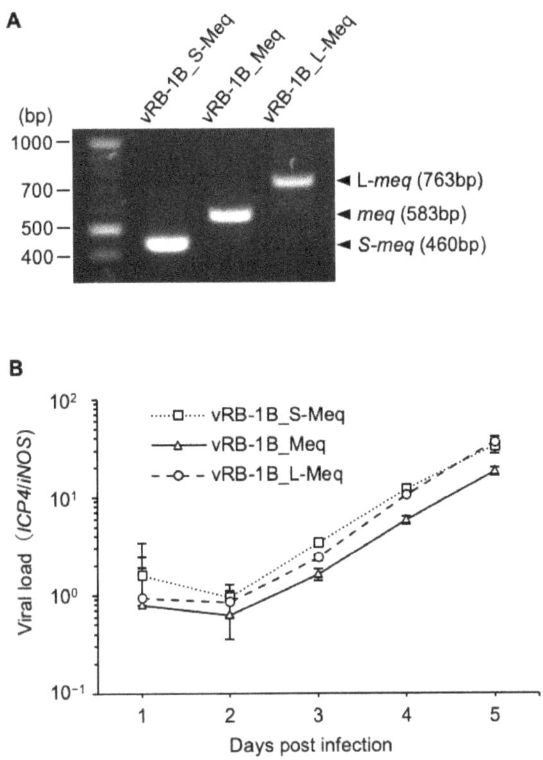

Figure 3. Expression analysis and replication of recombinant MDVs in cell culture. (**A**) mRNA expression of each *meq*-isoform in chicken embryo fibroblasts (CEFs) infected with vRB-1B_S-Meq, vRB-1B_Meq, and vRB-1B_L-Meq was confirmed by reverse transcription-polymerase chain reaction (PCR). (**B**) CEFs were infected with 50 plaque-forming units of recombinant Marek's disease viruses. The infected cells were collected daily for 6 days. The viral loads in the infected cells were analyzed by quantitative PCR. The growth kinetics among the groups were analyzed using the Kruskal–Wallis test.

3.4. Replication of rMDVs In Vivo

We then assessed viral replication in vivo. Day-old chicks were inoculated with 5000 PFU of each rMDV via the intraabdominal route, and the viral loads were analyzed in the PBMCs, spleens, and feather tips from four chickens per group at 7, 14, 28, and 35 dpi. Although all of the rMDVs efficiently replicated in infected birds, vRB-1B_L-Meq showed higher viral loads at 28 and 35 dpi than those of vRB-1B_S-Meq and vRB-1B_Meq (Figure 4).

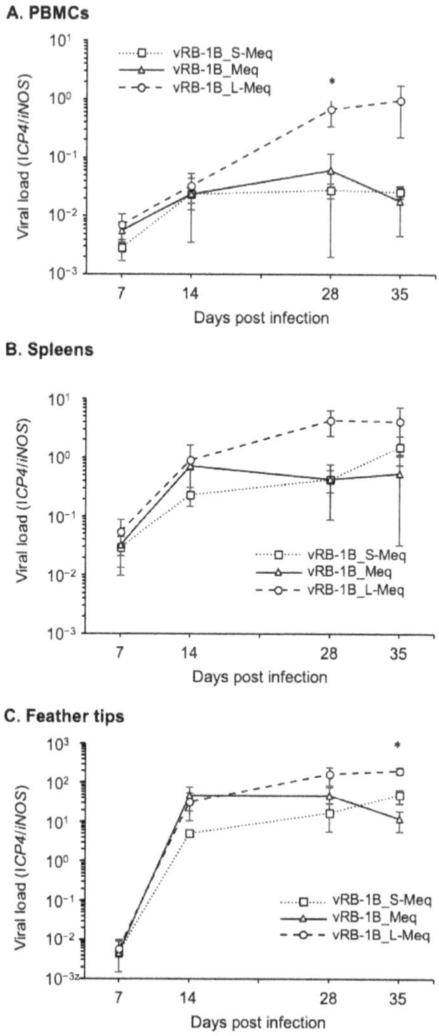

Figure 4. Replication of recombinant MDVs in vivo. The viral loads in the (**A**) peripheral blood mononuclear cells, (**B**) spleens, and (**C**) feather tips from chickens infected with vRB-1B_S-Meq, vRB-1B_Meq, and vRB-1B_L-Meq were determined by quantitative polymerase chain reaction. Asterisks indicate significant differences (* $p < 0.01$; Kruskal–Wallis test).

Remarkably, the viral loads in PBMCs at 28 dpi and in feather tips at 35 dpi from chickens infected with vRB-1B_L-Meq were significantly higher than those infected with vRB-1B_S-Meq and vRB-1B_Meq (Figure 4A,C). These data suggest that insertion in the transactivation domain does not affect the growth kinetics in the early phase of infection but resulted in a higher viral load during latency and in the transformation phases of infection. However, deletion in the transactivation domain did not influence the replication kinetics of rMDVs.

3.5. Pathogenicity of rMDVs In Vivo

To determine whether insertion and deletion in the transactivation domain affect pathogenicity, we monitored the survival rate and tumor incidence in chickens infected with the rMDVs. The vRB-1B_L-Meq was associated with the highest mortality and tumor incidence, whereas the mortality and tumor incidence of vRB-1B_S-Meq were lower than those of vRB-1B_L-Meq and vRB-1B_Meq (Table 3, Figure 5A,B). In addition, we compared viral loads in the PBMCs, spleens, and feather tips collected from chickens infected with each rMDV at the experimental endpoint (euthanized chickens with clinical signs during the experimental period and without clinical signs at 56 dpi). Higher viral loads were observed in the samples from chickens infected with vRB-1B_L-Meq, whereas the viral loads of vRB-1B_S-Meq were significantly lower in all samples (Figure 5C–E). Thus, the pathogenicity of rMDVs was apparently dependent on the length of the transactivation domain.

Table 3. Survival rate and tumor incidence at 56 days post-infection.

rMDV	Survival Rate	Tumor Incidence
vRB-1B_S-Meq	90.9% (10/11)	18.2% (2/11)
vRB-1B_Meq	41.7% (5/12)	58.3% (7/12)
vRB-1B_L-Meq	0% (11/11)	100% (11/11)

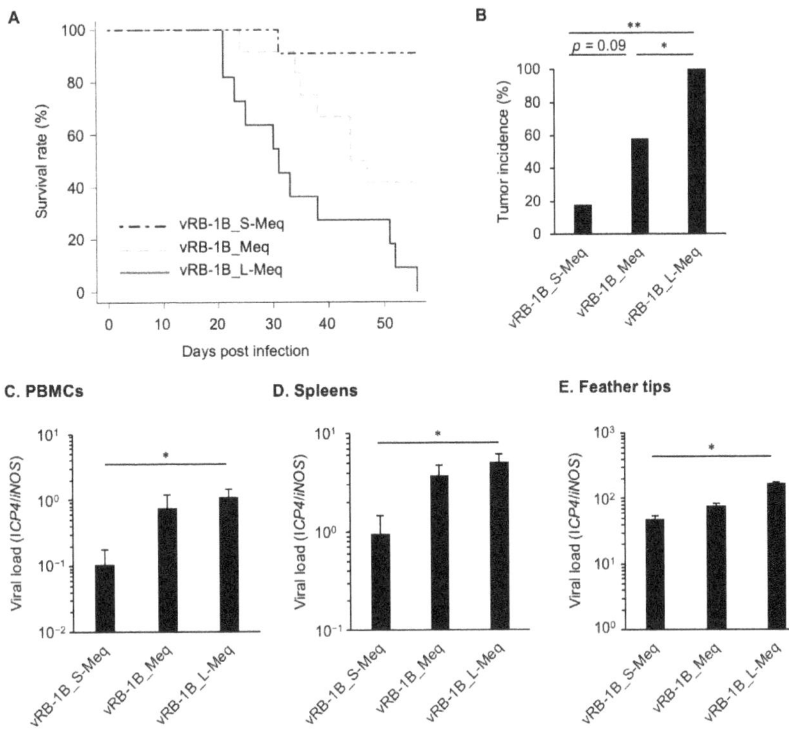

Figure 5. Mortality and tumor incidence in chickens infected with recombinant MDVs. (**A**) Survival rate in chickens infected with recombinant Marek's disease viruses (rMDVs). Asterisks indicate significant differences. (** $p < 0.01$, * $p < 0.05$; log-rank test). (**B**) Tumor incidence in chickens infected with rMDVs throughout the study period. Asterisks indicate significant differences (** $p < 0.01$, * $p < 0.05$; Fisher's exact test). The viral loads in (**C**) peripheral blood mononuclear cells, (**D**) spleens, and (**E**) feather tips from chickens infected with the rMDVs were determined by quantitative polymerase chain reaction. Asterisks indicate significant differences (** $p < 0.01$; Kruskal–Wallis test).

3.6. Pathogenicity of rMDVs in Vaccinated Chickens

Finally, we characterized the pathogenicity of each rMDV in HVT-vaccinated chickens to further characterize the pathogenicity enhanced by the insertion in the transactivation domain. At 28 dpi, no significant difference was observed in the viral loads in whole blood of the vaccinated chickens infected with the vRB-1B_L-Meq strain and other rMDV strains (Figure 6); however, the viral loads in the PBMCs of chickens infected with vRB-1B_L-Meq alone were higher than those of chickens infected with other rMDV strains at 28 dpi (Figure 5A), suggesting that HVT vaccination reduced the viral loads in chickens infected with the vRB-1B_L-Meq strain. However, the viral loads in chickens infected with the vRB-1B_L-Meq strain were higher than those of chickens infected with other rMDVs in the later phase of infection, although the difference in the growth kinetics of rMDVs was not statistically significant (Figure 6). Clinical signs were observed in 2 of the 12 chickens infected with vRB-1B_L-Meq during the experimental period, whereas chickens infected with vRB-1B_Meq or vRB-1B_S-Meq did not exhibit any clinical signs (Table 4). In addition, tumors were observed in 4 of the 12 chickens infected with vRB-1B_L-Meq at termination; however, in the other groups, a tumor was observed in only one chicken infected with vRB-1B_S-Meq (Table 4). These data support the observation that insertion in the transactivation domain of Meq contributes to the increased pathogenicity of this virus.

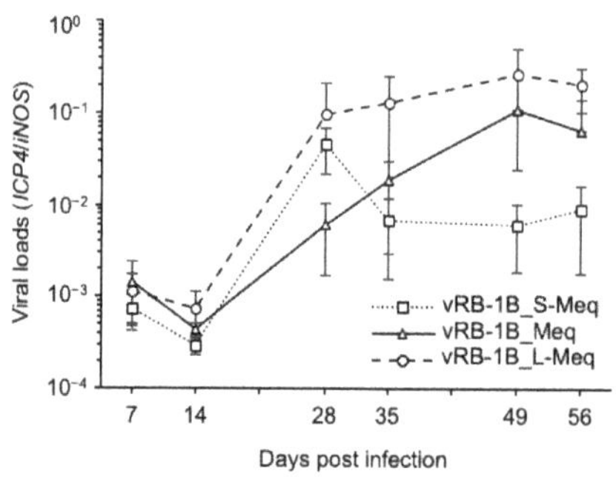

Figure 6. Replication of recombinant MDVs in vaccinated chickens. The chickens were vaccinated and superinfected with vRB-1B_S-Meq, vRB-1B_Meq, or vRB-1B_L-Meq. The viral loads in whole blood were monitored using quantitative polymerase chain reaction with primers specific to the *infected cell protein 4* (*ICP4*) gene of Marek's disease virus. The growth kinetics among the groups were analyzed using the Kruskal–Wallis test.

Table 4. Survival rate and tumor incidence in vaccinated chickens at 56 days post-infection.

rMDV	Survival Rate	Tumor Incidence
vRB-1B_S-Meq	100% (8/8)	12.5% (1/8)
vRB-1B_Meq	100% (9/9)	0% (0/9)
vRB-1B_L-Meq	83.3% (10/12)	33.3% (4/12)

4. Discussion

In this study, we investigated the pathogenicity of rMDVs encoding Meq with insertion or deletion in the transactivation domain of Meq in order to better understand the contribution of Meq to MDV virulence. We compared the transactivation activities among RB-1B-Meq, wild-type S-Meq (Kgw-c2), wild-type L-Meq (CVI988), S-Meq (RB-1B), and L-Meq (RB-1B). S-Meq (RB-1B) and L-Meq (RB-1B) showed higher activities than those of wild-type S-Meq and L-Meq, respectively. These data are consistent with the results that several amino acid substitutions in Meq, such as a glutamic acid-to-lysine substitution, a tyrosine-to-aspartate substitution, and a proline-to-alanine substitution at positions 77, 80, and 217, respectively, contributed to the higher transactivation activity of Meq [21,22] However, the transactivation activities of S-Meq (RB-1B) and L-Meq (RB-1B) were higher than that of the parental RB-1B-Meq. Insertion and deletion into the transactivation-associated domain caused an increase and decrease in the number of PRRs, respectively [19,20]. The PRR, minus the last 39 aa of Meq has been considered to exhibit a transrepressive effect [14], and therefore, an increase or decrease in the number of PRRs was predicted to induce a reduction or increase in the transactivation activity. We previously reported that deletion in the transactivation domain could enhance transactivation activity [20]. However, the L-Meq (RB-1B) exhibited the highest transactivation potential among the Meq constructs, contrary to the PRR theory. According to a previous study, an rMDV encoding CVI-988-L-Meq demonstrated higher pathogenicity than did CVI-988-Meq [30], and these data imply that the insertion in the transactivation domain enhanced the Meq functions related to tumorigenesis. Thus, the enhancement of the transactivation activity observed in L-Meq (RB-1B) seemed to be caused by an unknown mechanism(s) different from the previously suggested theory. As the PRR is theorized to be highly disordered in structure and contains likely binding sites for cellular proteins, alterations in this sequence could have myriad consequences for differential binding of factors. Further analyses are required to clarify the role of PRRs in the protein functions of Meq.

The rMDVs encoding RB-1B-Meq, S-Meq (RB-1B), or L-Meq (RB-1B) exhibited no difference in viral replication in vitro, because Meq was not involved in lytic replication in the infected chickens [18]. In vivo experiments revealed that the viral loads of rRB-1B_L-Meq were higher than those of other rMDVs in the later phases of infection, which could be explained by virus reactivation and/or an increased number of transformed cells upon disease progression. Indeed, vRB-1B_L-Meq caused the highest tumor incidence compared with the other rMDVs. In addition, the increase in viral loads in the feather tips from chickens infected with rRB-1B_L-Meq could facilitate more efficient viral shedding, because the feather follicle epithelium is a site for the production of cell-free viruses [40]. Thus, insertion within the transactivation domain could enhance MDV pathogenicity and efficient virus transmission.

In the present study, vRB-1B_L-Meq showed the highest mortality and tumor incidence rates. Conradie et al. previously reported that an RB-1B-based rMDV encoding wild-type L-Meq from CVI988 was more virulent than an rMDV encoding wild-type Meq from CVI988 [30]. Thus, insertion in the transactivation domain could enhance MDV pathogenicity due to the increase in transactivation activity. In contrast, vRB-1B_S-Meq exhibited the lowest mortality rate and tumor incidence, although the transactivation activity of S-Meq (RB-1B) was higher than that of RB-1B-Meq. Meq has been reported to have several functions, including inhibition of the cGAS–STING pathway [41], interaction with tumor suppressors [42], and transcriptional regulation [43,44], and deletion in the transactivation domain may reduce some of these protein functions or other unknown functions. For instance, Meq could inhibit the cGAS-STING DNA-sensing pathway, which plays a vital role in innate immunity in chickens [41] by interacting with STING through the C-terminal domain in Meq, thereby suppressing the expression of type 1 interferon [41]. Therefore, insertion and deletion in the transactivation domain in Meq may affect the immunosuppressive effects through the interaction of STING with the C-terminal domain of Meq. Further analyses are required to elucidate the molecular mechanisms by which insertion and deletion in the transactivation domain of Meq modulate MDV pathogenicity.

In the present study, we found that insertion in the transactivation domain causes enhanced pathogenicity, whereas deletion results in reduced pathogenicity. Indeed, virulent MDV strains circulating in Australia encode an L-Meq and exhibit the features of MDV strains with high virulence [24]. Therefore, insertion in the transactivation domain may be responsible for the high virulence of MDV strains in Australia, in addition to the polymorphisms in Meq. However, there is a discrepancy between the emergence of MDV strains with enhanced virulence and the years in which the MDV strains with insertions and deletions were detected when we considered the timeline of MDV evolution. MDV strains encoding L-Meq were originally found in field strains from the 1960s and 1970s with low virulence in the US and an attenuated vaccine strain, CVI988 [19], whereas MDV strains with S-Meq have been reported in the field since the 2010s. Moreover, a very short isoform of Meq, which encodes one copy of PRR in the transactivation domain, was reported in a virulent MDV strain circulating in Iran [45]. Although the pathogenicity of such field strains having the S-Meq is uncertain, MDV strains with short isoforms of Meq have emerged more recently than MDV strains with L-Meq and Meq along the timeline.

In the present study, however, MDV pathogenicity was reduced by deletion in the transactivation domain, which was inconsistent with the tendency of MDV strains to exhibit enhanced virulence in the field. The reduced virulence following the loss of sequences in Meq may result in the prolonged survival of chickens infected with MDV, thereby leading to the efficient virus shedding and maintenance of MDV in the poultry houses; it may indicate an aspect of virus evolution aimed at symbiosis with the host. In addition, although the polymorphisms and insertion/deletion in Meq could be factors contributing to MDV virulence, other viral factors may be involved in the evolution of MDV. This study could improve our understanding of the mechanisms by which MDV alters the virulence and survival strategy of MDV in the environment.

As limitations in the present study, we could not identify why the deletion in the transactivation domain causes the reduced pathogenicity of Meq, despite the enhanced transactivation activity. On the other hand, the insertion in the transactivation domain enhanced the transactivation activity, contrary to the previously suggested theory. Thus, the PRR domain in Meq appears to have unknown functions separate from a role in transcriptional regulation. Therefore, to better understand the molecular basis of MDV pathogenicity and its enhanced virulence, it is necessary to further investigate the functions of Meq and the proteins that could interact with the various Meq isoforms. In addition, in the present study, we used the rMDVs encoding S-Meq that the amino acid mutations introduced, and we did not evaluate the pathogenicity of rMDV encoding wild-type Kgw-c2 S-Meq by comparing with that of rMDV encoding S-Meq (RB-1B). Therefore, to evaluate the risks of future outbreaks caused by MDV strains encoding S-Meq, the pathogenicity of field strains encoding S-Meq with some amino acid substitutions toward higher virulence should be investigated. Likewise, to investigate the contribution of amino acid substitutions at positions other than the insertion in the transactivation domain to the MDV pathogenicity, it is necessary to compare the pathogenicity of rMDV encoding wild-type L-Meq with that of rMDV encoding L-Meq (RB). Additionally, we used the RB-1B-based rMDVs to determine whether the insertion or deletion contribute to the pathogenicity, and then, the mutations were introduced to L-*meq* and S-*meq* genes to match the sequences with that of the *meq* gene of RB-1B. Therefore, the possibility that the artificially introduced mutations affected the viral biology should be considered. The use of rMDVs from other strains, such as the CVI988-based rMDVs, should be tested to determine their effect on pathogenicity.

Supplementary Materials: The following supporting information can be downloaded at: https://www.mdpi.com/article/10.3390/v14020382/s1, Figure S1: The original uncropped image related to Figure 2B; Figure S2: The original uncropped image related to Figure 3A.

Author Contributions: Conceptualization: J.S., S.M., B.B.K. and M.S.P. Data curation: J.S. and S.M. Formal analysis: J.S. and S.M. Funding Acquisition: S.M., B.B.K. and K.O. Investigation: J.S., S.M., Z.Y., S.F. and H.S. Project administration: S.M. Resources: S.M., B.B.K., N.M., T.O., N.O., S.K. and K.O. Supervision: S.M. Writing—original draft: J.S. and S.M. Writing—review and editing: all authors. All authors have read and agreed to the published version of the manuscript.

Funding: This work was supported in part by Grants-in-Aid for Scientific Research (B: 18H02332 and B: 20H03137), Grant-in-Aid for Challenging Research (Exploratory) (20K21357) from the Japan Society for the Promotion of Science and the Volkswagen Foundation Lichtenberg grant A112662 awarded to B.B.K., and was partially supported by the Training Program for Asian Veterinarians from the Japan Veterinary Medical Association.

Institutional Review Board Statement: All animal experiments were approved by the Institutional Animal Care and Use Committee of Hokkaido University (approval number: 19-0081). All experiments were performed in accordance with the relevant guidelines and regulations of the Faculty of Veterinary Medicine of Hokkaido University, which is fully accredited by the Association for Assessment and Accreditation of Laboratory Animal Care International.

Informed Consent Statement: Not applicable.

Data Availability Statement: Data are contained within the article.

Conflicts of Interest: The authors declare no conflict of interest.

References

1. Osterrieder, N.; Kamil, J.P.; Schumacher, D.; Tischer, B.K.; Trapp, S. Marek's disease virus: From miasma to model. *Nat. Rev. Microbiol.* **2006**, *4*, 283–294. [CrossRef]
2. Schat, K.A. History of the first-generation Marek's disease vaccines: The science and little-known facts. *Avian Dis.* **2016**, *60*, 715–724. [CrossRef]
3. Othman, I.; Aklilu, E. Marek's disease herpesvirus serotype 1 in broiler breeder and layer chickens in Malaysia. *Vet. World* **2019**, *12*, 472–476. [CrossRef]
4. Raja, A.; Raj, G.D.; Bhuvaneswari, P.; Balachandran, C.; Kumanan, K. Detection of virulent Marek's disease virus in poultry in India. *Acta Virol.* **2009**, *53*, 255–260. [CrossRef] [PubMed]
5. Shi, M.Y.; Li, M.; Wang, W.W.; Deng, Q.M.; Li, Q.H.; Gao, Y.L.; Wang, P.K.; Huang, T.; Wei, P. The emergence of a vv + MDV can break through the protections provided by the current vaccines. *Viruses* **2020**, *12*, 1048. [CrossRef] [PubMed]
6. Sun, G.R.; Zhang, Y.P.; Lv, H.C.; Zhou, L.Y.; Cui, H.Y.; Gao, Y.L.; Qi, X.L.; Wang, Y.Q.; Li, K.; Gao, L.; et al. A Chinese variant Marek's disease virus strain with divergence between virulence and vaccine resistance. *Viruses* **2017**, *9*, 71. [CrossRef]
7. Sung, H.W. Recent increase of Marek's disease in Korea related to the virulence increase of the virus. *Avian Dis.* **2002**, *46*, 517–524. [CrossRef]
8. Zhuang, X.; Zou, H.; Shi, H.; Shao, H.; Ye, J.; Miao, J.; Wu, G.; Qin, A. Outbreak of Marek's disease in a vaccinated broiler breeding flock during its peak egg-laying period in China. *BMC Vet. Res.* **2015**, *11*, 157. [CrossRef] [PubMed]
9. Morrow, C.; Fehler, F. Marek's disease: A worldwide problem. In *Marek's Disease: An Evolving Problem*; Davison, F., Nair, V., Eds.; Elsevier: Amsterdam, The Netherlands, 2004; pp. 49–61.
10. Trimpert, J.; Groenke, N.; Jenckel, M.; He, S.; Kunec, D.; Szpara, M.L.; Spatz, S.J.; Osterrieder, N.; McMahon, D.P. A phylogenomic analysis of Marek's disease virus reveals independent paths to virulence in Eurasia and North America. *Evol. Appl.* **2017**, *10*, 1091–1101. [CrossRef] [PubMed]
11. Padhi, A.; Parcells, M.S. Positive selection drives rapid evolution of the meq oncogene of Marek's disease virus. *PLoS ONE* **2016**, *11*, e0162180.
12. Murata, S.; Machida, Y.; Isezaki, M.; Maekawa, N.; Okagawa, T.; Konnai, S.; Ohashi, K. Genetic characterization of a Marek's disease virus strain isolated in Japan. *Virol. J.* **2020**, *17*, 186. [CrossRef]
13. Witter, R. Increased virulence of Marek's disease virus field isolates. *Avian Dis.* **1997**, *41*, 149–163. [CrossRef]
14. Liu, J.L.; Kung, H.J. Marek's disease herpesvirus transforming protein MEQ: A c-Jun analogue with an alternative life style. *Virus Genes* **2000**, *21*, 51–64. [CrossRef] [PubMed]
15. Qian, Z.; Brunovskis, P.; Lee, L.; Vogt, P.K.; Kung, H. Novel DNA binding specificities of a putative herpesvirus bZIP oncoprotein. *J. Virol.* **1996**, *70*, 7161–7170. [CrossRef] [PubMed]
16. Levy, A.M.; Gilad, O.; Xia, L.; Izumiya, Y.; Choi, J.; Tsalenko, A.; Yakhini, Z.; Witter, R.; Lee, L.; Cardona, C.J.; et al. Marek's disease virus Meq transforms chicken cells via the v-Jun transcriptional cascade: A converging transforming pathway for avian oncoviruses. *Proc. Natl. Acad. Sci. USA* **2005**, *102*, 14831–14836. [CrossRef] [PubMed]
17. Liu, J.-L.; Ye, Y.; Lee, L.F.; Kung, H.-J. Transforming potential of the herpesvirus oncoprotein MEQ: Morphological transformation, serum-independent growth, and inhibition of apoptosis. *J. Virol.* **1998**, *71*, 388–395. [CrossRef]
18. Lupiani, B.; Lee, L.F.; Cui, X.; Gimeno, I.; Anderson, A.; Morgan, R.W.; Silva, R.F.; Witter, R.L.; Kung, H.-J.; Reddy, S.M. Marek's disease virus-encoded Meq gene is involved in transformation of lymphocytes but is dispensable for replication. *Proc. Natl. Acad. Sci. USA* **2004**, *101*, 11815–11820. [CrossRef]
19. Shamblin, C.E.; Greene, N.; Arumugaswami, V.; Dienglewicz, R.L.; Parcells, M.S. Comparative analysis of Marek's disease virus (MDV) glycoprotein-, lytic antigen pp38- and transformation antigen Meq-encoding genes: Association of meq mutations with MDVs of high virulence. *Vet. Microbiol.* **2004**, *102*, 147–167. [CrossRef]
20. Murata, S.; Yamamoto, E.; Sakashita, N.; Maekawa, N.; Okagawa, T.; Konnai, S.; Ohashi, K. Research Note: Characterization of S-Meq containing the deletion in Meq protein's transactivation domain in a Marek's disease virus strain in Japan. *Poult. Sci.* **2021**, *100*, 101461. [CrossRef]
21. Murata, S.; Okada, T.; Kano, R.; Hayashi, Y.; Hashiguchi, T.; Onuma, M.; Konnai, S.; Ohashi, K. Analysis of transcriptional activities of the Meq proteins present in highly virulent Marek's disease virus strains, RB1B and Md5. *Virus Genes* **2011**, *43*, 66–71. [CrossRef]

22. Murata, S.; Hashiguchi, T.; Hayashi, Y.; Yamamoto, Y.; Matsuyama-Kato, A.; Takasaki, S.; Isezaki, M.; Onuma, M.; Konnai, S.; Ohashi, K. Characterization of Meq proteins from field isolates of Marek's disease virus in Japan. *Infect. Genet. Evol.* **2013**, *16*, 137–143. [CrossRef] [PubMed]
23. Conradie, A.M.; Bertzbach, L.D.; Trimpert, J.; Patria, J.N.; Murata, S.; Parcells, M.S.; Kaufer, B.B. Distinct polymorphisms in a single herpesvirus gene are capable of enhancing virulence and mediating vaccinal resistance. *PLoS Pathog.* **2020**, *16*, e1009104. [CrossRef] [PubMed]
24. Renz, K.G.; Cooke, J.; Clarke, N.; Cheetham, B.F.; Hussain, Z.; Islam, A.F.M.F.; Tannock, G.A.; Walkden-Brown, S.W. Pathotyping of Australian isolates of Marek's disease virus and association of pathogenicity with meq gene polymorphism. *Avian Pathol.* **2012**, *41*, 161–176. [CrossRef] [PubMed]
25. Mescolini, G.; Lupini, C.; Felice, V.; Guerrini, A.; Silveira, F.; Cecchinato, M.; Catelli, E. Molecular characterization of the meq gene of Marek's disease viruses detected in unvaccinated backyard chickens reveals the circulation of low- and high-virulence strains. *Poult. Sci.* **2019**, *98*, 3130–3137. [CrossRef] [PubMed]
26. Deng, Q.; Shi, M.; Li, Q.; Wang, P.; Li, M.; Wang, W.; Gao, Y.; Li, H.; Lin, L.; Huang, T.; et al. Analysis of the evolution and transmission dynamics of the field MDV in China during the years 1995–2020, indicating the emergence of a unique cluster with the molecular characteristics of vv+ MDV that has become endemic in southern China. *Transbound. Emerg. Dis.* **2021**, *68*, 3574–3587. [CrossRef]
27. Spatz, S.J.; Petherbridge, L.; Zhao, Y.; Nair, V. Comparative full-length sequence analysis of oncogenic and vaccine (Rispens) strains of Marek's disease virus. *J. Gen. Virol.* **2007**, *88*, 1080–1096. [CrossRef]
28. Wajid, S.J.; Katz, M.E.; Renz, K.G.; Walkden-Brown, S.W. Prevalence of Marek's disease virus in different chicken populations in Iraq and indicative virulence based on sequence variation in the ecoRI-q (meq) gene. *Avian Dis.* **2013**, *57*, 562–568. [CrossRef]
29. Lee, S.I.; Takagi, M.; Ohashi, K.; Sugimoto, C.; Onuma, M. Difference in the meq gene between oncogenic and attenuated strains of Marek's disease virus serotype 1. *J. Vet. Med. Sci.* **2000**, *62*, 287–292. [CrossRef]
30. Conradie, A.M.; Bertzbach, L.D.; Bhandari, N.; Parcells, M.; Kaufer, B.B. A common live-attenuated avian herpesvirus vaccine expresses a very potent oncogene. *mSphere* **2019**, *4*, e00658-19. [CrossRef]
31. Mohamed, M.H.A.; El-Sabagh, I.M.; Al-Habeeb, M.A.; Al-Hammady, Y.M. Diversity of Meq gene from clinical Marek's disease virus infection in Saudi Arabia. *Vet. World* **2016**, *9*, 572–578. [CrossRef]
32. Osterrieder, N. Sequence and initial characterization of the U(L)10 (glycoprotein M) and U(L)11 homologous genes of serotype 1 Marek's Disease Virus. *Arch. Virol.* **1999**, *144*, 1853–1863. [CrossRef] [PubMed]
33. Schumacher, D.; Tischer, B.K.; Fuchs, W.; Osterrieder, N. Reconstitution of Marek's disease virus serotype 1 (MDV-1) from DNA cloned as a bacterial artificial chromosome and characterization of a glycoprotein B-negative MDV-1 mutant. *J. Virol.* **2000**, *74*, 11088–11098. [CrossRef] [PubMed]
34. Jarosinski, K.W.; Schat, K.A. Multiple alternative splicing to exons II and III of viral interleukin-8 (vIL-8) in the Marek's disease virus genome: The importance of vIL-8 exon I. *Virus Genes* **2007**, *34*, 9–22. [CrossRef] [PubMed]
35. Engel, A.T.; Selvaraj, R.K.; Kamil, J.P.; Osterrieder, N.; Kaufer, B.B. Marek's disease viral interleukin-8 promotes lymphoma formation through targeted recruitment of B cells and CD^{4+} CD^{25+} T cells. *J. Virol.* **2012**, *86*, 8536–8545. [CrossRef]
36. Tischer, B.K.; Kaufer, B.B. Viral bacterial artificial chromosomes: Generation, mutagenesis, and removal of mini-F sequences. *J. Biomed. Biotechnol.* **2012**, *2012*, 472537. [CrossRef]
37. Tischer, B.K.; von Einem, J.; Kaufer, B.; Osterrieder, N. Two-step red-mediated recombination for versatile high-efficiency markerless DNA manipulation in Escherichia coli. *BioTechniques* **2006**, *40*, 191–197.
38. Borenshtain, R.; Davidson, I. Marek's disease virus genome separation from feather tip extracts by pulsed field gel electrophoresis. *J. Virol. Methods* **2002**, *101*, 169–174. [CrossRef]
39. Bello, N.; Francino, O.; Sánchez, A. Isolation of genomic DNA from feathers. *J. Vet. Diagn. Investig.* **2001**, *13*, 162–164. [CrossRef]
40. Baigent, S.J.; Kgosana, L.B.; Gamawa, A.A.; Smith, L.P.; Read, A.F.; Nair, V.K. Relationship between levels of very virulent MDV in poultry dust and in feather tips from vaccinated chickens. *Avian Dis.* **2013**, *57*, 440–447. [CrossRef]
41. Li, K.; Liu, Y.; Xu, Z.; Zhang, Y.; Luo, D.; Gao, Y.; Qian, Y.; Bao, C.; Liu, C.; Zhang, Y.; et al. Avian oncogenic herpesvirus antagonizes the cGAS-STING DNA-sensing pathway to mediate immune evasion. *PLoS Pathog.* **2019**, *15*, e1007999. [CrossRef]
42. Deng, X.; Li, X.; Shen, Y.; Qiu, Y.; Shi, Z.; Shao, D.; Jin, Y.; Chen, H.; Ding, C.; Li, L.; et al. The Meq oncoprotein of Marek's disease virus interacts with p53 and inhibits its transcriptional and apoptotic activities. *Virol. J.* **2010**, *7*, 348. [CrossRef]
43. Subramaniam, S.; Johnston, J.; Preeyanon, L.; Brown, C.T.; Kung, H.-J.; Cheng, H.H. Integrated analyses of genome-wide DNA occupancy and expression profiling identify key genes and pathways involved in cellular transformation by a Marek's disease virus oncoprotein, Meq. *J. Virol.* **2013**, *87*, 9016–9029. [CrossRef] [PubMed]
44. Qian, Z.; Brunovskis, P.; Rauscher, F., III; Lee, L.; Kung, H. Transactivation activity of Meq, a Marek's disease herpesvirus bZIP protein persistently expressed in latently infected transformed T cells. *J. Virol.* **1995**, *69*, 4037–4044. [CrossRef] [PubMed]
45. Molouki, A.; Ghalyanchilangeroudi, A.; Abdoshah, M.; Shoushtari, A.; Abtin, A.; Eshtartabadi, F.; Akhijahani, M.M.; Ziafatikafi, Z.; Babaeimarzango, S.S.; Allahyari, E.; et al. Report of a new meq gene size: The first study on genetic characterisation of Marek's disease viruses circulating in Iranian commercial layer and backyard chicken. *Br. Poult. Sci.* **2021**, *20*, 1–8. [CrossRef] [PubMed]

MDPI
St. Alban-Anlage 66
4052 Basel
Switzerland
Tel. +41 61 683 77 34
Fax +41 61 302 89 18
www.mdpi.com

Viruses Editorial Office
E-mail: viruses@mdpi.com
www.mdpi.com/journal/viruses

www.ingramcontent.com/pod-product-compliance
Lightning Source LLC
Chambersburg PA
CBHW041948260926
51850CB00018B/2049
* 9 7 8 3 0 3 6 5 6 6 2 3 8 *

The Teaching of Ethics X

– PROPERTY OF –
MUNICIPAL POLICE
TRAINING COUNCIL

Ethics in Criminal Justice Education

170.7
SHE
C.2

Lawrence W. Sherman

INSTITUTE OF
SOCIETY, ETHICS AND
THE LIFE
SCIENCES
THE
HASTINGS
CENTER

Copyright © 1982 by the Institute of Society, Ethics and the Life Sciences

All rights reserved. No part of this book may be reproduced or transmitted in any form or by any means, electronic or mechanical, including photocopying, recording or by any information storage and retrieval system, without permission in writing from the Publisher.

The Hastings Center
Institute of Society, Ethics and the Life Sciences
360 Broadway
Hastings-on-Hudson, New York 10706

Library of Congress Cataloging in Publication Data

Sherman, Lawrence W.
Ethics in criminal justice education.

(The Teaching of ethics ; 10)
Bibliography: p.
1. Criminal justice, Administration of—Study and teaching—United States—Moral and ethical aspects. 2. Police—Training of—United States—Moral and ethical aspects. I. Title. II. Series.
HV8143.S425 170'.7'1173 82-978
ISBN 0-916558-14-2 AACR2

Printed in the United States of America